International Conference on Computer Information and Automation Engineering (ICCIAE 2018)

IOP Conference Series: Materials Science and Engineering
Volume 359

Beijing, China
16 – 18 March 2018

ISBN: 978-1-5108-6314-9
ISSN: 1757-8981

Printed from e-media with permission by:

Curran Associates, Inc.
57 Morehouse Lane
Red Hook, NY 12571

Some format issues inherent in the e-media version may also appear in this print version.

Printed by Curran Associates, Inc. (2018)

For permission requests, please contact the Institute of Physics at the address below.

Institute of Physics
Dirac House, Temple Back
Bristol BS1 6BE UK

Phone: 44 1 17 929 7481
Fax: 44 1 17 920 0979

techtracking@iop.org

Additional copies of this publication are available from:

Curran Associates, Inc.
57 Morehouse Lane
Red Hook, NY 12571 USA
Phone: 845-758-0400
Fax: 845-758-2633
Email: curran@proceedings.com
Web: www.proceedings.com

Table of contents

Volume 359

International Conference on Computer Information and Automation Engineering

17–19 November 2017, Yichang, China

Accepted papers received: 1 May 2018
Published online: 16 May 2018

Preface

Papers

IOP Conf. Series: Materials Science and Engineering **359** (2018) 011001 doi:10.1088/1757-899X/359/1/011001

Preface

The 2018 International Conference on Computer Information and Automation Engineering (ICCIAE 2018) has been successfully held during March 16-18, 2018 in Beijing, China. The ICCIAE 2018 serves as an platform for researchers, engineers, academicians as well as industrial professionals from all over the world to present their research results in Computer Information and Automation Engineering. This conference provides opportunities for the delegates to exchange new ideas and application experiences face to face, to establish business or research relations and to find global partners for future collaboration.

There are 3 keynote talks and 54 contributed talks presented in this conference. All papers included in the conference proceedings were peer reviewed according to IOP Publishing standards.

We would like to deeply acknowledge all the parties involved in making this conference successfully held: the conference session chairs, organizing committee, authors, reviewers and IOP Publishing.

General chair

Dr. Xiaolong Li

Peer review statement

All papers published in this volume of *IOP Conference Series: Materials Science and Engineering* have been peer reviewed through processes administered by the proceedings Editors. Reviews were conducted by expert referees to the professional and scientific standards expected of a proceedings journal published by IOP Publishing.

IOP Conf. Series: Materials Science and Engineering **359** (2018) 012001 doi:10.1088/1757-899X/359/1/012001

Study on Hybrid Image Search Technology Based on Texts and Contents

H.T. Wang[1], F.L. Ma[2], C Yan[2] and H Pan[2]

[1]Network Information Center, Wuhan University of Technology, Wuhan 430070,PR China

[2]School of Computer Science and Technology, Wuhan University of Technology, Wuhan 430070,PR China.

Abstract. Image search was studied first here based on texts and contents, respectively. The text-based image feature extraction was put forward by integrating the statistical and topic features in view of the limitation of extraction of keywords only by means of statistical features of words. On the other hand, a search-by-image method was put forward based on multi-feature fusion in view of the imprecision of the content-based image search by means of a single feature. The layered-searching method depended on primarily the text-based image search method and additionally the content-based image search was then put forward in view of differences between the text-based and content-based methods and their difficult direct fusion. The feasibility and effectiveness of the hybrid search algorithm were experimentally verified.

1.Introduction

Text-based image search has been being further studied by many scholars so far. References indicate that many scholars focused on types of image correlation information in the stage for annotation of most of the current text-based image search; on the other hand, less study was carried out to the weight related to the context information[1,2]. The content-based image search existed in some ready-made systems abroad; for example, Web SEEK and Visual SEEK have image search tools with color features (such as texture features based on the wavelet transform and visual features based on the color set) for applications[3,5]. On the other hand, the domestic research on the content-based image search started relatively late, which always focused on some single features not enough to express information of entire images; moreover, those search algorithms based on features of single contents often present unsatisfactory effects. Thus, multi-feature fusion technology of the content-based image search is still to be further studied. The fusion issues focuses on mutual compensation of different factors and integration of their advantages to get a better effect. The traditional fusion methods include the linear fusion, fuzzy analytical hierarchy process[6-9], factor analysis and so on. The fusion issue in the image field is generally to bridge the "gap" between image text and visual information and discuss how to perform their fusion for better description of images[10,11].

2.Text-based image feature extraction

2.1 Processing information related to images. First, the text information in the page including an image is preprocessed. The processing flow diagram is shown in Figure 1.

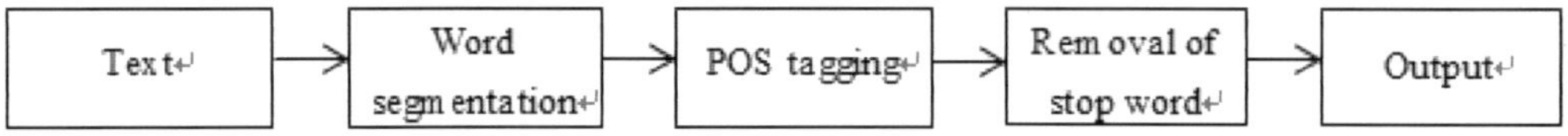

Figure 1. Text information processing flow diagram

NLPIR was selected here as the word segmentation tool and Sogou scel was introduced to deal with new words. After completion of word segmentation of a document, stop words may be removed by means of screening of part-of-speech (POS) in accordance with NLPIR Chinese (POS) annotation set.

2.2 Weight-based TF-ID selection of candidate phrases. AHP was selected here as the method for analysis of the weights of information related to images[12].

2.2.1 Establishment of the hierarchy structure model. Text information was selected in 3 positions, namely: webpage title (B1), image title (B2) and image context information (B3). The specific hierarchy structure is schematically shown in Figure 2.

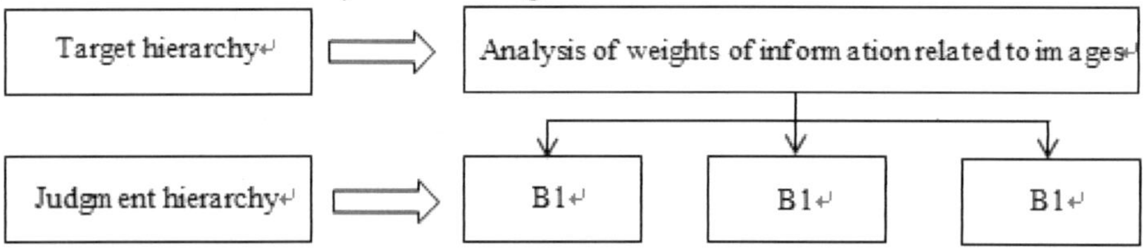

Figure 2. Hierarchy structural representation for analysis of weights related to information

2.2.2 Establishment of the judgment matrix (M). The method for establishment of the judgment matrix is described as follows: its overall objective is analysis of weights of information related images and the scoring mode is adopted for judgment of the relative importance of any two factors among 3 factors.

2.2.3 Calculation of eigenvectors of the judgment matrix by means of Asymptotic Normalization Coefficient.

2.2.4 Consistency checking. During the experiment, the indexes for checking results are generally the C.I. (consistency index) and C.R. (consistency ratio). The specific steps are as follows:

Step 1: calculation of C.I. is performed by the following equation:

$$C.I. = \frac{\lambda_{\max} - n}{n - 1} \tag{1}$$

Where: $\lambda_{\max}$ represents the maximum eigenwert of the matrix.

Step 2: checking the mean random C.I. in the lookup table.

Step 3: calculation of C.R. is performed by the following equation:

$$C.R. = \frac{C.I.}{R.I.} \tag{2}$$

It is gained from Steps 1 and 3 that: C.I. and C.R. are 0; obviously, both of them are less than 0.1; thus, this experiment passed the consistency checking; and the estimated weights of 3 kinds of text information related to images are presented in Table 1.

Table 1. Types and weights of text information related to images

Factor	Type	English	Weight
B1	Webpage title	Page Title	0.35
B2	Image title	Title	0.5
B3	Webpage text	Text	0.15

Weights of context information (Table 1) and the TF- algorithm were here integrated for operation; and the specific equations are as follows:

Weighted term frequency (TF):
$$wtf = \sum_{i=1}^{m} \frac{n_i \times w_i}{n} \quad (3)$$

Inverse document frequency (IDF):
$$idf = \log\left(\frac{N}{N_i + 1}\right) \quad (4)$$

$$wtf - idf = wtf \times idf \quad (5)$$

Where: m represents the total number of types of context information related images (m=3 here);n_i represents the occurrence frequency of the keywords in the ith type of context information; w_i represents the weight of the ith type of context information; n represents the total number of candidate phrases in the context information related images; N represents the size of the corpus; N_i represents the number of texts including this term; and *wtf-idf* represents the weight-based TF-IDF value of keywords.

2.3Extraction of text keywords by integrating topical features

2.3.1 Selection of subject elements. POS: Nouns and verbs shall be selected to lower the calculation complexity.

TF: those terms closely related to the theme generally occur more frequently in a paper. A TF calculation method was utilized here for normalization of TF. The specific equation is as follows:

$$tf\left(w_{ij}\right) = f_j\left(w_i\right) / n\left(d_j\right) \quad (6)$$

Where:$tf(w_{ij})$ represents the relative TF of Word w_i in Document d_j; $f(w_i)$ represents the occurrence frequency of Word w_i in Document d_j; and $n(d_j)$ represents the number of notional words in the document.

Document frequency (DF): DF refers to the number of documents including a certain word; and DF is calculated by:

$$df(w_i) = -\frac{n(w_i)}{N} \times \log\frac{n(w_i)}{N} - (1 - \frac{n(w_i)}{N})\log\frac{n(w_i)}{N} \quad (7)$$

Where: $df(w_i)$ represents the DF theme expression of Word w_i; $n(w_i)$ represents the number of documents including Word w_i; and N represents the total number of documents.

Position of a word: the weight of the position of a word is calculated by:

As for Word w_i in a title: $f_p(w_i) = 1 + 4 \times \frac{f_t(w_i)}{s_t(d_j)}$ (8)

As for Word w_i in texts: $f_p\left(w_i\right) = 1$ (9)

Where: $f_p(w_i)$ represents the weight of the position of Word w_i; $f_t(w_i)$ represents the occurrence frequency of Word w_i in a title; and $s_t(d_j)$ represents the number of notional words in the title of Document d_j.

2.3.2 Co-occurrence analysis of phrases. The co-occurrence phenomenon of phrases refers to the statistical relationship between phrases or words. The following keywords were extracted here. The specific weight is calculated by:

$$g\left(w_i\right) = n(w_i) / (n(w_i) + 1) \quad (10)$$

Where: $n(w_i)$ represents the number of fixed collocations of Word w_i in the document space; and $g(w_i)$ represents the theme expression of Word w_i.

2.3.3 Calculation of theme semantic features. After word segmentation and removal of stop words, some significant nominal words remains; moreover, evaluation of expression weights of words primarily depends on their TFs and positions; and co-occurrence information centralized in texts may have a reference value. In general, the theme expression of information (such as TF) is dominant to reflect the correlation between words or phrases and the theme to a great extent. The co-occurrence phenomenon of phrases reflects whether those words centralized in texts are keywords. The scoring equation is as follows:

$$W(w_i) = tf(w_i) \times f_p(w_i) \times d\,f(w_i) \times (1 + g(w_i)) \tag{11}$$

Where: $tf(w_{ij})$ represents the relative TF of Word w_i in Document d_j; $d\,f(w_i)$ represents the DF weight of Word w_i; $f_p(w_i)$ represents the weight of Word w_i; and $g(w_i)$ represents the theme expression of Word w_i.

The total of the weight of a candidate centralized word with statistical features and its theme weights is regarded as the final scores of the candidate word.

$$p(d, w_j) = wtf - idf + W(w_j) \tag{12}$$

Where: $wtf\text{-}idf$ represents the weight-based TF-IDF value of the keyword; and $w(w_j)$ represents the theme semantic feature value of the word.

3.Content-based Image Search

3.1 Content-based image feature selection. The following factors were taken into account here for selection of the image feature set:

How to select the feature from various features for expression of the image contents as can as possible;

How to efficiently extract the image features.

As for experimentally extraction of image features, an image set including 110 images (resolution: 1024*740; and bit depth: 24 bits) acted as the test objects and each image feature was extracted for 15 times; and then the measured data were averaged. Figure 3 show comparison of extraction rates for various features, where the vertical axis represents various features and the horizontal axis represents time whose unit in second for extraction of a feature.

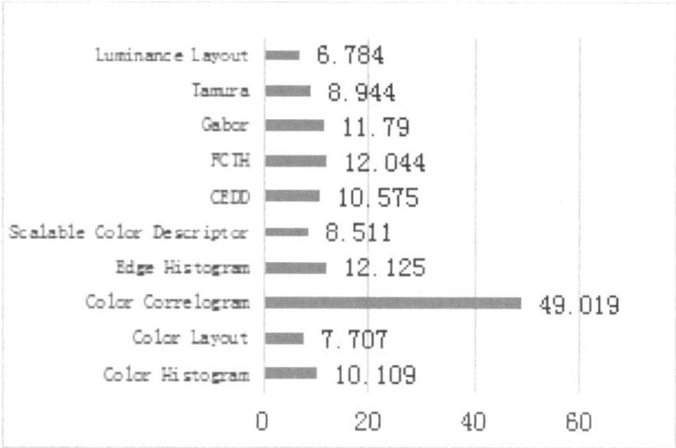

Figure 3. Comparison of extraction rates for image features

Figure 3 indicates that extraction of features of the Color Correlogram takes a long time and extraction of features of other items spend equivalent time. Considering the generation efficiency of the eigenvalues of images, fusion of features may occur for those items except the Color Correlogram.

3.2 Content-based multi-feature fusion scheme. Our fusion scheme is calculated by:

$$w = \sum k_i \times s_i \tag{13}$$

Where: w represents the weight of images; k_i represents the weight of the ith image feature solved by means of the genetic algorithm; s_i represents the degree of similarity based on the ith image feature.

To determine the individuals and genes: the weights vector of our genes (image features) is (k0k1k2k3k4k5k6k7k8); and individuals include genes and their presenting fitness.

To determine the criterion for judgment of the fitness of individuals: the fitness is calculated by:

$$fitness = \sum \left(\frac{1}{2} \right)^{index} + 1 \tag{14}$$

Where: index is the subscript of a similar image in the search results (the subscript of the similar image in the first place is defined as 0).

To determine the elimination criteria: while our applying the genetic algorithm, only some individuals with the maximum fitness remain to improve the calculation efficiency and appropriately lower the computational complexity.

To determine the gene recombination rules: the specific calculation equation is as follows:

$$k_{child} = (k_{father} + k_{mother}) / 2 \tag{15}$$

To determine the gene mutation criteria: the weights after gene recombination are as follows:

$$k_1 = k_1 - random \tag{16}$$

$$k_2 = k_2 + random \tag{17}$$

In summary, the method for solving the weights of image features based on the genetic algorithm is schematically in Figure 4.

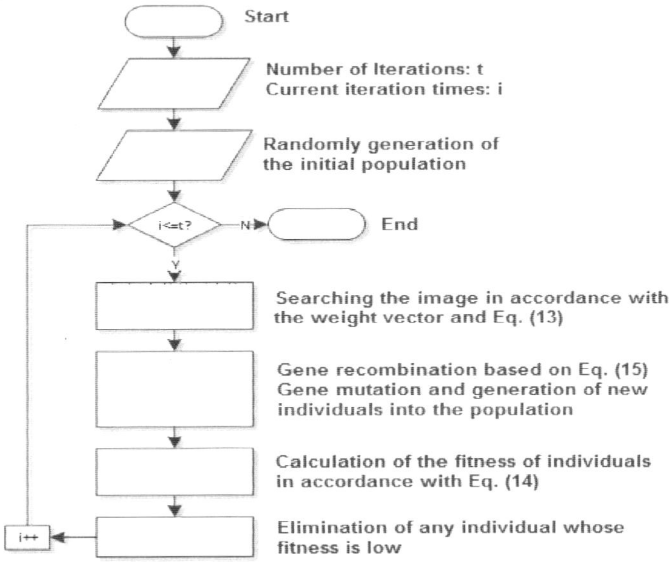

Figure 4. Flow diagram of the scheme for solving weights of image features

The actual implementation is a continuous cyclic process. While the iterative process has been completed, the individual whose fitness peaks is selected from those survival individuals, whose genes are selected as the weights vector of image features; and such vector is integrated into the equation for solving the fusion similarity.

4.Hybrid image search scheme based on texts and contents and application

Considering the algorithms for processing texts and images and their time and space complexities, the text-based search is dominant and the content-based image search is subsidiary for sufficient fusion of their advantages and further improvement of the search efficiency; moreover, duplicate and precision checking rates of the image search engine shall be guaranteed.

4.1Experiments for extraction of text features of images and results analysis

4.1.1 Experimental setup. As for our experiments for extraction of text features of images, the experimental data are from the Palace Museum Network (http://www.dpm.org.cn/). The specific experimental steps are as follows:
Step 1: Crawling and analysis of 1,000 ceramics images and related text information: they are divided into 5 categories (200 images per category), which are "Bottles", "Bowls", "Pots", "Flagons" and "Dishes"; and the related information is preprocessed.
Step 2: the weight of each keyword is calculated by means of the weight-based TF-IDF algorithm, ten of which are extracted as the text feature candidate set of images.
Step 3: Calculation of topical eigenvalues of words.
Step 4: To determine the regulatory factor (η) for mixing the statistical and topical features.
η is among 0-1 for our experiment and its step size is 0.1. 100 documents are selected at random. While 5 keywords are extracted per document in accordance with the algorithm, the average annotation precision (avg_P), recall (avg_R) and comprehensive evaluation index (F) changes along with the growth of η (Figure 5: x axis － η and y axis - avg_P, avg_R and F). Figure 5 indicates that fusion has the best effects while η =0.6.

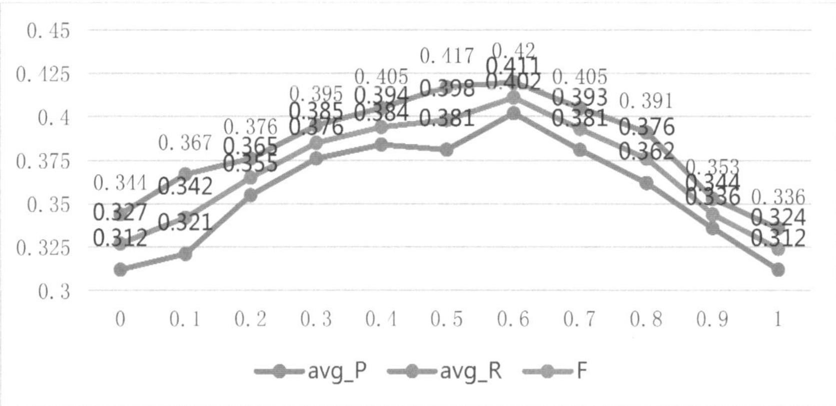

Figure 5. avg_P, avg_R and F vs. η

Step 5: calculation of overall scores of words based statistical and topical features in accordance with Eq. (16); and these scores are sorted from small to big. The top 5 scores are extracted as text features of images.

4.1.2 Experimental evaluation indexes. avg_P, avg_R and F are utilized for measurement of the precision of extraction of text features of images, which are calculated by:

$$avg\_P = \frac{1}{N}\sum_{i=1}^{N} P(I_i) \tag{18}$$

$$avg\_R = \frac{1}{N}\sum_{i=1}^{N} R(I_i) \tag{19}$$

$$F = \frac{2 \times avg\_P \times avg\_R}{avg\_P + avg\_R} \tag{20}$$

Where: N represents the size of the measurement image set; and $P(I_i)$ and $R(I_i)$ represents the precision and recall of annotation of the Measurement Image I_i, respectively, whose calculation equations are as follows:

$$P(I_i) = \frac{|A_i \cap G_i|}{|A_i|} \tag{21}$$

$$R(I_i) = \frac{|A_i \cap G_i|}{|G_i|} \tag{22}$$

IOP Conf. Series: Materials Science and Engineering **359** (2018) 012001 doi:10.1088/1757-899X/359/1/012001

Where: A_i represents the annotation gained based on the above algorithm; G_i represents the standard keyword of the image; $|A_i \cap G_i|$ represents the number of correct annotations; $P(I_i)$ represents the annotation precision of a certain image; and $R(I_i)$ represents the integrity for annotation of a certain image.

4.1.3 Check experiment settings and summary. For verification of feasibility and effectiveness of the algorithm for capturing text features of images by integrating topical and statistical features of words, the following check experiments are set currently; and the corresponding comparison results are as follows.

Table 2. Comparison of results for various algorithms

Algorithm	avg_P	avg_R	F
TF/IDF	0.325	0.367	0.345
Weight-based TF/IDF	0.376	0.405	0.39
Integration of topical features	0.402	0.420	0.411

Given all that, not only extraction of text-based image features but also utilization of statistical and topical features of words may improve the precision of extraction.

4.2Content-based image search check experiment and results analysis

4.2.1 Experimental evaluation indexes. Precision-Recall criteria are utilized for evaluation of performances of the algorithm. Precision and Recall are calculated by:

$$\text{Precision} = B / A \tag{23}$$

$$\text{Recall} = B / C \tag{24}$$

Where: represents the number of returned images after searching; represents the number of images related to searching images in the search results; and represents the total number of images related to the search images in the search database.

4.2.2 Experimental design and summary. The experiments were carried out in Corel 1000 of the public data set. This image database is made up of 10 types of subsets; and 90 images were selected at random per subset to form the image set (including 900 images) and solve the optimized weight for various features by means of the genetic algorithm; moreover, the remaining 100 images as the measurement image set were used for searching and calculation of the mean precision of search results of the two experiments as follows.

Experiment 1: the multi-feature fusion scheme was used based on the weight of the same feature; namely, it is thought that the contribution of each type of features is the same for the image contents. For example, while searching Image 611.jpg, the search results are represented in Figure 6.

Experiment 2: the above genetic algorithm was utilized for calculation of the weight of each feature; moreover, the multi-feature fusion scheme was utilized based on various weights. For example, while searching Image 611.jpg and using this algorithm, the search results are represented in Figure 7.

Figure 6. Search results of Experiment 1

Figure 7. Search results of Experiment 2

The evaluation indexes are listed in Table 3.

Table 3. Various searching precisions for the both algorithms
(Number of returned images: 50)

Type	Experiment 1	Experiment 2 (the algorithm of this section)
Africa	0.35	0.47
Beach	0.26	0.31
Buildings	0.25	0.38
Bus	0.51	0.65
Dinosaurs	0.36	0.51
Elephants	0.34	0.48
Flowers	0.37	0.42
Horses	0.51	0.58
Mountains	0.16	0.22
Food	0.32	0.42

The above search results indicate that the degree of similarity of search results is higher and the images are positioned centrally in the front while the genetic algorithm is applied. Table 3 indicates that the algorithm utilized here leads remarkable improvement of Precision and Recall. Experiments were carried out to the public data set to gain that the mean precisions for the both algorithms are 0.343 and 0.444, respectively. Thus, the multi-feature fusion algorithm shall be feasible and effective based on calculation of weights of various features by means of the genetic algorithm.

4.3Hybrid image search scheme. The hybrid search scheme schematically shown in Figure 8 will be described in detail.

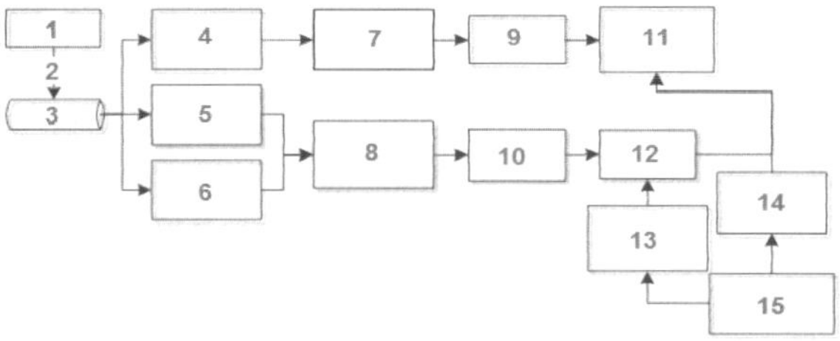

1 – Crawler; 2 - html parser; 3 – Database; 4- Extraction of content features of images; 5 - Extraction of topical features of words; 6 - Extraction of the candidate set of keywords by means of the weight-based TF/IDF; 7 – Calculation of the weight per feature by means of the genetic algorithm; 8 - Extraction of keywords by integrating topical and statistical features; 9 – Establishment of the image indexes library; 10 - Establishment of the inverted indexes of keywords; 11 – Display of results; 12 – Matching; 13 – Inputting keywords; 14 – Inputting images; 15 - User Inquiry

Figure 8. Flow diagram of the hybrid search scheme

As for the text-based image search, the AHP-based weight analysis method is applied for analysis of context information of images. On the other hand, the genetic algorithm is utilized as for the content-based image search to gain the new solution of weights of visual features of images; finally, more precise image search scheme may be achieved. Based on the hierarchy search thought, preliminary search is first carried out by means of keywords and the secondary fined search is then performed by means of Search By Image from the primary searched result set.
While a user inputs a keyword or image, search results are returned after layered-searching.

4.4Hybrid image search scheme

4.4.1 Experimental data set and evaluation indexes. The experiment data set is the same as that in Section 4.1 and the indexes for evaluation of performances of the algorithm are the same as those mentioned in Section 4.2.

4.4.2 Experiment settings and results analysis.
Experiment preprocessing:

- Establishment of the image library
- Establishment of the image indexes

An indexer sub-thread is established for each image folder; at the same time, while the start method of Indexer is called in the thread, an Indexer file will be established for each image in the folder for searching. The indexer sub-thread is under control of the Thread Pool in the outermost layer of the program.

- Calculation of weights of features of images

Prior to operation of the genetic algorithm, a group of similar image sets may be necessarily selected by people as the search standards. Our image set met such requirement during the experimental period. Some images in the image set are shown in Figure 9.

Figure 9. Some images in the image set

After selecting the similar image set, a group of source images shall be selected as inputs to start the program for solution of weights of features of images based on the genetic algorithm.
Weights are calculated by means of Eqs. (16) and (17) given in Section 3.2 for this experiment.
All weights of features of images shall be written in the configuration file as the standard weights for searching images. At this point, all basic search work is OK; next, the image search is necessarily created and the search is ready for inputting an image.

After completion of java and html parser, 3 search scenes were established for measurement of performances of the hybrid algorithm, which are as follows:
Experiment 1: Using the text-based image search method
Experiment 2: Using the content-based image search method
Experiment 3: Hybrid image search method based on texts and contents: the addresses for searching keywords and images are input to the search bars at the same time and the multi-tiered search method is utilized for getting similar images.
The examples of sampling results for experiments are shown in Table 4.

Table 4. Experimentally sampling presentation

Keyword	Input image	Number of related images (C)	Number of search results (A)	Number of related images in the search results (B)	Recall (B/A)	Recall (B/C)
Bottle	N/A	180	196	110	0.561	0.611
Bowl	N/A	180	192	118	0.615	0.656
N/A	1-36.jpg	180	138	83	0.601	0.461
N/A	5-5.jpg	180	149	86	0.577	0.478
Pots	3-8.jpg	180	158	112	0.709	0.622
Flagons	4-7.jpg	180	148	116	0.784	0.644

Experiments 1, 2 and 3 were carried out for 50 times, respectively; and the Precision - Recall results are shown in Figure 10 (horizontal ordinate: Recall; and vertical axis: Precision). While Recall is 0, Precision is the mean precision while the first image and 50 search images in the same type have been searched; and while Recall is 20%, Precision is the mean precision of 50 search images while 20% of B (number of related images in the search results) images have been searched; similarly, while Recall is 100%, Precision is the mean precision while B (number of related images in the search results) images have been searched.

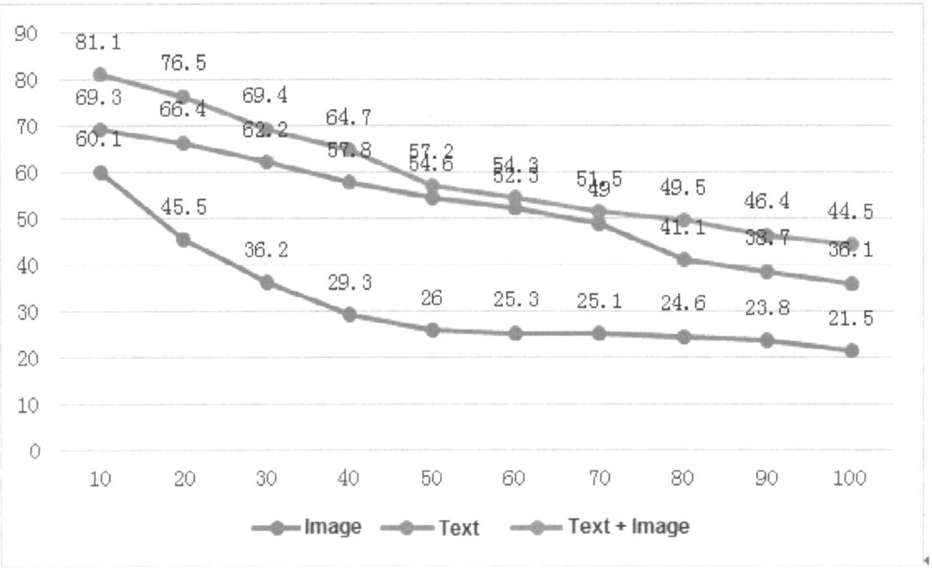

Figure 10 Precision vs. Recall

IOP Conf. Series: Materials Science and Engineering **359** (2018) 012001 doi:10.1088/1757-899X/359/1/012001

The above figures and tables indicate that the precision for the hybrid image search based texts and contents is higher than that for the text-based or content-based image search; moreover, the feasibility and effectiveness of the hybrid image search algorithm are verified; in addition, such case also indicates text and visual features of any image are able to reinforce each other and complementary.

5.Conclusions

The feasibility and effectiveness of the hybrid image search algorithm were verified here and the hybrid searching was performed based on improvement of performances of the text-based and content-based image search methods; and the primary work is described in detail as follows:

1). Extraction of features of images based on texts was put forward by integrating statistical and topical features to overcome the limitation (extraction of keywords only by means of statistical features of words).

2). Search By Image was put forward by means of the multi-feature fusion method to solve the imprecision for the single-feature content-based image search.

3). The genetic algorithm was used to determine the weights of features to clarify the contributions of various features to contents of any image.

4). The layered-searching method was applied to overcome the difference between the text-based and content-based image search methods; for example, "Semantic Gap" exits so that direct fusion shall be very difficult.

1. References

[1] Wang X J, Zhang L and Ma W Y *Duplicate-Search-Based Image Annotation Using Web-Scale Data [J]*

[2] roceedings of the IEEE, 2012, 100(9):2705-2721

[3] Huang Peng 2008 *Research on Web Image Retrieval based on the Fusion of Textual and Visual Information [D]* (Zhejiang University)

[4] Qiu Zhaowen 2009 *Research on Key Techniques of User-oriented Web Image Retrieval [D]* (Harbin Institute of Technology)

[5] Xu Hongtao 2009 *Study of Webimage Semantic Analysis and Automatic Labeling [D]* (Fudan University)

[6] Zhong Yi, Deng Hui and Zhang Hui 2015 *Basic Theory and comparison of Content-Based Image Retrieval [J]* (China New Telecommunications) chapter 14 pp 106-106

[7] Eakins J and Graham M 1999 *Content-based Image Retrieval [R]* A Report to the JISC Technology Applications Programme

[8] Datta R, Li J and Wang J Z 2005 *Content-based image search: approaches and trends of the new age [J]* (Proceedings of MIR'05. 2005), chapter3 pp 36-43

[9] Yuan Yunli 2013 *The Research on the Construction Engineering Project Risk Management based on Fuzzy Analytic Hierarchy Process (FAHP) [D]* (Chongqing University)

[10] Tian L, Zheng D and Zhu C 2011 *Research on image classification based on a combination of text and visual features.[C]// Eighth International Conference on Fuzzy Systems and Knowledge Discovery* (Shanghai, China.) pp 1869-1873

[11] Wu Wei 2014 *Image Classification and Retrieval Based on Semantic Distance and Features Combination [D]* (Inner Mongolia University)

[12] Deng Xue, Li Jiaming, Zeng Haojian, et al 2012 *Research on Computation Methods of AHP Wight Vector and Its Applications [J] (* Mathematics in Practice and Theory) chapter 7 pp 93-100

IOP Conf. Series: Materials Science and Engineering **359** (2018) 012002 doi:10.1088/1757-899X/359/1/012002

Multi exposure image fusion algorithm based on YCbCr space

T T Yang[1] and P Y Fang[2]

[1]Computer Science, University of Posts and Telecommunications, Beijing, China

[2]Computer Science, University of Posts and Telecommunications, Beijing, China.

E-mail:muyi1024b@163.com

Abstract. To solve the problem that scene details and visual effects are difficult to be optimized in high dynamic image synthesis, we proposes a multi exposure image fusion algorithm for processing low dynamic range images in YCbCr space, and weighted blending of luminance and chromatic aberration components respectively. The experimental results show that the method can retain color effect of the fused image while balancing details of the bright and dark regions of the high dynamic image.

1. Introduction

In real life, the intensity of the daily scenes is far greater than the dynamic range recorded by image acquisition device. Generally, when the intensity of the scene changes too much, it is inevitable that some details will be lost due to overexposure and underexposure. By adjusting the exposure time, the multiple exposure images of the same scene are processed to expand the dynamic range of low dynamic range images. Debevec[1] restores the camera response curve and tone mapping, which is complex and the fusion image produces a clear halo. Mertens[2] calculates contrast, saturation and exposure as fusion guidance indicators, and complete image fusion in Laplacian Pyramid. Due to the lack of considering the correlation of RGB color channels, fusion image color degradation and more details are lost in bright and dark areas. Paul[3-5] deals with the luminance and chromatic aberration domains of the image respectively in YCbCr color space. The fused image has higher saturation and the details of the darkness are promoted, but the overall image is darker and with a smaller contrast. This paper proposes a multi exposure image fusion algorithm based on YCbCr space, processing in luminance and chromatic aberration components respectively. Fusion image retains details in bright and dark areas, at the same time, obtains good color expression.

The rest of the paper is organized as follows. In Sec. 2, the proposed multi exposure image fusion algorithm is presented. In Sec. 3, the high dynamic images synthesized by three methods are evaluated and compared. Finally, In Sec. 4, the main conclusions are drawn.

2. Multi exposure image fusion algorithm in YCbCr space

In Mertens and Paul methods, the details and saturation of the image cannot be optimized at the same time when dealing with the multi exposure images. We propose to transform the low dynamic range images into YCbCr space. In luminance component Y, Firstly, the weight map W is calculated according to the contrast, saturation, and proper exposure. Then, W and Y are decomposed in Laplacian pyramid and weighted blending on multiresolution scale. Finally, operating the gamma correction and histogram equalization for the fused luminance component Y'. In chromatic aberration components Cb and Cr, take the offset maximum absolute values corresponding to the chromatic aberration value as Cb' and Cr'.

Figure 1. Flow chart of fusion algorithm

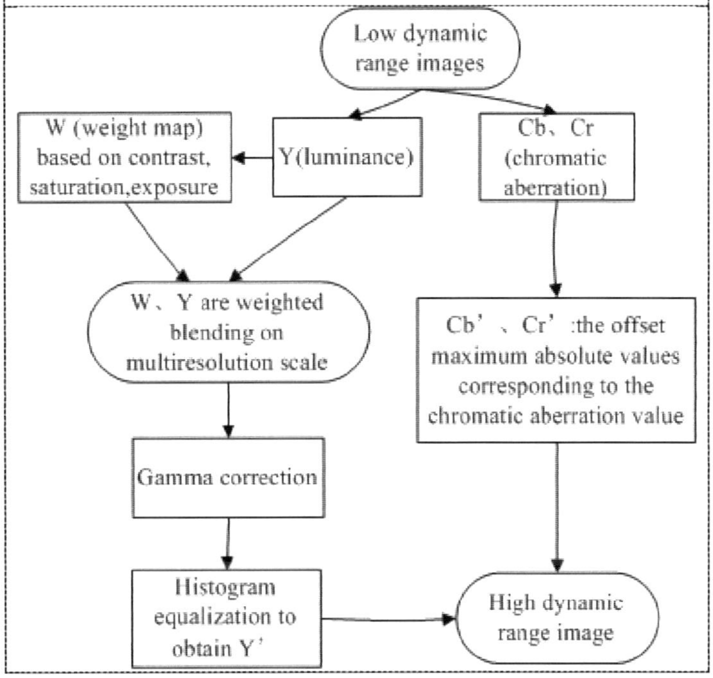

2.1YCbCr color space. The YCbCr color space is the offset version of YUV color space, where Y is the luminance component, Cb and Cr are chromatic components. Cb and Cr reflect the difference between the blue and red part of the RGB input signal and the intensity value.The advantage of YCbCr color space is that its luminance components Y and chromatic components Cb and Cr are separated. It can be operated separately in each channel without mutual effect. The conversions between YCbCr color space and RGB color space is as follows:

$$\begin{pmatrix} Y \\ Cb \\ Cr \end{pmatrix} = \begin{bmatrix} 0.257 & 0.564 & 0.098 \\ -0.148 & -0.291 & 0.439 \\ 0.439 & -0.368 & -0.071 \end{bmatrix} \begin{pmatrix} R \\ G \\ B \end{pmatrix} + \begin{bmatrix} 16 \\ 128 \\ 128 \end{bmatrix} \quad (1)$$

$$\begin{pmatrix} R \\ G \\ B \end{pmatrix} = \begin{bmatrix} 1.164 & 0 & 1.596 \\ 1.164 & -0.392 & -0.813 \\ 1.164 & 2.017 & 0 \end{bmatrix} \begin{pmatrix} Y-16 \\ Cb-128 \\ Cr-128 \end{pmatrix} \quad (2)$$

Where Y~[16 235], Cb、Cr~[16 240].

2.2Luminance domain processing.

2.2.1. Weight map. The fused luminance component is obtained by weighting the corresponding position luminance value of source images. The weight coefficient determines the quality of the fusion image. The weight map for each luminance image is determined by three factors: contrast, saturation, and exposure.

- Contrast $C_k(i, j)$.There are distinctions in details at the same scene with different exposure times in which the area with a larger contrast remains more edges and outlines. In the process of calculating contrast, Y is convolved by Laplacian operator. After convolution, the absolute value of coefficient is used as the contrast weight of corresponding image pixels.

- Saturation $S_k(i, j)$. Under the appropriate exposure, the color expression of the fusion image is the strongest, which should be given a greater weight. When the chromatic components Cb and Cr are all 128, this image loses color and becomes a grayscale map. The larger the deviation from 128, the more colorful the color is. The standard deviation of |Cb-128|/240 and |Cr-128|/240 is taken as the saturation weight.
- Exposure $E_k(i, j)$. Overexposure and underexposure will lose the details of the scene, which is not consistent with the visual characteristics of the human eye, and should be given a smaller weight. When the exposure is moderate, the intensity is mostly distributed in the middle region, so more weights are expected to be allocated to the pixels in the middle brightness area. Using the method of Mertens[2], we calculate how close each intensity is to 0.5 with a Gauss curve, and the result coefficient is used as the saturation weight. Its calculation is shown in the following formula:

$$E_k(i, j) = \exp(-\frac{(I_k(i,j)-0.5)^2}{2\sigma^2}) \tag{3}$$

$I_k(i, j)$ is the pixel value at the position of (i, j) in the k-th image; σ refers to the variance, which equals 0.2 in our implementation.

2.2.2. Gamma correction. Converting luminance component blended at multiresolution scale to uint8 type, Y~[0 255] exceeds the luminance range in the YCbCr color space. thus the Y is corrected by the following gamma correction, after correction Y' ~[16 235]:

$$Y'(i,j) = \left(\frac{Y(i,j)-Y_{\min}(i,j)}{Y_{\max}(i,j)-Y_{\min}(i,j)}\right)^{\gamma} \times R_C + L \tag{4}$$

Where $\gamma = \log_e(R_C)/\log_e(R_I)$, R_I is the range of luminance threshold to be corrected, R_C is the Y channel luminance range. H and L are the maximum and minimum values of the Y channel, R_C=H-L, H=235, L=16.

2.2.3. The processing steps of the luminance component Y.

- Compute the weight of contrast, saturation, and exposure.
- Fusion weight map W is obtained according to the following formula:

$$W_k(i, j) = C_k(i, j) \times S_k(i, j) \times E_k(i, j) \tag{5}$$

- The fusion weight map W and the luminance component Y are weighted blending on the multiresolution scale in Laplacian Pyramid which is given in paper[2].
- Fusion brightness is converted to uint8 type and performed gamma correction.
- The corrected luminance is balanced by histogram equalization to Y'.

2.3Chromatic aberration domains processing. When the value of Cb and Cr are all equal 128, the image loses color and is a grayscale map. The higher the degree of deviation from 128, the more obvious the color effect is, and the more suitable for human visual system to observe. Therefore, The chromatic value of the corresponding position of the maximum absolute value of deviation from 128 is used as the chromatic component of the fused image:

$$Cb' = Cb_n, \max_{0<n\leq k}|Cb_n\text{-}128| \tag{6}$$

$$Cr' = Cr_n, \max_{0<n\leq k}|Cr_n\text{-}128| \tag{7}$$

k represents the number of fused low dynamic images.

3. Analysis of experimental results

In this paper,we select two groups of multi exposure images. The fused results are compared with the methods proposed by Mertens[2] and Paul[3]. The experiment is verified by MATLAB R2013a, and the experimental system is windows7 64bits.

3.1Subjective evaluation. We can see the "Floor" and "Table" series of fusion images. The Mertens[2] method shows color degradation and details loss in the light and dark areas, such as "Floor" to the right window and the lamp filament of "Table". The method of Paul[3] shows the best performance of the details of the scene, but the integrated image is dark and not conducive to visual observation. The result images of our method can see more scene details than Mertens and better color expressiveness than Paul.

3.2Objective evaluation. The entropy, mean variance contrast and saturation of the two groups of high dynamic images in figure 2 and figure 3 are respectively evaluated. Entropy is calculated on the corresponding grayscale map. The larger the entropy is, the more information is contained in the image. Its calculation formula is as follows:

$$H = -\sum p(x_i)\log p(x_i) \tag{8}$$

$p(x_i)$ is the frequency of greyscale x_i.

Evaluation methods in literature[6] are used to calculate the mean variance contrast and saturation. The high dynamic range image is transferred to HSI space, and the mean value of S component represents the global saturation of the image. The mean variance contrast refers to the dynamic range of the image, which is computed in the I component:

$$C = \{\frac{1}{rc}\sum_{x=0}^{r-1}\sum_{y=0}^{c-1}[I(x,y)-I_{mean}]\}^{\frac{1}{2}} \tag{9}$$

Where I_{mean} is the mean value of the I component in a $r \times c$ sizes image. The larger the value of the three evaluation criteria, the more information, color, and details the fusion image has.

Figure 2. Comparison of Floor scenes in different methods.

(a)Mertens method (b)Paul method (c)Our method

IOP Conf. Series: Materials Science and Engineering **359** (2018) 012002 doi:10.1088/1757-899X/359/1/012002

Figure 3. Comparison of Table scenes in different methods.

(a)Mertens method (b)Paul method (c)Our method

Table 1. Fusion image evaluation indicators.

Image sequence	Mean variance contrast			Entropy			Saturation		
	Mertens	Paul	Ours	Mertens	Paul	Ours	Mertens	Paul	Ours
Floor	0.1359	0.1116	**0.1361**	6.7170	6.6135	**6.8807**	0.5129	0.5587	**0.6250**
Table	0.2095	0.1737	**0.2136**	7.5651	7.3228	**7.5950**	0.1931	0.2219	**0.2823**

From table 1, we can see that the three evaluation indicators of our algorithm are better than those of the other two methods. The mean variance contrast is slightly improved than Mertens, and the saturation is larger than Paul. It shows that the proposed method has been improved in both image details and color expressiveness compared with the existing methods.

4. Conclusion

For Mertens and Paul methods, the details of light and dark areas and image saturation can not be simultaneously optimized when expanding the dynamic range of low dynamic images. This paper deals with luminance and chromatic aberration components in YCbCr space respectively. The experiment shows that our method can achieve better visual effect while balancing the light and dark area details of low dynamic images.

References

[1] Debevec, P.E, and Malik. Recovering High Range Radiance Maps from Photographs[J]. in: ACM SIGGRAPH 1997 Proceedings 1997, pp.369-378.

[2] Mertens T, Kautz J, Reeth F V. Exposure Fusion: A Simple and Practical Alternative to High Dynamic Range Photography [J]. Computer Graphics Forum, 2009, 28(1):161-171.

[3] S. Paul, I.S. Sevenco, P. Agathoklis. Multi-Exposure and Multi-Focus Image Fusion in Gradient Domain. In: Journal of Circuits, Systems, and Comuters Vol. 25, No.10(2016) 1650123

[4] Q Xie, XZ Fang. Research on indoor video monitoring based on wide dynamic technology [J]. Information technology, 2013, 37(05):115-117+120.

[5] GZ Xu, XL Huang. Synthesis of high dynamic range images based on YUV space [J]. Journal of China media university (natural science edition), 2017, 24(03):11-13.

[6] Vanmali A V, Deshmukh S S, Gadre V M. Low complexity detail preserving multi-exposure image fusion for images with balanced exposure [C] // National Conference on Communications. New Delhi: IEEE, 2013: 1-5.

IOP Conf. Series: Materials Science and Engineering **359** (2018) 012003 doi:10.1088/1757-899X/359/1/012003

Research on Signature Verification Method Based on Discrete Fréchet Distance

J L Fang, W Wu

Shen Yang JianZhu University Liaoning, China

E-mail: 1328767406@qq.com

Abstract. This paper proposes a multi-feature signature template based on discrete Fréchet distance, which breaks through the limitation of traditional signature authentication using a single signature feature. It solves the online handwritten signature authentication signature global feature template extraction calculation workload, signature feature selection unreasonable problem. In this experiment, the false recognition rate (FAR) and false rejection rate (FRR) of the statistical signature are calculated and the average equal error rate (AEER) is calculated. The feasibility of the combined template scheme is verified by comparing the average equal error rate of the combination template and the original template.

1.INTRODUCTION

With the rapid development of electronic transactions and the increasing demand for social security and identity authentication, biometrics-based identification technology has been rapidly developing. Among them, hand-written signature authentication is one of the most promising technologies in biometric identification technology. Handwritten signatures have different uniqueness between people and constant stability over a period of time, making signatures difficult to forge and fake. At present, the Hausdorff distance as a measure of distance is widely used to determine the similarity between two point sets, while the continuous Fréchet distance [1] is used to study the continuity of two consecutive curves. Zhang Zhigang[2], Qiu Yiming[3] have done to determine the similarity of the two curves of the study; HOLM et al.[4] applied the continuous Fréchet distance to the similarity of the protein structure arrangement; C. Bargetz.J[5] proposed the definition of discrete Fréchet distance on the basis of continuous Fréchet distance, and Jiang Minghui et al.[6] applied it to determine the structure of protein, and achieved some effect, but discrete Fréchet Distance in the signature certification has not been a comprehensive application; You Qingcheng et al.[7] considered the signature of the location, pressure and grip pen tilt, the use of HMM-SVM hybrid model for signature verification. Faundez-Zanuy[8] proposed a DTW algorithm based on feature victimization for online signature verification. Huang Chengjie et al [9] proposed an improved DTW online signature authentication method, the use of special points on the signature segmentation, and then use the sub-signature for DTW certification.

Based on the discrete Fréchet distance, this paper compares the template features with the signatures to be tested, and uses a reasonable threshold to sign the authentication within the range of tolerance. Finally, a multi-feature combination template is generated to break through the single signature authentication in traditional signature authentication limitation. In this paper, the SVC2004 signature database is used to ensure the reliability of the experimental results.

2.SIGNATURE AUTHENTICATION ALGORITHM

A. Abbreviations and Acronyms. Definition 1 ①Given a polygon chain with n highest points $P =< p_1, p_2, ..., p_n >$, a k-step along P, the highest point of the division P becomes k non-intersecting nonempty subsets $\{P_i\}_{i=1,...,}$,meet $P_i =< p_{n_{i-1}+1}, L, p_{n_i} >$ and $1 = n_0 < n_1 < L < n_k = n$.

②Given two polygon chains A and B, AB combination step consists of k steps $\{A_i\}_{i=1...k}$ along A and k step $\{Bi\}_{i=1...k}$ along B, Satisfy formula $1 \le i \le k$, and there are only one high point in A_i and B_i.

③The cost of a combination of $W = \{(Ai, Bi)\}$ along the chains A and B is:

$$d_F^w(A,B) = \max_i \max_{(a,b)\in A_i \times B_i} dist(a,b) \qquad (1)$$

Where $dist()$ is the Euclidean distance between A and B, The discrete Fréchet distance between chains A and B is:

$$d_F^w(A,B) = \min_W d_F^W(a,b) \qquad (2)$$

This combination is called the Fréchet arrangement of chains A and B. Since the discrete Fréchet distance is only able to examine the distance between the peak points, it is far from enough to judge the similarity of the curve. Therefore, this paper introduces the definition of the new judgment curve similarity.

Definition 2 Let $A =< a_1,...,a_m >$ and $B = \langle b_1,...,b_n \rangle$ be two curves of discrete points, $d_F^1(A,B)$ is the discrete Fr-échet distance between their highest points, and $d_F^2(A,B)$ is the Fréchet distance between their lowest points. Suppose there is ε, if $|d_F^1(A,B) - d_F^2(A,B)| \le \varepsilon$ then called A, B similar, otherwise they are not similar.

As shown in Fig. 1, a peak point and a trough point are taken as an example in which a and m are a peak point and a trough point of curve A, respectively, b and n are a peak point and a trough point of curve B, d1 is the distance between their crests a and b, and $d2$ is the distance between their troughs m and n. If | $d1$-$d2$ | <ε is true for a given ε, then curves A and B are similar, otherwise they are not similar.

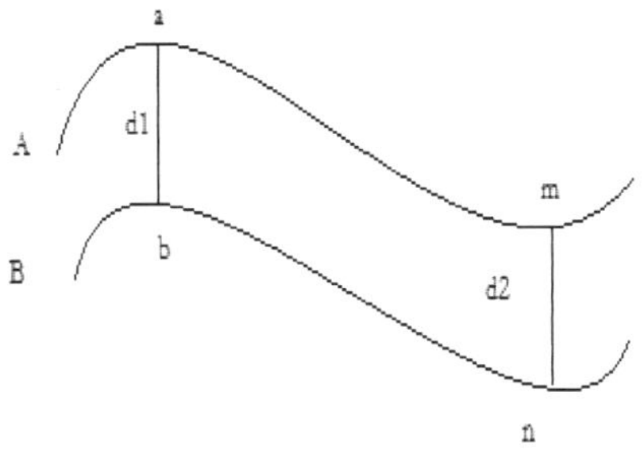

Fig.1 Two curves similar schematic diagram

The same is true for the two curves of the discrete points. When the discrete points are enough, find out all the peaks and valleys, and the number of extremes of the similar curves is not much different, and the Fréchet permutation is always the best arrangement of the arrangement, the arrangement of the calculated discrete Fréchet distance used as a judgment curve similarity can reduce the error.

B. Algorithm steps. ①*A*, *B* is a finite curve composed of two curves. First find the highest point and the lowest point of the curve, E.g. $A = \langle A_1, \ldots, A_m \rangle$, $a = \langle a_1, \ldots, a_n \rangle$, $B = \langle B_1, \ldots, B_m \rangle$ and $b = \langle b_1, \ldots, b_n \rangle$ where m ≤ n (A_1, ..., A_m and a_1, ..., a_n are *m* highest points and *n* lowest points of the chain A, B_1, ..., B_m and b_1, ..., b_n are *m* highest points and *n* lowest points of the chain *B*). If $n - m \geq 5$ (since the total number of signatures is between 200 and 300 points, the total of the high and low points is between 20 and 30, and the high and 20% Similarity, so the number of differences set here is limited to five), then they are not similar, the algorithm is over, otherwise the next step.

②Based on the chain *A* with a small peak point, the chain *B* with the peak point is divided into *m* steps using the definition of Fréchet permutation and the above constraint, and then find all the conditions to meet the division. Suppose there are *k* species ($k \in$ N), each division $W_j = \{(A_i, B_i)\}$, $1 \leq i \leq m$ and $1 \leq j \leq k$.

③In each division of Wj, first calculate the maximum distance between all the corresponding points in each step, and then find the maximum of the maximum distance in all the steps of this division. As follows:

$$d_F^W(A,B) = \max_i \max_{(a,b)\in A_i \times B_i} dist(a,b) \quad (3)$$

④The best way to find a division is to find the minimum of the distance in all the division methods, which is the discrete Fréchet distance between chains *A* and *B*.

$$d_F(A,B) = \max_W d_F^W(A,B) \quad (4)$$

⑤In this way, we can get the minimum discrete Fréchet distance $d_F^1(A,B)$ between the two peaks and the minimum discrete Fréchet distance $d_F^2(A,B)$ between the two troughs. Finally, with $d_F^1(A,B) - d_F^2(A,B)$, if the final result is less than a specific threshold ε, it is judged that the two curves are similar, otherwise they are not similar and the algorithm ends.

3. PARAMETER

A. *Characteristic curve.* Different signature features, parameter acquisition is different, Such as the signature of the horizontal, vertical, pressure signal signature characteristics can be directly through the handwriting device to collect data, and other signature features are calculated by the formula, the calculation formula as shown in Table 1.

TABLE 1 ONLINE SIGNATURE AUTHENTICATION FEATURE SET

Num	Feature	Description
1	X direction displacement	X(i)
2	Y direction displacement	Y(i)
3	Pressure	P(i)
4	XY Absolute displacement	$XY=sqrt(X(i)^2+Y(i)^2)$
5	X linear velocity	Vx(i)=(X(i+1)-X(i-1))/2
6	Y linear velocity	Vy(i)=(Y(i+1)-Y(i-1))/2
7	linear velocity	$V(i)= sqrt(Vx(i)^2+Vy(i)^2)$
8	X Acceleration	aX(i)=(Vx(i+1)-Vx(i-1))/2
9	Y Acceleration	aY(i)=(Vy(i+1)-Vy(i-1))/2
10	Centrifugal acceleration	a(i)=(Vx(i)aY-Vy(i)aX)/V(i)
11	Speed and X axis angle cosine value	cosV(i)=Vx(i)/V(i)

Taking the linear velocity as an example, the point *Vx(i)* in the X direction and the point *Vy* (*i*) in the Y direction are connected to the line by curve to obtain the linear velocity curve of *XY*. In combination with definition 1, the maximum value set *A* and the minimum value set *B* of the linear velocity curve are obtained. Calculate the wave troughs of discrete Fréchet distance $d_F^1(A,B)$ and $d_F^2(A,B)$. Figure 2 shows the true signed X-direction linear velocity curve and Y-direction linear velocity curve.

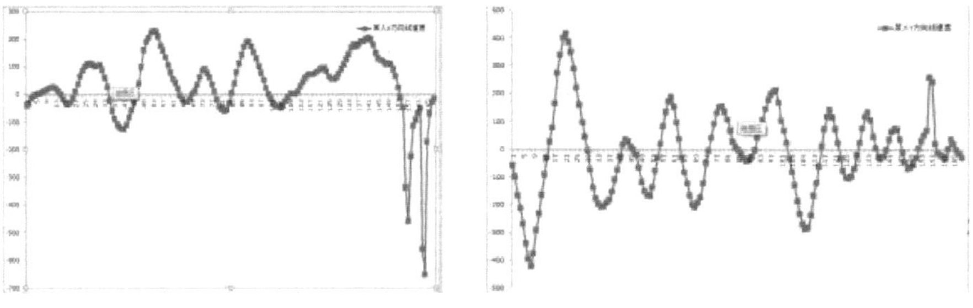

(a) X linear velocity (b) Y linear velocity

Fig.2 True signature X, Y axis direction linear velocity

IOP Conf. Series: Materials Science and Engineering **359** (2018) 012003 doi:10.1088/1757-899X/359/1/012003

B. Extreme Point Judgment Threshold Analysis

Use the data in the SVC2004 database Task1 user1, user2, user3, user4, user5 five true signatures, and skilled forged signature test, through experiments to determine their threshold. In the signature sample library to take each person's real signature sample 10, skilled forged signature samples to match the 10, so as to observe the changes in parameters on the rate of error and the impact of the rate of rejection.

The initial matching threshold of the extreme points of the coordinate curve is analyzed by taking the X axis displacement signed by user1 as an example. By experimenting with the signatures of real signatures and others forged in the signature sample library, we define the discrete Ferchet distances $d_F^1(A,B)$ and $d_F^2(A,B)$ of true and false signatures in conjunction with definition 1. Then, in conjunction with definition 2, the value of $d_F^1(A,B)$ - $d_F^2(A,B)$ is obtained and all discrete Ferchet distance differences are obtained as shown in Table 2.

TABLE 2 TRUE AND FALSE SIGNATURE DISCRETE FRÉCHET DISTANCE

User1	1	2	3	…	8	9	10
True	980	84	138	…	302	454	480
False	1306	780	975	…	1959	663	1006

The matching result is observed by setting the gradient value of ε to 250, 370, 490, 610, 730, and 850. So as to get the best threshold X displacement ε = 490, the matching results shown in Figure 3:

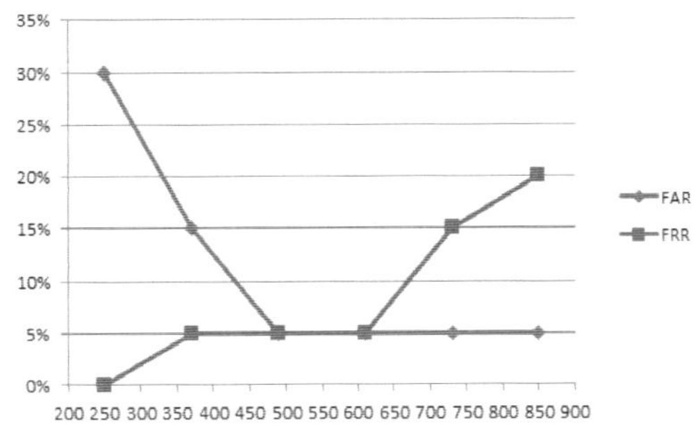

Fig.3 FAR and FRR curves with different values of ε

It can be quickly seen from Figure 3 the optimum threshold values ε of displacement X is 490, the calculated threshold value ε Similarly this experiment, the other characteristic parameters 10, as shown in Table3 (in mm) :

TABLE 3 OTHER PARAMETER THRESHOLDS E

Feature	*Y*(*i*)	*P*(*i*)	*XY*(*i*)	*Vx*(*i*)	*Vy*(*i*)
ε	200	600	800	300	300
Feature	*V*(*i*)	*aX*(*i*)	*aY*(*i*)	*a*(*i*)	*cosV*(*i*)
ε	400	200	110	520	1.3

4. SIGNATURE TEMPLATE SELECTION

A. Orthogonal test design. In order to reduce the number of experiments, this paper introduces the experimental scheme of orthogonal experiment design. The essence of orthogonal design is to use the method of orthogonal experiment to arrange the experiment and analyze the data. The survey index of this experiment is the average error rate (AEER) of the signature authentication. There are 11 factors, and the number of levels is 2, which means that the signature data entered by the user is selected twice. From this analysis can be found, this test to meet the orthogonal table L12(112)of the design, the orthogonal table as shown in Table 4, with the number 1 and 2 represent the number of horizontal, A1-A11 on behalf of the test factors, 1-12 On behalf of the experimental serial number.

TABLE 4 SIGNATURE VERIFICATION ORTHOGRAPHIC FORM

	A1	A2	A3	…	A9	A10	A11
1	1	1	1	…	1	1	1
2	1	1	1	…	2	2	2
3	1	1	2	…	2	2	2
4	1	2	1	…	1	1	2
5	1	2	2	…	1	2	1
6	1	2	2	…	2	1	1
7	2	1	2	…	1	2	1
8	2	1	2	…	1	1	2
9	2	1	1	…	2	1	1
10	2	2	2	…	2	1	2
11	2	2	1	…	1	2	2
12	2	2	1	…	2	2	1

As can be seen from Table 4, the orthogonal table gives a combination of the number of experiments and the number of factors, this feature coincides with the signature designation of the signature template and signature number of the combination of design. Therefore, we can use the orthogonal design table design signature template.

B. Signature template generation. According to the orthogonal Table 4, 11 signature parameters are selected according to the order of factors: $X(i)$、$Y(i)$、$P(i)$、$XY(i)$、$Vx(i)$、$Vy(i)$、$V(i)$、$aX(i)$、$aY(i)$、$a(i)$、$cosV(i)$, corresponding to $A1$-$A11$ in Table 4. Select the user two signature information as the experimental level, with the number 1, 2, which can be designed to generate the signature experiment of the 12 templates (m_1, m_2 ... m_{12} corresponding to Table 4, 1-12) program combination table. Where number 1 indicates that all the signature parameters of the user's first signature are selected in the database and number 2 is the characteristic parameter of the second signature of the user in the selected database.

C. *Experimental data and analysis.* For example, take user1's third signature as a template, written with u3.Select 10 real signatures and 10 false signatures to test u3, and the whole experiment is divided into the following.

The first step, by definition 3 for template and 20 signatures of 11 sets of stable characteristic curve of the minimum set $d_F^1(A,B)$ and maximum set $d_F^2(A,B)$, calculating the discrete Fréchet distance $d_F^1(A,B)$ - $d_F^2(A,B)$, finally get the difference of Fréchet distance set parameters.

The second step, according to the threshold epsilon of each characteristic value in Table 3, statistics FAR and FRR, and obtains the values of FAR and FRR of the u3 template.

The third step, according to the second step results, can be summed up u3 template average FAR value of 7.72% and the average FRR value of 8.18%. (Average FAR and FRR value is all the characteristics of FAR and FRR after the sum of the value).

And so on, you can calculate the average of the remaining twelve FAR templates and average FRR value. Through the above calculation, the average FAR and FRR all template values, the end of the experiment, the original signature template will mean FAR and FRR 12 template obtained respectively and untreated (user with a signature machine the average FAR and FRR) for comparison, ultimately determine the feasibility of combination of the template. In this experiment, orthogonal design experiment using the user first and second signature data, therefore also need to calculate the second signatures as the average FAR and FRR template (U2 table). Through the calculation of second times the average signature as FAR template the value is 12.27%, the average FRR value is 11.82%.

The average error rate (AEER) after averaging FAR and FRR is shown in Table 5 below:

TABLE 5 THE TEMPLATE OF THE AEER

template	m1	m2	m3	m4	m5
AEER	6.82%	7.5%	10.23%	8.41%	9.55%
template	m6	m7	m8	m9	m10
AEER	11.36%	12.5%	7.95%	9.55%	11.14%
template	m11	m12	u2	u3	
AEER	10%	8.18%	12.05%	7.95%	

5.Conclusion.

As it can be seen from table 5, the initial template and with second and third overall signatures as signature authentication templates, the average error rate is 12.05% and 7.95%, while the highest in the composite template is 6.82%.According to table 4, you can quickly get the combination template generation method : take the first signature *Y*(i) *XY*(*i*), *aX*(*i*), *aY*(*i*), *a*(*i*)and second signature of the *X*(*i*), *P*(*i*), *Vx*(*i*), *Vy*(*i*), *V*(*i*).Therefore, combination of *m*1 template is the user1 template signature.

And so on, it can be analyzed that the signature authentication template combination of 4 other users. Through the experiment, the signature of the user2 template using the third combinations in table 4, the average error rate of template AEER reached 7.95%, while the original template average error rate of AEER is 9.55%;The signature template of user3 uses eleventh combinations in Table 4, which makes the average error rate of AEER equal to 9.55%, while the average error rate of the original template AEER is 10%;the signature template of user4 uses eighth combinations in Table 4, which makes the average error rate of

AEER equal to 11.14%, while the average error rate of the original template AEER is 12.05%;the signature template of user5 uses fourth combinations in Table 4, which makes the average error rate of AEER equal to 10%, while the average error rate of the original template AEER is 11.14%.

6.SUMMARY

This paper mainly analyzes the characteristics of the Fréchet discrete distance of curves, combined with online signature verification, and presented a signature template combination of multiple features the signature, and gives the analysis of the experimental data and the corresponding experimental data. The acquisitions of signature characteristic curve, signature feature data calculation, extraction of curve extreme value are introduced. The progress of signature verification in the current environment is also pointed out. In the SVC2004 data set, the experimental part of this paper will be combined with multi signature template features traditional template to compare the results found using the traditional signature information the average error rate of AEER compared with the signature composition template average error rate of AEER is higher, indicating signature template combination is proposed to optimize the feasible.

REFERENCES

[1] Efrat A and Fan Q 2007 Curve Matching, Time Warping, and Light Fields: *New Algorithms for Computing Similari-ty between Curves.* J Math Imaging, 27:203-216.

[2] Zhang Z G, Zhou M Q and Geng G H 2008 Study on similarity of contours of skull contours. *Journal of Chinese Computer Systems*, (07): 1304-1307.

[3] Qiu Y M, Hu H C, Zhang J B, Chen Q H 2014 *An online signature verification method based on curve similarity*, (05):1016- 1020.

[4] Holm L and Sander C 1996 *Mapping the protein universe Science*, 273(5275): 595 — 602.

[5] Bargetz C, Kąkol J and Kubiś W 2016 A separable Fréchet space of almost universal disposition. *Journal of Functional Analysis*.

[6] Jiang Mi H, Xu Y and Zhu B H 2008 Protein Structure- Structure Alignment with *Discrete Fréchet Distance*[DB/OL]. 01-20.http:// www.cs.mon-tana.edu/bhz/2007apbc.pdf.

[7] Zhu Y C, Zhou J, Deng C G and Du H W 2010 Improved HMM-SVM hybrid model handwritten Chinese signature verification method. *Journal of Chinese Computer Systems*, (09):1869-1872.

[8] Faundez-Z M 2007 On-Line Signature Recognition Based on VQ-DTW. *Pattern Recognition*, 40(3): 981-992.

[9] Huang C J, Shen F and Wu Z C 2009 Online signature verification method based on dynamic time planning . *Computer Systems & Applications*, (02):49-52.

IOP Conf. Series: Materials Science and Engineering **359** (2018) 012004 doi:10.1088/1757-899X/359/1/012004

Research on Performance of Wire-controlled Hydraulic Steering System Based on Four-wheel Steering

P Tao[1] and X H Jin[1]
[1] Wuhan University of Science and Technology, Wuhan, Hubei 430081, PR China
E-mail:taopin_88@163.com

Abstract. In this paper, the steering stability and control strategy of forklift are put forward. Drive based on yawing moment distribution of rotary torque coordination control method, through analyzing the linear two degree of freedom model of forklift truck, forklift yawing angular velocity and mass center side-slip Angle of expectations, as the control target parameters system, using fuzzy controller output driving forklift steering the yawing moment, to drive rotary torque distribution, make the forklift truck to drive horizontal pendulum angular velocity and side-slip Angle tracking reference model very well. In this paper, the lateral stability control system were designed, the joint simulation in MATLAB/Simulink, the simulation results show that under the different partial load, the control system can effectively to control side forklift lateral stability, enhanced the forklift driving safety, for the side forklift steering stability study provides a theoretical basis.

1.Introduction

The steer-by-wire system is a new way to turn over the limitations of traditional steering systems due to mechanical connections[1].At present, the engineering vehicle mainly adopts two-wheel steering （Two-wheel steering, 2WS）, and the ordinary two-wheel steering vehicle has large turning radius, and has the disadvantages of slow steering response and poor stability of high-speed steering[2].The four-wheel steering[2]（Four-wheel steering, 4WS）can independently control the movement trajectory and attitude of the vehicle. It has a small turning radius, effectively reduces the lag of the steering force and obtains better steering performance. Therefore, it is necessary to combine the steer-by-wire control technology with the four wheel steering technology to improve the driving performance of the engineering vehicles. Liang Chunming and others[3] designed the four-wheel synchronous steering system for the universal forklift, and added four-wheel steering device on the ordinary two-wheel steering system, so that the forklift could travel in any direction. Mu Xihui[4] proposed a hydraulic four-wheel steering system with double steering gear and double steering cylinder interlocking by controlling electromagnetic commutation to achieve two-wheel independent steering and four-wheel combined steering. Bekheira Tabbache[5] studies the stability of the four wheel independent driving vehicle under the extreme conditions such as braking and cornering, and concludes that the four-wheel drive system is superior to the two-wheel drive system through comparative analysis.

With the side forklift turning frequently in the narrow space, this paper adopts some control strategy, established a wire-controlled hydraulic steering system based on four-wheel steering mode; Through modeling and simulation analysis, the dynamic characteristics of the system are obtained to provide theoretical guidance for the promotion of the wire-controlled hydraulic steering system in the vehicle engineering.

2. System working principle

As shown in Figure 1, the control block diagram of wire-controlled hydraulic steering system based on four-wheel steering mode is given. It works by transmitting the steering signal of steering wheel(steering wheel angle) and speed to the electronic control unit(ECU), and the electronic control unit performs some calculation according to the corresponding control strategy and control target to obtain the desired angle of each wheel. Then, the electronic control unit converts these desired angles into voltage signals and transmits them to the electro-hydraulic steering systems of each wheel. The voltage signal drives the electro-hydraulic steering system of each wheel to control the steering of each wheel, thus obtaining the actual steering angle of each wheel(In this paper, δ is the wheel angle, subscript *if*、*of*、*ir*、*or* represent inner front wheel, outer front wheel, inner rear wheel and outer rear wheel respectively).

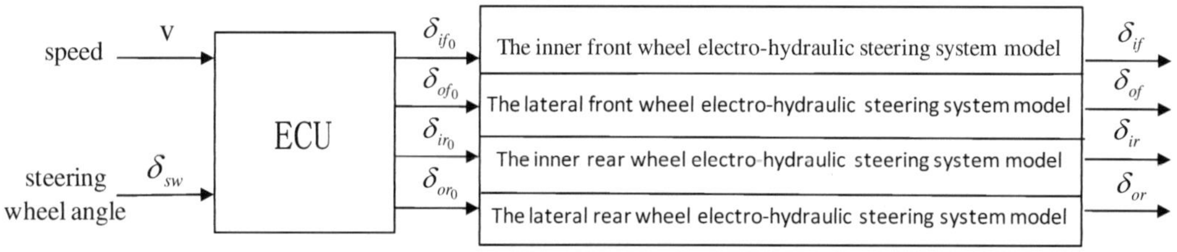

Figure 1. the control block diagram of wire-controlled hydraulic

The electro-hydraulic steering system of each wheel, which is in exactly the same form, relatively independent, is closed loop proportional position control system. The independent control of each wheel steering is ensured by parallel connection, and the coordinated control of each wheel movement is carried out by ECU.

3. System control strategy and mathematical model

*3.1 Control strategy. I*n order to improve the flexibility of engineering vehicle steering and the driving stability, a certain control strategy is needed; Since the main object of our research is the side forklift, it needs to change frequently in the narrow space, and the steering response speed is higher than that of ordinary vehicles, so as to reduce the turning radius.

Through the analysis to determine the control objectives[6] are as follows: to reduce the centroid of the sideslip angle; Low-speed turning radius is small , and the high-speed operation stability is good.; When the tire reaches its attachment limit, the response frequency is still within the operative limit.

In the four-wheel steering analysis, the vehicle is generally simplified into two-wheel model[6], that is, the two-degree-of-freedom linear model of vehicle[6].

$$\begin{bmatrix} \beta' \\ r' \end{bmatrix} = \begin{bmatrix} -\dfrac{C_f + C_r}{mv} & -(1 + \dfrac{l_f C_f - l_r C_r}{mv^2}) \\ -\dfrac{l_f C_f - l_r C_r}{I} & -\dfrac{l_f^2 C_f + l_r^2 C_r}{Iv} \end{bmatrix} \begin{bmatrix} \beta \\ r \end{bmatrix} + \begin{bmatrix} \dfrac{C_f}{mv} & \dfrac{C_r}{mv} \\ \dfrac{l_f C_f}{I} & -\dfrac{l_r C_r}{I} \end{bmatrix} \begin{bmatrix} \delta_f \\ \delta_r \end{bmatrix} \tag{1}$$

In the formula, m is the whole vehicle mass,*Kg*; *I* is the yaw moment of inertia around the center of mass, $Kg \cdot m^2$; v is the speed of the vehicle, m/s; β is the sideslip angle of center of mass, rad; r is the yaw velocity, rad/s; δ_f and δ_r are the virtual front and rear wheel angles respectively, rad; l_f and l_r are the distance from the vehicle centroid to the front and rear axis respectively, *m;* C_f and C_r are virtual front and rear wheel side stiffness(Take positive value), *N*/rad;

The meaning of each parameter in the formula is shown in [6];

According to the above control objectives, make the sideslip angle β=0, and in the steady state β′= 0 and r'= 0 ,plug that into the equation (1) ,and then simplify it:

$$\frac{\delta_r}{\delta_f} = \frac{-l_r + \dfrac{ml_f}{C_r l}v^2}{l_f + \dfrac{ml_r}{C_f l}v^2} = K_r \tag{2}$$

From the formula (2), the ratio of the virtual rear wheel angle to the virtual front wheel angle K_r is the function of the speed v, and K_r determines whether the four-wheel steering is the same phase or reverse phase; When the speed v is low, Kr < 0 indicates that the front and rear wheels turn in the opposite direction, reducing the turning radius and improving operation flexibility; When the speed v is high, Kr >0 indicates that the front and rear wheels turn in the same direction ,and it can improve stability and anti-lateral interference ability.

3.2 Relation model between wheel angle and electronic steering wheel angle. Through the dynamic analysis and calculation of the vehicle[6], the relationship between the four wheel angles and the virtual front and rear wheel angles δ_f and δ_r can be obtained:

$$\delta_{if} = \frac{1}{1-(1-K_r)w/(2l)\delta_f}\delta_f \qquad \delta_{ir} = \frac{1}{1-(1-K_r)/K_r \cdot w/(2l)\delta_r}\delta_r$$

$$\delta_{of} = \frac{1}{1+(1-K_r)w/(2l)\delta_f}\delta_f \qquad \delta_{or} = \frac{1}{1+(1-K_r)/K_r \cdot w/(2l)\delta_r}\delta_r \tag{3}$$

The virtual front wheel angle and steering wheel angle meet the following relationship[7]:

$$\delta_f = K_f^{'}\delta_{sw} \tag{4}$$

In the formula, K'_f is the ratio of the steering wheel to the virtual front wheel; δ_{sw} is the electronic steering wheel angle, °;

According to the previous formula (2), (3) and (4), in the four-wheel steering mode, the steering angle (desired steering angle) of each wheel is related to the electronic steering wheel angle, as shown in Figure 2:

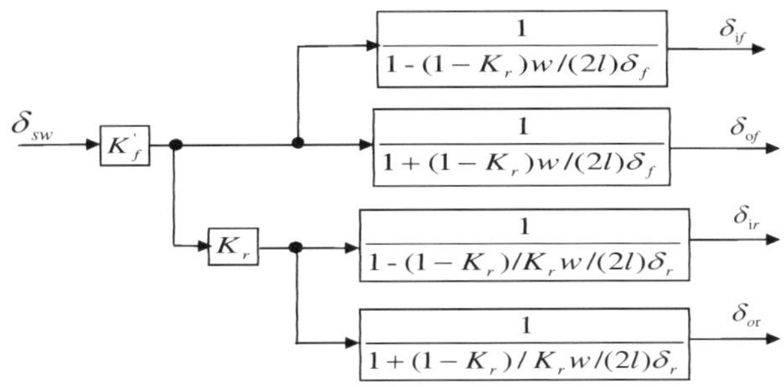

Figure 2. the relationship between desired wheel angle and electronic

3.3 The mathematical model of electro-hydraulic proportional steering system. The control block diagram of electro-hydraulic proportional steering system is shown in Figure 3, and the system consists of comparison element, controller, proportional directional valve, hydraulic cylinder, wheel and corner sensor etc. The angle sensor feedback the actual deflection angle of the current wheel, and the system compares and calculates the angle deviation between actual deflection angle and desired deflection angle of wheel. The system translates the angle deviation into the electrical signal and transmits it to the proportional amplifier to obtain the voltage signal after power amplification so as to achieve closed-loop control.

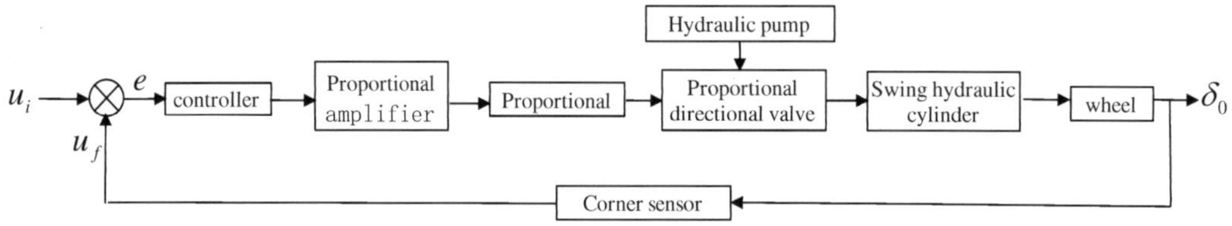

Figure 3. control block diagram of the system

Through the analysis of the working principle and characteristics concerned of the system, the transfer function of the main modules of the system is obtained. Finally, the block diagram of the whole system's transfer function model is given, as shown in Figure 4.

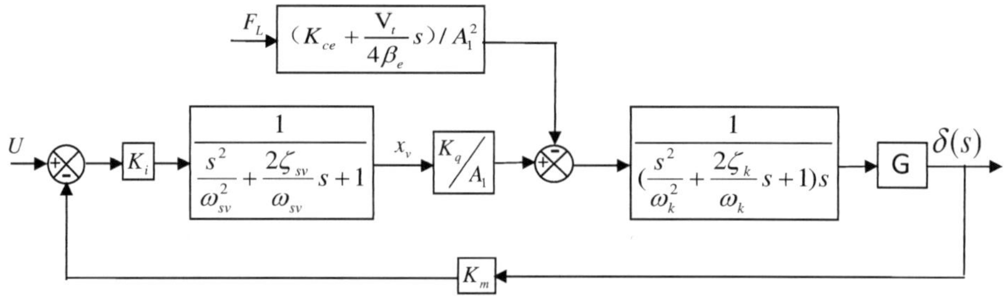

Figure 4. the block diagram of the whole system's transfer function model

4. Joint simulation of steering system

The simulation model of each wheel corner control in ECU is established by the relation between the wheel angle and the electronic steering wheel angle, as shown in Figure 5.

Based on the principle of wire-controlled hydraulic steering system, simulation mode of single-wheel electro-hydraulic proportional steering system was established by AMESim simulation software[8]. After the figure 5, PID controller and comparison element are encapsulated, the model of ECU can be obtained. Combined with the simulation model of single-wheel electro-hydraulic proportional steering system, the simulation model of four-wheel independent electro-hydraulic steering system is set up, as shown in Figure 6.

5. Simulation result analysis

In order to fully reflect the performance of the steering system and the performance advantages of the four-wheel steering system, we have simulated ordinary two-wheel steering and four-wheel steering of forklift truck these two steering modes, and the simulation results of the two are compared and analyzed.

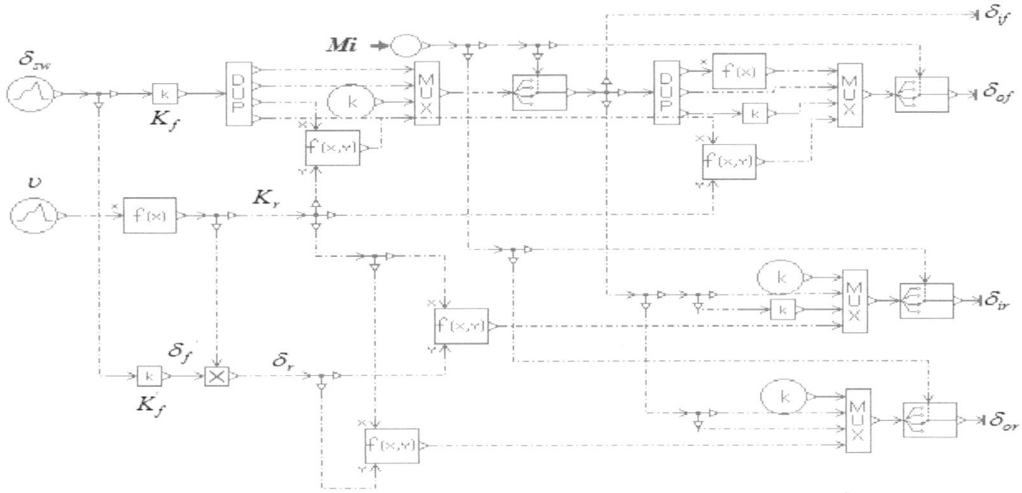

Figure 5. simulation model of each wheel angle control

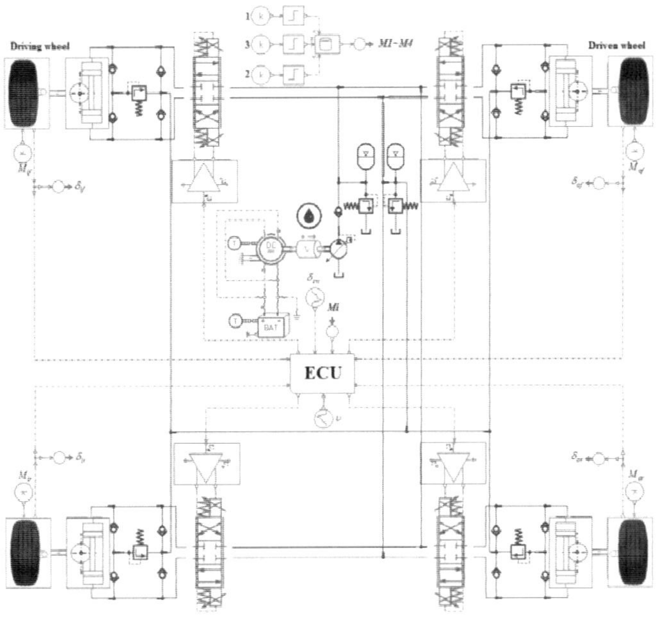

Figure 6. simulation model of the steering system

Select the angle input of the steering wheel as a symmetrical trapezoid, as shown in Figure 7. Since the steering wheel angle directly controls the virtual front wheel when the four-wheel is turned, the left front wheel angle of the two steering modes is unequal in the case of the same steering wheel angle input. For a better analysis and comparison of the two steering performance, we set steady state steering wheel angle of 410 degrees at the four-wheel steering and 360 degrees at the ordinary two-wheel steering to ensure that the left front wheel angle of the two steering modes is equal.

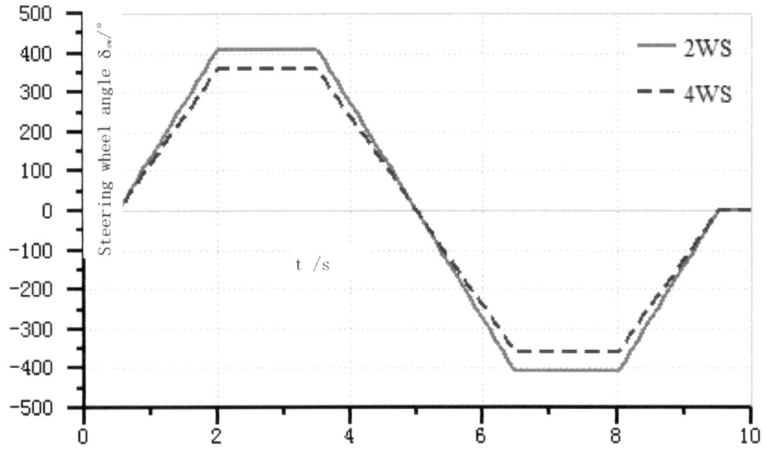

Figure 7. steering wheel angle input

5.1 Transient response and tracking error of wheel angle. The steering system under four-wheel steering mode with the above steering wheel angle input is simulated, and the transient response and tracking error of wheel angle are obtained, as shown in Figure 8.In the figure the solid line represents the theoretical value of each wheel angle, and the dotted line represents the actual value of each wheel angle.

From the simulation results, it can be seen that each wheel angles can better track the theoretical values, and the angle tracking error of the steering wheel constant angle is no more than 0.5 degrees, and the steering accuracy is higher. However, when the steering wheel angle input is zero, there is still an error of about 1 degrees, i.e., the wheel failed to return to zero. This is due to the electromagnetic hysteresis and mechanical hysteresis, and it is necessary to eliminate the repetition error by setting up the central locking mechanism to ensure that the vehicle returns to the straight driving state.

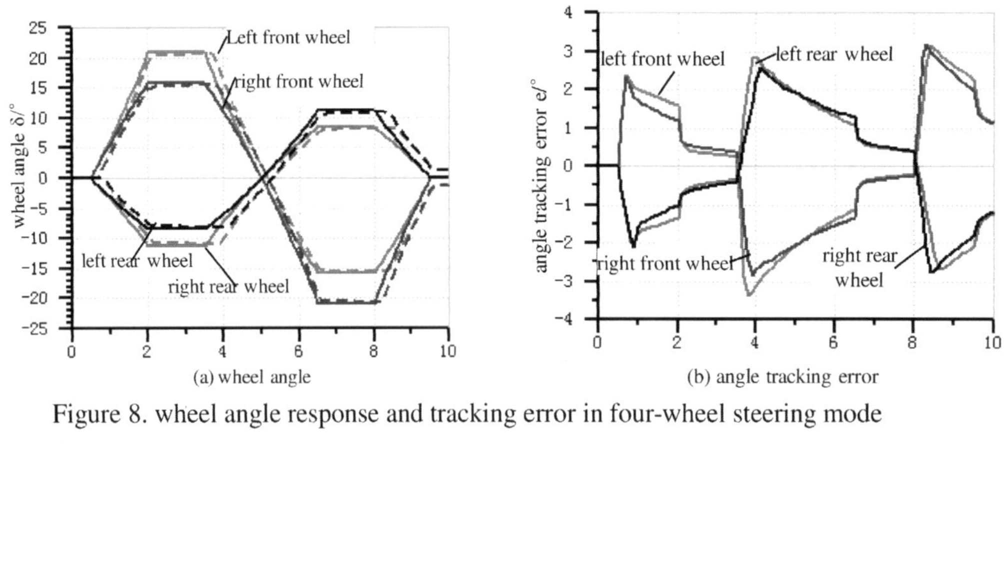

Figure 8. wheel angle response and tracking error in four-wheel steering mode

5.2 The transient response of the sideslip angle of the center of mass and the yaw velocity. Figure 9 shows the transient response of the sideslip angle and yaw velocity of the ordinary two-wheel steering mode and the four wheel steering modes under low speed respectively. It can be seen from Figure 9 (a) that, at low speed, the inverse phase four-wheel steering can be used to obtain a smaller sideslip angle of the center of mass, and the vehicle's stability is better. The forklift is low-speed vehicle, which needs frequent steering in narrow space, and its requirements for steering response speed are higher than that of ordinary vehicles. Therefore, the yaw velocity feedback is introduced[9]. From Figure 9, the yaw velocity of the four-wheel steering is greater than that of the ordinary two-wheel steering, and the sensitivity is high, and the overall rotation amplitude of the vehicle is larger, and a smaller turning radius can be obtained.

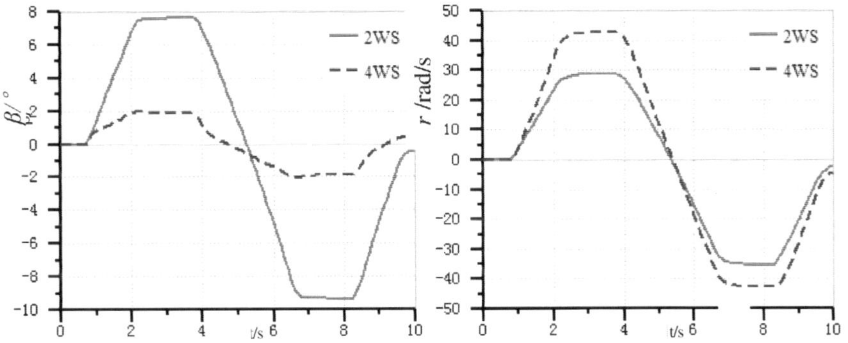

(a) the sideslip angle (b) the yaw velocity

Figure 9. the transient response of the sideslip angle of the center of mass and the yaw velocity

5.3 Stability and response speed. Figure 10 shows the stability and response velocity analysis of the steering system in the four-wheel steering mode. Figure 10 (a) show the system center of mass side-slip angle frequency response. Figure 10 (b) is the frequency response of the yaw velocity. The more the amplitude frequency curve is, the smaller the steering distortion is; and the wider the frequency band is, the better the steering stability is. The limit of the driver's steering wheel is 2Hz, and the amplitude frequency is effective at 0 ~ 2Hz. It can be seen from the figure that the bandwidth is about 1Hz and the stability is better. The smaller the phase lag in phase frequency characteristics, the faster the response speed of the vehicle steering. It can be seen from the figure that, before 0.1 Hz, there is basically no lag, and the steering system has a better following and faster response.

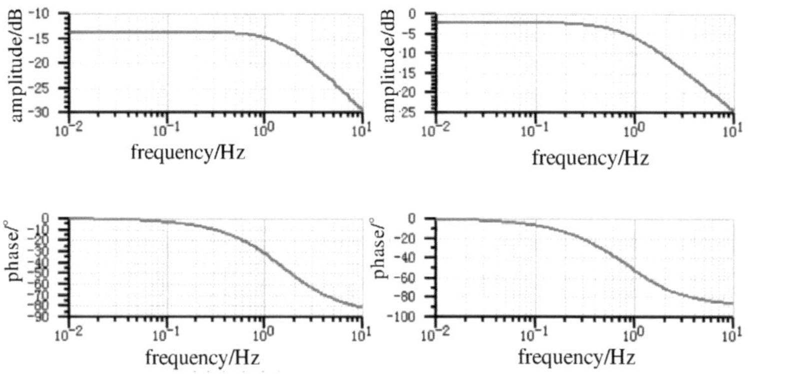

(a) the frequency response of sideslip angle (b) The frequency response of the yaw velocity

Figure 10. Steady-state characteristics of the sideslip angle of the center of mass and yaw velocity

6. Conclusion

Aiming at the working characteristics of the side forklift, wire-controlled hydraulic steering system based on four-wheel steering mode has better steering performance , and the simulation results show that: The steering system has high steering accuracy, fast response time, low system pressure and flow fluctuation, better static and dynamic performance, and meets the engineering practical requirements of side forklift truck.

Reference

[1] Zhu Z C 2012 Research Summary and Technical Summary for Steer-by-wire System. *Light Vehicle Technology*, (6):7-12

[2] Hu J Q 2013 Structure and Control Research for Steer-by-wire System. *Beijing automotive*, (5):11-15

[3] Zhang Q L 2015 Research on Fuzzy Control Strategy for Steer-by-wire System of Forklift. Control and Instruments in *Chemical Industry*, (8):134-136

[4] Mu X H 2010 Omni-directional Explosion-proof Forklift Side Key Technology Research. *Papers of Peak BBS on Logistics Engineering*,2010.

[5] Bekheira T, Abdelaziz K, Mohamed E and Hachemi B 2010 An adaptive electric differential for electric vehicles motion stabilization. *IEEE Transactions on Vehicular Technology*, 1(60):104-110.

[6] Zhang X Z 2006 Basis and Application of Vehicle Control Theory. Beijing: *Chemical Industry Press*.

[7] Chen G Y 2012 Research on the Design and Integration Control Strategy of Four-wheel Independent Electric Vehicle Testing Platform. *Changchun: University of Automotive Engineering, Jilin University*.

[8] Tian S J and Hu Q Y 2012 Dynamic Characteristics of Hydraulic System Digital Simulation. *Da lian: Dalian University of Technology Press*.

[9] Li X S 2013 Research on the Performance of the Electric Power Steering System for Electric Forklift. *Heilongjiang: Harbin Institute of Technology*.

IOP Conf. Series: Materials Science and Engineering **359** (2018) 012005 doi:10.1088/1757-899X/359/1/012005

A scheme of hidden-structure attribute-based encryption with multiple authorities

J Ling[1] and A X Weng[1*]

[1]Faculty of Computer, Guangdong University of Technology, Guangzhou 510006, PR. China

E-mail:wenganxiang@163.com

Abstract. In the most of the CP-ABE schemes with hidden access structure, both all the user attributes and the key generation are managed by only one authority. The key generation efficiency will decrease as the number of user increases, and the data will encounter security issues as the only authority is attacked. We proposed a scheme of hidden-structure attribute-based encryption with multiple authorities, which introduces multiple semi-trusted attribute authorities, avoiding the threat even though one or more authorities are attacked. We also realized user revocation by managing a revocation list. Based on DBDH assumption, we proved that our scheme is of IND-CMA security. The analysis shows that our scheme improves the key generation efficiency.

1.Introduction

The cipher policy attribute-based encryption (CP-ABE)[1] has been used more and more widely since it was first proposed because of its flexible access control policy. Under CP-ABE, the encryptor uses public key to encrypt the message, and the decryptor gets decryption key from a trustable authority according to his attribute set. If the attribute set of decryptor is dissatisfied to the access structure embedded into the ciphertext, the decryptor cannot decrypt the ciphertext.
However, there are two main problems existing in the traditional CP-ABE schemes: (1) There is only one authority in most of CP-ABE schemes. The key generation efficiency will reduce as the number of attribute grows, and this will reduce the efficiency of the whole system. Besides, once the authority is attacked, the user's key will be compromised, leading to the threat of user's privacy and data security. (2) Both access structure and ciphertext are sent to the user in most of CP-ABE schemes, and the access structure probably includes information about the message. If the ciphertext is intercepted, the privacy of encryptor will be under the risk of being compromised.
To solved the first defect described above, Chase[2] first proposed multi-authority attribute-based encryption(MA-ABE). Gao et al.[3] introduced privacy homomorphism and proposed a scheme of MA-ABE for protecting user privacy. Li et al.[4, 5] proposed an adaptive secure ABE scheme with multiple authorities, and the attributes could be repeated. An ABE scheme was proposed by Wu[6], and it introduced proxy re-encryption to increase encryption efficiency and realized policy update. Neethu et al.[7] proposed a scheme, which deletes the central authority, and all the authorities generate secret key independently with each other. Besides, it enables users to decrypt more than one file by key aggregation concept.
To guarantee the confidentiality of access policy, Song et al.[8] used composite order bilinear groups to realize the policy-hidden scheme based on access tree. Based on policy-hidden ABE, Du et al.[9] and Lei et al.[10] proposed an access control scheme which is applicable in the cloud. Wang et al.[11] and Weng et al.[12] proposed the ABE schemes with hidden policy, which uses access tree to express access structure and realized anonymous access structure. Xie et al.[13] used asymmetric pairings to realize hidden access policy, but this scheme increases the computational overhead of bilinear pairings. Ying et al.[14] propose an ABE scheme, which supports dynamic policy update, but it the policy is only partially hidden.

The schemes mentioned above either adopted multiple authorities or realized hidden policy, but there exists rare scheme which combines these two features. Zhong et al.[15] realized hidden policy with multiple authorities as well as user revocation by the linear secret sharing scheme(LSSS), but neither encryption nor decryption are not efficient. Fan et al.[16] adopted key fragmenting to realize multi-authority attribute-based encryption with policy hidden, but the data owner needs to keep online. Aiming at the problems mentioned above, based on the scheme in [5], we propose a scheme of hidden-structure attribute-based encryption with multiple authorities, where the access structure is transform to access tree, the user private key is generated by multiple authorities, increasing the key generation efficiency, and avoiding the security issues brought by the attacked authorities. We realized the hidden access structure and the user privacy could not be leaked by the attacked authorities with the condition that the authorities are not completely trustworthy. Besides, we allocated a global unique identifier, using which to generate a user revocation list which is used to update private key for legal users.

2.Prelininaries

2.1Bilinear map. Let G_1 and G_2 be two multiplicative cyclic groups with prime order p, and g be a generator of G_1.A map $e:G_1 \times G_1 \rightarrow G_2$ is a bilinear map if the following properties can be satisfied:
(1) Bilinearity: $\forall a,b \in Z_p$, $\forall f,h \in G_1$, we have $e(f^a,h^b) = e(f,h)^{ab}$.
(2) Non-degeneracy: $\exists f \in G_1$, $e(f,f) \neq 1$.
(3) Computability: For all $f,h \in G_1$, there exists an valid algorithm to compute $e(f,h)$.

2.2Decisional bilinear Diffie-Hellman assumption(DHDH assumption). Choose some random values $a,b,c,z \in Z_p$. The DBDH assumption says that, there is no probabilistic polynomial time algorithm that can differentiate the tuples $[g^a,g^b,g^c,e(g,g)^{abc}]$ and $[g^a,g^b,g^c,e(g,g)^z]$.

2.3Access Structure. We assume that there are n kinds of attributes in the system. Let $U=\{U_1,U_2,\ldots,U_n\}$ be the set of system attributes and $U_i=\{u_{i,1},u_{i,2},\ldots,u_{i,m}\}$(m≥1), $S=\{S_1,S_2,\ldots,S_n\}$ be the set of user attributes, where $S_i=\{att_{i,1},att_{i,2},\ldots,att_{i,m}\}$. $\mathbf{A}=\{A_1,A_2,\ldots,A_n\}$ indicates the access structure and $A_i \subseteq U_i$. For $\forall i \in [1,n]$, $S_i \subseteq A_i$, we say that the user attributes satisfies the access structure.

In this paper, the access structure adopts the 'AND-OR' gates with multi-valued attributes. The access structure is transformed into an access tree before encrypting. The leaf node in the tree represents the attribute value, and the non-leaf nod represents 'AND' gate or 'OR' gate. Each attribute name of leaf node is replaced by a value chosen by data owner when encrypting. A user will not obtain the information about the attributes and data owner. As is shown in the fig 1, '∧'indicates the 'AND' gate, and '∨' indicates the 'OR' gate. Assuming that the access structure $\mathbf{A}=[\{v_{1,2}\},\{v_{2,1},v_{2,4}\},\{v_{3,1}\},\{v_{4,2},v_{4,3},v_{4,4}\}]$ indicates the access structure, we say that the attribute set $S=\{v_{1,2},v_{2,1},v_{3,1}, v_{4,2}\}$ or $S=\{v_{1,2},v_{2,4}, v_{3,1}, v_{4,3}\}$ satisfies the access structure.

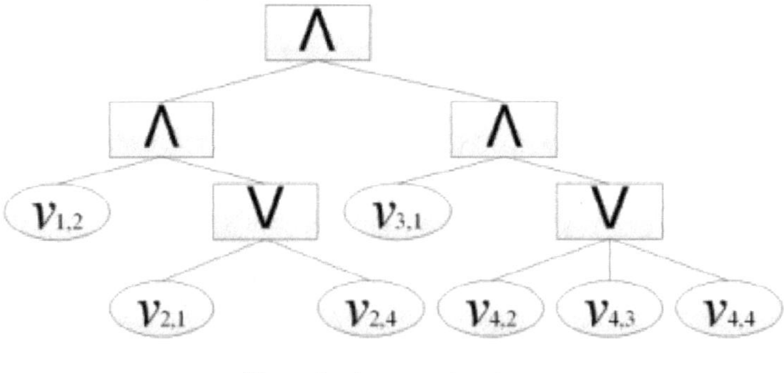

Figure1. Access structure

2.4System composition. In this paper, there are five participants: cloud service provider(CSP), central authority(CA), attribute authority(AA), data owner(DO) and user(UID).

CSP: Store encrypted data for DO, and it is not trusted.
CA: Generate public parameters and confirm the legality of UID.
AA: Each AA generates a part of private key for UID. In this paper, an AA can manage multiple attributes and each attribute is managed by only one AA.
DO: Decide access structure and encrypt data, and upload the encrypted data to the cloud.
UID: Download data from encrypted data. UID can decrypt the ciphertext if and only if its attribute set satisfies the access structure.

2.5Security model. The scheme proposed in this paper satisfies indistinguishability of ciphertext under chosen-message(IND-CMA), the security model based on which is described as an interactive game between adversary A and challenger B as follows:

Step 1: The adversary A submits the access structure W to challenge to the challenger B. After selecting a security parameter k, B runs initialization algorithm and get the system public key PK, the system main key MK, attribute authority public key (APK) and private key of attribute authority(ASK). B preserves MK and ASK, and sends PK and APK to A.
Step 2: A queries B about attribute list S, but B cannot run the generation algorithm of private key and cannot send the private key SK to A, unless S does not satisfy the access structure W. A can query for polynomial times.
Step 3: A submits two groups of plaintext M_0 and M_1, which are different but have the same length. B chooses $\varphi(\varphi \in \{0,1\})$ randomly by tossing a coin, then runs the encryption algorithm to encrypt M_φ, and returns CT(the corresponding ciphertext of M_φ) to A.
Step 4: Repeat step 2 and step 3, but the query from A must be limited, that is, S does not satisfy W.
Step 5: A outputs a guess φ' of $\varphi(\varphi' \in \{0,1\})$. If $\varphi'=\varphi$, A wins the game, otherwise loses the game.

Suppose the advantage that A wins the game is $Ad_A = \Pr[\varphi = \varphi'] - \frac{1}{2}$.

Within polynomial time, if there is not a polynomial-time adversary who can win the game with an unnegligible advantage, we can say that the scheme of multiple-authority attribute-based encryption with hidden policy is of IND-CMA security.

3.Our scheme

(1) Initialization

CA setup: Input a security parameter k, then output parameter as follows:

1) Generate a bilinear group G and a bilinear map $G \times G \rightarrow G_T$, G is a multiplicative cyclic group with prime order N.

2) CA selects $\alpha \in Z_N^*$ randomly, and outputs the public key of system

$$PK = \{e, g, e(g,g)^\alpha, G, G_T\} \qquad g \text{ is the generator of } G$$

The main key of system is

$$MK = \{g^\alpha\}$$

CA sends MK to all the AAs through a secure channel.

AA setup: CA allocates a unique global indentifier *aid* for each legal AA. Let $U=\{U_1, U_2, \ldots, U_n\}=\{att_1, att_2, \ldots, att_n\}$ indicate a complete attribute set. Divide U into $k(k \leqslant n)$ subsets and distribute these k subsets to k AAs to manage. AA randomly selects $a_{i,j} \in Z_N^*(i \in [1,n], j \in [1,n_x]$, where n_x represents the number of att_x. Each AA calculates its own public key

$$APK = \{g^{a_{i,j}}\}$$

And saves its own private key

$$ASK = \{a_{i,j}\}$$

User registration: CA allocates a unique global indentifier *uid* for each user, and uses *uid* to generate identify code Certificate(*uid*) , which is stored in the database of CA. Certificate(*uid*) is a hash algorithm which uses *uid* to generate a hash code with fixed length.

(2) Key generation

After receiving decryption request from user, CA uses *uid* to search Certificate(*uid*) in database, to confirm the legality of this user. If this user is legal, CA will send a message to all the AAs that they can generate private key for this user, otherwise, CA will reject decryption request.

Let $S_{uid}=\{att_{1,uid},att_{2,uid},\ldots,att_{m,uid}\}$ be the attribute set of UID, where $S_{uid}\subseteq U, m\in[1,n]$. $S_{uid,aid}=\{att_{1,aid}, att_{2,aid},\ldots, att_{q,aid}\}$ represents the attributes of UID which AA_{aid} manages, where $S_{uid,aid}\subseteq S_{uid}, q\in[1,m]$. AA_{aid} selects a value $\lambda_i \in Z_N^*$ for each $att_{i,aid}$, and uses *MK* sent by CA to calculate

$$D_1 = g^{\alpha} \cdot g^{\lambda_i a_{i,j}} = g^{\lambda_i a_{i,j}+\alpha}$$

$$D_2 = g^{\lambda_i}$$

The output of the private key is

$$SK = \{D_1, D_2\}$$

(3) Encryption

1) Input the access structure $\mathbf{A}$, public key of the system(*PK*) and the plaintext(*M*).

2) The encryptor transforms $\mathbf{A}$ to the corresponding access tree τ with the rule described in section 2.3. Let s be the root of τ, which is set to be the root node of the access tree and marked as read status. All the child nodes are marked as unread status. The encryptor conducts recursive computation for each unread non-leaf node.

Suppose t is the number of the child node of the current node. If current node is '∧'and all of its child nodes are unread, the encryptor randomly chooses $s_i \in Z_N^*$ for each child node, and calculates the value of the last leaf node as $s_t = s - \sum_{i=1}^{t-1} s_i \bmod N$, then sets these nodes as read status.

If current node is '∨' and its all child nodes are under unread status, the encryptor randomly chooses $s' \in Z_N^*$, assigns all the child nodes as s', then all the chilled nodes as read status. The ciphertext is calculated as follows.

$$C_0 = M \cdot e(g, g)^{\alpha s}$$

$$C_1 = g^{s_i}$$

$$C_2 = g^{s_i a_{i,j}}$$

Finally, the output of the ciphertext is

$$CT = \{C_0, C_1, C_2\}$$

(4) Decryption

After the UID downloads ciphertext from CSP, UID can calculate as follows to decrypt if its attribute set satisfies the access structure.

$$D' = \frac{\prod e(D_1, C_1)}{\prod e(D_2, C_2)} = \frac{\prod e(g^{\alpha+\lambda_i a_{i,j}}, g^{s_i})}{\prod e(g^{\lambda_i}, g^{s_i a_{i,j}})} = \frac{\prod e(g, g)^{\lambda_i s_i a_{i,j}+\alpha s_i}}{\prod e(g, g)^{\lambda_i s_i a_{i,j}}}$$

$$= \frac{e(g, g)^{\alpha s} \cdot \prod e(g, g)^{\lambda_i s_i a_{i,j}}}{\prod e(g, g)^{\lambda_i s_i a_{i,j}}} = e(g, g)^{\alpha s}$$

$$M = \frac{C_0}{D'} = \frac{M \cdot e(g,g)^{\alpha s}}{e(g,g)^{\alpha s}}$$

(5) User revocation
In this paper, the first step is that UID submits decryption request. If UID is legal, AA will allocate private key to UID. When CA finds malicious user, it will delete the corresponding Certificate(*uid*) from its database, then generate a user revocation list, update *MK* and send them to all the AAs with secure channel. After receiving new *MK* and revocation list, AA will regenerate *SK* for the users who are not in the revocation list. The legality of malicious user cannot be confirmed when applying decrypt new file. Even though this illegal user jumps over the CA, it cannot get the new *SK* from AA.

4.Security proof

Based on DBDH assumption, our scheme is of IND-CMA security under the security model described in section 2.5.
The proof is as follows.
Suppose G_1 and G_2 are two multiplicative cyclic groups with prime order p, and g be a generator of G_1. Map $e:G_1 \times G_1 \rightarrow G_2$ is a bilinear map. The challenger B tosses a coin to choose a number $\xi \in \{0,1\}$. If $\xi=0$, B chooses $(g,g^a,g^b,X)=(g,g^a,g^b,e(g,g)^{ab})$, otherwise, B chooses $(g,g^a,g^b,X)=(g,g^a,g^b,e(g,g)^z)$, where $a,b \in Z_N^*$.
(1) The adversary A sends an access structure he wants to challenge to B. B calculates $Y=e(g^a,g^b)=e(g,g)^{ab}$, then randomly chooses $a_{i,j} \in Z_N^* (i \in [1,n], j \in [1,n_i])$ and executes the initialization algorithm to generate *PK*, *MK* and *APK*, where $PK=\{e,g,e(g,g)^\alpha,G,G_T\}$, $MK=\{g^\alpha\}$, $APK=\{g^{a_{i,j}}\}$, $ASK=\{a_{i,j}\}$. B sends *PK* and *APK* to A and preserves *MK* and *ASK*.
(2) A submits his attribute set *S* to B for private key query, but *S* does not satisfy *W*. After receiving private key query, B randomly chooses $\lambda_x \in Z_N^*$ for each attribute of A, and calculates *SK* as follows.

$$D_1 = g^\alpha \cdot g^{\lambda_x a_{i,j}} = g^{\lambda_x a_{i,j}+\alpha}$$

$$D_2 = g^{\lambda_x}$$

(3)B tosses a coin an outputs a random value $b \in \{0,1\}$. Let r be the root of τ, which is set to be the root node of the access tree and marked as read status. All the child nodes are marked as unread status. B conducts recursive computation for each unread non-leaf node.
Suppose v is the number of the child node of the current node. If current node is '∧' and all of its child nodes are unread, B randomly chooses $r_i \in Z_N^*$ for each child node, and calculates the value of the last leaf node as $r_v = r - \sum_{i=1}^{v-1} r_i \bmod N$, then sets these nodes as read status.

If current node is '∨' and its all child nodes are under unread status, B randomly chooses $r' \in Z_N^*$, assigns all the child nodes as r', then all the chilled nodes as read status. The ciphertext is calculated as follows.

$$C_0 = M \cdot e(g,g)^{ar}$$

$$C_1 = g^{r_i}$$

$$C_2 = g^{r_i a_{i,j}}$$

Finally, the output of the ciphertext is

$$CT = \{C_0, C_1, C_2\}$$

B sends *CT* to A.
(4) Execute step(2).
(5) A outputs a guess φ' of φ. If $\varphi'=\varphi$, then B outputs the guess result $\xi'=0$, otherwise, the guess result is $\xi'=1$. So there are two situations as follows:

1) When ξ'=0, $X=e(g^a,g^b)=e(g,g)^{as}$, and the ciphertext is valid. A can guess the correct ξ' with an unnegligible advantage. Under this condition, we have $\Pr[\varphi'=\varphi \mid \xi'=0]=\frac{1}{2}+\varepsilon$.

2) When ξ'=1, $X=e(g,g)^z$. Because z is a random number from Z_N^*, the ciphertext cannot be identified. Under this condition, we have $\Pr[\varphi'\neq\varphi \mid \xi'=1]=\frac{1}{2}$.

Therefore, the probability that B successfully breaks the DBDH assumption is:

$$Ad_A=\frac{1}{2}(\Pr[\varphi'=\varphi \mid \xi'=0])+\frac{1}{2}(\Pr[\varphi'\neq\varphi \mid \xi'=1])-\frac{1}{2}=\frac{\varepsilon}{2}$$

As we can know according to the proof above, If the advantage of A is ε within polynomial time, the probability that B breaks the scheme proposed in this paper is $\frac{\varepsilon}{2}$. Therefore, if ε is unnegligible, the advantage that B breaks our scheme is also unnegligible.

5.Efficiency analysis

Let $|G|,|G_T|,|Z_N^*|$ respectively represent the bit length of G,G_T,Z_N^*. Let t_G , t_{GT} be the time for calculating on the groups G and G_T, and t_b be the time for calculating bilinear pairs.

Let n be the total amount of system attribute. $L=\sum_{i=1}^{n} n_i$ represents the total amount of possible attribute, where n_i is the amount of attribute i.

Table 1. Comparison of parameter length

Scheme	PK	MK	SK	CT
[11]	$(L+1)\|G\|+\|G_T\|$	$L\|Z_N^*\|+\|G_T\|$	$(2n+1)\|G\|$	$(2n+1)\|G\|+\|G_T\|$
[16]	$\|G\|+\|G_T\|$	$\|Z_N^*\|+\|G\|$	$(2n+1)\|G\|$	$(2n+1)\|G\|+\|G_T\|$
Our scheme	$\|G\|+\|G_T\|$	$\|Z_N^*\|+\|G\|$	$2n\|G\|$	$2n\|G\|+\|G_T\|$

As is shown in table 1, both *PK* and *MK* have the same length as the scheme in [16], but are shorter than the scheme in [11]. For the reason that the DO does not take part in private key generation, so the length of private key and ciphertext are shorter than the scheme in [16].

Table 2. Comparison of time overhead

Scheme	Encryption	Decryption
[11]	$(2n+1)t_G+t_{GT}$	$(2n+1)t_b+3t_{GT}$
[16]	$(2n+1)t_G+t_{GT}$	$(2n+2)t_b+4t_{GT}$
Our scheme	$2n*t_G+t_{GT}$	$2n*t_b+2t_{GT}$

As can be seen from table 2, our scheme spends less time on both encryption and decryption than scheme in [11] as well as scheme in [16]. Because in the scheme in [16], DO participates generation of private key, our scheme spends more time. But for the scheme in [16], the cost for increasing efficiency of private key generation is the DO needs to keep online, which greatly consumes resources. Based on the above analysis, our scheme is better than both the scheme in [11] and the scheme in [16].

6.Conclusion

This paper improves the scheme of attribute-based encryption with only one authority, proposing a scheme of hidden-structure attribute-based encryption with multiple authorities. In our scheme, the private key of user is generated by multiple authorities, improving encryption efficiency as well as eliminating the threat brought by one or more attacked authorities. Besides, we realized the user revocation by updating private key for legal users. We proved that our scheme is of IND-CMA security under DBDH assumption. Lastly, comparing with other schemes, we analyzed the efficiency of ours. But the prerequisite for realizing our scheme is that the center authority is completely credible. Once successfully attacks the center authority, the attacker can register any amount of legal user and get private key from the attribute authorities, so that the data security will be threaten. Our next research is multi-authority attribute-based encryption without center authority.

7.Acknowledgement

This work is supported by the science and technology project of Guangdong Province (No.2015B010128014, 2016B090918039, 2016B090918058), and the project of Guangzhou Science and Technology (No.201508010026, 201604016067).

References

[1] Bethencourt J, Sahai A and Waters B 2007 Ciphertext-Policy attribute-based encryption. *Proc. of the 2007 IEEE Symp. on Security and Privacy. Washington: IEEE Computer Society*.321-334.
[2] Chase M. 2007 Multi-authority attribute based encryption *Theory of cryptography. Springer Berlin Heidelberg*. 515-534.
[3] Gao A and Li Z Z 2013 Improving user's privacy for multi-authority ABE using privacy homomorphism *J. IEICE Transactions on Fundamentals of Electronics Communications and Computer Sciences*. **96**(3):724-727.
[4] Li Z H, Yang M M and Chen X Y 2015 Adaptive Secure and Attribute Repeatable Multi-authority ABE scheme *J. Computer Engineering*. **41**(12):21-25.
[5] Li Z H and Chen X Y 2016 Research and application of multi-authority attribute-based encryption scheme *J. COMPUTER ENGINEERING AND DESIGN*. **37**(8):1981-1985.
[6] Wu G Q 2016 Multi-Authority CP-ABE with Policy Update in Cloud Storage *J. Journal of Computer Research and Development*. **53**(10):2393-2399.
[7] Neethu A N and Prem K 2016 Efficient Security in Decentralized Cipher text-Policy Attribute-Based Encryption. *IJSART*. **2**(5):682-686.
[8] Song Y, Han Z, Liu F M and Liu L 2015 Attribute-based encryption with hidden policies in the access tree *J. Journal on Communications*. **36**(9):119-126.
[9] Du R Y, Shen J, Chen J and Zhou S G 2016 An Access Control Scheme in cloud Storage with Policy Hiding ABE *J. Wuhan Univ.(Nat. Sci.Ed.)*. **62**(3):242-248.
[10] Lei L, Cai Q W, Jing J W, Lin J Q, Wang Z and Chen B 2016 Enforcing access controls on encrypted cloud storage with policy hiding *J. Journal of Software*. **27**(6):1432-1590.
[11] Wang H P and Zhao J J 2016 Ciphertext-policy Attribute-based Encryption with Anonymous Access Structure *J. Computer Science*. **43**(2):175-178,198.
[12] Weng A X and Ling J 2017 Improved scheme of CP-ABE with Hidden Access Structure *J. Computer Science*. **44**(11A):377-380.
[13] Xie L and Ren Y L 2015 Efficient attribute-based encryption with hidden access structures *J. JOURNAL OF XIDIAN UNIVERSITY*. **42**(3):97-102.
[14] Ying Z B, Ma J F and Cui J T 2015 Partially policy hidden CP-ABE supporting dynamic policy updating *J. Journal on Communication*. **36**(12):178-189.
[15] Zhong H, Zhu W L , Xu Y and Cui J 2016 Multi-authority attribute-based encryption access control scheme with policy hidden for cloud storage *J. Soft Couputing*.1-9.
[16] Fan Y D and Wu XP 2017 A Multi-authority Attribute-based Encryption Scheme with Hidden Policies *J. Computer Engineering*. http://kns.cnki.net/kcms/detail/31.1289.TP.20170421.1439.010.html

IOP Conf. Series: Materials Science and Engineering **359** (2018) 012006 doi:10.1088/1757-899X/359/1/012006

Research on SOC Calibration of Large Capacity Lead Acid Battery

W Q Ye and Y X Guo
1 School of computer science, Hunan University of Technology, Zhuzhou,China

Abstract: Large capacity lead-acid battery is used in track electric locomotive, and State of Charge(SOC) is an important quantitative parameter of locomotive power output and operating mileage of power emergency recovery vehicle. But State of Charge estimation has been a difficult part in the battery management system.In order to reduce the SOC estimation error better, this paper uses the linear relationship of Open Circuit Voltage(OCV)and State of Charge to fit the SOC-OCV curve equation by MATLAB.The method proposed in this paper is small, easy to implement and can be used in the battery non-working state SOC estimation correction, improve the estimation accuracy of SOC.

1. Introduction

This paper is based on the power battery management system of CRRC power repair locomotive. The sample used is 2V/160Ah's NM-160 valve controlled sealed lead-acid battery. The purpose of this thesis is to provide a more efficient SOC correction method for battery management system to improve the accuracy of SOC estimation in battery management system[1].BMS (Battery Management System)can prolong the service life of batteries and prevent overcharge and over discharge in the process of using batteries. In this paper, the correction of SOC estimation is studied by exploring the corresponding relationship between the open circuit voltage and the SOC in stationary state.In order to better fit the linear relationship between OCV and SOC, a large number of charging and discharging experiments are carried out to collect data.

2. SOC Principle and Measurement

The SOC represents the battery's current state of charge, that is, the remaining battery power.The United States Advanced Battery Consortium (USABC) defines SOC as the ratio of the remaining capacity of the battery [2]to the nominal capacity under the same conditions at a certain discharge rate.According to this definition is derived from the definition of the ampere hour method:

$$SoC = SOC_0 - \int_0^t Ki\mathrm{d}t / Q_c \tag{1}$$

Among them, Q_C is rated capacity, SOC_0 is the initial battery capacity, i I is the discharge current at a certain time, and K is the influence factor.Ampere hour method principle is simple and easy to implement. In addition, the computational complexity of the algorithm, but this algorithm has accumulated error and the accuracy depends on the current sampling frequency.The open circuit voltage method uses the linear relationship between the open circuit voltage[3]-[5] and the SOC to estimate SOC through the voltage when the battery is not working.This method is simple, but it needs to be done in the non working state of the battery.

In addition, the estimation error of the algorithm is large when the battery runs.In this paper, the corresponding relation between the stationary voltage and the SOC of the battery is tested, and the estimation of SOC is corrected by the corresponding relation when the battery is not working.

3. Open circuit voltage

Open circuit voltage is an important parameter in the fitting of OCV-SOC relation[6].In order to obtain the true correspondence of OCV-SOC, a lot of experiments have been carried out.The picture is a voltage change curve during the discharge of a lead acid battery.The region marked by the letter A represents the discharge process.The voltage rises at the end of the discharge.The voltage change is caused by the internal resistance of the battery.In addition, the voltage tends to be stable in a short time.The voltage at the stationary stage represents the voltage at which the battery is not working[7].The steady change of the open circuit voltage is beneficial to fit the corresponding relation between OCV and SOC.

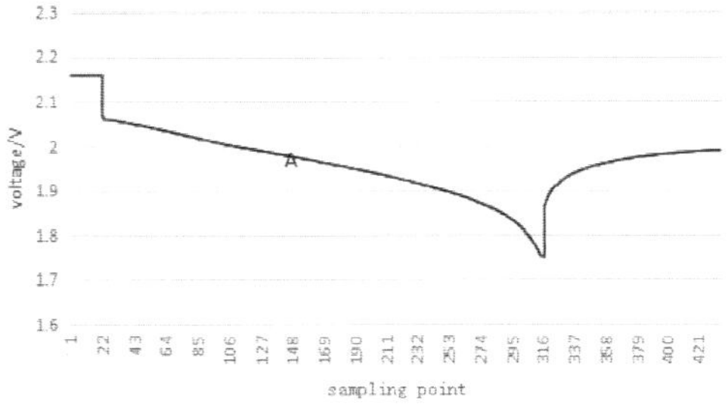

Fig. 1 voltage change during the discharge of lead-acid battery

4. Experimental analysis and simulation

The automatic test equipment of BTS-M battery parameters is used to control the experiment process in order to obtain more accurate data. The device supports programmable charging current, programmable charging time, programmable discharge current, and discharge time.The data collected by the device is sent to the computer terminal through the serial port.The experiment process has realized automation and digitization.The temperature is controlled by a constant temperature device to obtain experimental data at different temperatures.The experimental equipment and batteries are shown in Figure 2.

Figure 2 experimental battery and charge and discharge test chamber

4.1 Experimental design

In the experiment, the NM-160 valve controlled sealed lead-acid battery was used as the experimental object.Each individual cell voltage is 2V.The 3 individual batteries are connected in series to form a battery pack,which the voltage of battery pack voltage is 6V.Experiments were carried out at different temperatures and data were collected.The experimental steps are:

- The battery is charged by 0.2C constant current until the voltage reaches the cut-off voltage.When the cut-off voltage is reached, the constant voltage charge is applied until

IOP Conf. Series: Materials Science and Engineering **359** (2018) 012006 doi:10.1088/1757-899X/359/1/012006

the battery cutoff current is reached.The battery is stationary for 3 hours after charging.

- When the battery does not work, collect the voltage data for 10 minutes through the BTS-M device before discharging.Then,The battery is discharged in 0.1C constant current.The battery discharges to different depths each time.For example,0.9SOC, 0.8SOC ... 0.1SOC.
- When the battery discharge is over, the device collects the battery voltage at standstill every minute.The acquisition process lasted three hours.
- The battery continues to discharge until the battery voltage reaches the cut-off voltage.Then, the battery is stationary for 5 hours.Experiment cycle execution step 1.But the discharge of the battery will be discharged at different currents.For instance, 0.2C, 0.3C ... 0.9C and 1C The experimental data will be recorded that discharge currents is different .

4.2 Data analysis

The static voltage data at the same temperature, the same discharge depth and the discharge current are selected to analyze.Static voltage data at the same temperature, at the same discharge depth and at different discharge currents are used to analyze voltage variation trends.The voltage change curve is shown in Figure 3.In the figure, when the battery discharges at different currents to the same SOC, the static voltage tends to stabilize with the increase of the settling time.Therefore, the conclusion is that the final static voltage is related to the final charge of the battery.

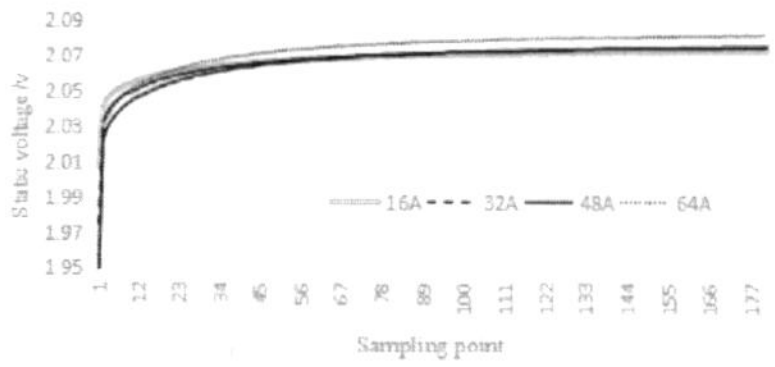

Figure 3 different current discharge to the same SOC voltage curve.

During the experiment, the battery discharges to different SOC at different discharge rates(0.1C,0.2C,0.3C...1C).Figure 4 shows the trend of the open-circuit voltage when the battery is discharged to a different SOC.When the remaining battery power is not the same, the open circuit voltage of the battery is obviously different.However, the open circuit voltage of the battery with the same remaining power remains stable after about 20 sampling periods.When the discharge is completed, the battery has more residual energy, the higher the voltage of the battery.

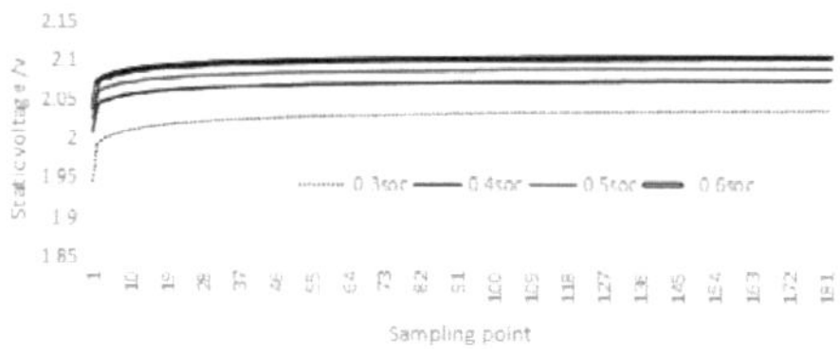

Figure 4 Battery packs are different from the SOC corresponding to the static voltage

The constant current discharge experiment was carried out on a NM-160 valve-regulated sealed

IOP Conf. Series: Materials Science and Engineering **359** (2018) 012006 doi:10.1088/1757-899X/359/1/012006

lead-acid battery (160 Ah, 2 V) at a discharge rate of 0.1 ° C at 30 ° C. According to the experimental data, the corresponding curves of OCV and SOC are obtained by processing and analyzing in MATLAB software[8].Figure 5 shows that there is a good linear relationship between SOC and OCV.This good linear relationship facilitates the estimation of the current SOC in the stationary state by measuring the OCV.In order to better correct the SOC and exclude the influence of temperature on the SOC estimation, this paper adopts different methods to deal with the data at different temperatures.

In MATLAB, the cftool tool is used to fit the linear formula. Where Uocv represents open circuit voltage. The formula for fitting at 30 degrees Celsius is:

$$SOC = -0.1022 \times U_{ocv}^{\ 2} + 3.729U_{ocv} - 10.1 \tag{2}$$

Similarly, the formula at 25 degrees Celsius is:

$$SOC = -2.949U^{2}_{ocv} - 21.55U_{ocv} - 38.07 \tag{3}$$

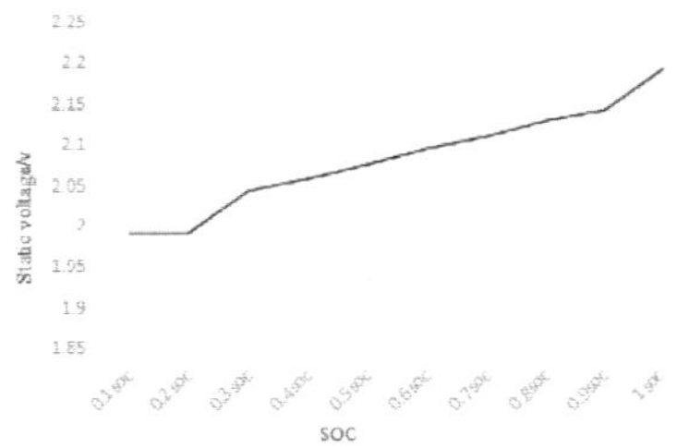

Figure 5 OCV and SOC correspondence between

The charge and discharge data of the NM-160 VRLA battery are collected as simulation and simulation data. The charge discharge data of NM-160 valve controlled sealed lead-acid batteries are used as the simulation data to analyze errors.Through the data comparison of Table 1, it is found that the error range is not more than 5%. Therefore, the fitting formula is very helpful to the correction of SOC estimation. And the accuracy is relatively high.

Table 1 Comparison of actual SOC with predicted SOC data

expected value	Measuring voltage	Estimated SOC	Deviation
0.1	1.988	0.1048	0.048
0.2	1.989	0.2078	0.039
0.3	2.042	0.3081	0.027
0.4	2.057	0.4144	0.036
0.5	2.074	0.5215	0.043
0.6	2.094	0.6192	0.032
0.7	2.109	0.7287	0.041
0.8	2.128	0.8248	0.031
0.9	2.141	0.9252	0.028
1	2.193	1.05	0.05

5. Conclusion

The discharge experiments of lead-acid batteries are carried out at different depths to obtain the open circuit voltage at standstill.In order to eliminate the influence of temperature on SOC, experiments were carried out at different temperatures.Then,with the aid of MATLAB software, a large amount of data is used to fit the corresponding relationship about $soc = f(U_{ocv})$.This relationship will be used to correct BMS's estimate of SOC at rest[9].What's more, this method is high precision and reliability and easy to implement.But the fitting formula also has some limitations.For complex battery environments, there may be unknown errors.

References

[1] SHAO Chun-sheng, LI Bei, CAI Ji-he. SOC estimation of manganese acid lithium battery using rebound voltage[J]. *Power technology*,2017,(02):211-213+269.

[2] Zhang Ting,Du Shejiao. SOC estimation method of lithium battery based on improved Thevenin model[J]. *Power technology*,2015,(11):2400-2402+2496.

[3] LI Zheng,ZHI Ruodon,SUN Honwan. SOC estimation method based on the prediction of *open*-circuit voltage[J]. *Hebei Journal of Industrial Science & Technology*,2017,(01):36-40.

[4] HAN Xue-bing, OUYANG Ming-gao. Characteristics analysis of open circuit voltage of aged LiFePO 4 battery for electric vehicle[J]. *Power technology*,2015,(09):1876-1878.

[5] DENG Ye,HU Yue-li,TENG Hua-qiang. Open-circuit Voltage Prediction and SOC Estimation of Li-ion Battery [J]. *Instrumentation Technology*,2015,(02):21-24.

[6] Deng Kun,Wang Fuzhong,Wang Zhaosheng. Open experiment research for the mea suring precision of the voltage - ampere hour method effect [J]. *Energy Conservation*, 2013,(08):25-28+2.

[7] XU Xin-ge ,YANG Song ,LI Yan-fang. A method of SOC-estimate based on forecast of open-circuit voltage[J]. *International Electronic Elements* ,2011,(14):127-129.

[8] Wu Jiajia,Zhao Youqun. Parameter Identification of Ni / MH Battery Used in Hybrid Electric Vehicles Based on Thevenin Mode [J]. *Agricultural equipment and vehicle engineering*,2014,(01):1-5.

[9]Hu Xiuzhi,Jin Peng,CHANG Rui-feng. Ni-mh battery simulation based on Thevenin model [J]. *Power technology*,2014,(11):2063-2065.

IOP Conf. Series: Materials Science and Engineering **359** (2018) 012007 doi:10.1088/1757-899X/359/1/012007

Analysis Balance Parameter of Optimal Ramp metering

Y Li[1*] N Duan[2] X Yang[1]

[1] School of Information Engineering, Chang'an University, ShaanXi Province, R.P.China

[2] Institute of Foundation and Structure Technologies, Zhejiang Sci-Tech University, Zhenjiang, Province, R.P.China.

yingli@chd.edu.cn

Abstract. Ramp metering is a motorway control method to avoid onset congestion through limiting the access of ramp inflows into the main road of the motorway. The optimization model of ramp metering is developed based upon cell transmission model (CTM). With the piecewise linear structure of CTM, the corresponding motorway traffic optimization problem can be formulated as a linear programming (LP) problem. It is known that LP problem can be solved by established solution algorithms such as SIMPLEX or interior-point methods for the global optimal solution. The commercial software (CPLEX) is adopted in this study to solve the LP problem within reasonable computational time. The concept is illustrated through a case study of the United Kingdom M25 Motorway. The optimal solution provides useful insights and guidances on how to manage motorway traffic in order to maximize the corresponding efficiency.

1.Introduction

The performance of transport infrastructure is closely linked to the associated society's prosperity, economic growth and quality of life. Sustained growth in demand in virtually all modes of transport places tremendous pressure on the infrastructure. Continuous construction of new roads will not be a sustainable solution to traffic congestion due to the increasingly constrained financial, physical, and environmental conditions. Consequently, governments, businesses, and research teams around the world want to explore alternative ways to effectively utilize and manage existing road infrastructure. In fact, adding new infrastructure may induce more demand for travel and hence is not an effective option. Therefore, a sustainable solution for mitigating congestion calls for the effective management of the existing infrastructure through appropriate control methods. This paper presents an optimization framework for motorways through ramp metering maximizing.

This paper is organized as follows: Section 2 reviews the first order macroscopic traffic model. Section 3 presents the formulation that is used to derive the optimal ramp metering for maximizing the efficiency of the motorway traffic. Section 4 explores the sensitivity analysis of the balance parameter between the main road delay and ramp delay. Section 5 presents the optimal ramp metering through a case study of the United Kingdom M25 Motorway. Finally, Section 6 provides some concluding remarks.

2.Cell Transmission Model

The cell transmission model is a finite difference approximation of the first order LWR model proposed by Daganzo[1], the road section is discretized into a collection of sub-sections or 'cells'. In the cell transmission formulation, traffic dynamics are characterized by flow and density. The evolution of traffic flow and density is governed by the principles of flow conservation and propagation. Define f_i (k) the traffic outflow from cell i during time step k, and hence f_(i-1) (k) (outflow from upstream cell i-1) will be the inflow to cell i during the same time step k. It is convenient (e.g. Gomes and Horowitz[5]) to specify the exit flow si(k) through a split ratio β i, where 0 ⩽ β i ⩽ 1, to represent the proportion of traffic leaving the system through the sink link i during time step k. The density in cell i at the following time step k+1 can then be updated by the conservation equation as follows:

$$\rho_i(k+1) = \rho_i(k) + \frac{\Delta t}{\Delta x_i}\left(f_{i-1}(k) - f_i(k)/\beta_i + r_i(k)\right) \quad (1)$$

where Δt denotes the length of simulation time step, and Δx_i represents the length of cell i. $f_i(k)$ is the traffic outflow from cell i during time step k, and $\rho_i(k)$ is the density in the cell i during time step k. r_i(k) and s_i(k) represent the actual on-ramp inflow and off-ramp outgoing flow during time step k respectively.

By assuming a piecewise linear fundamental diagram, the cell transmission rule calculates the outflow from cell i within time step k, which can be updated under the given cell density as follows:

$$f_i(k) = min\{\, v_i\rho_i(k),\quad Q_i,\quad Q_{i+1},\quad w_{i+1}(\,\overline{\rho}_{i+1} - \rho_{i+1}(k))\,\} \quad (2)$$

where $f_i(k)$ is the traffic outflow from cell i during time step k, and $\rho_i(k)$ is the density in the cell i during time step k. The notations v_i, w_{i+1}, and $\overline{\rho}_{i+1}$ are free flow speed, shockwave speed and jam density respectively. The notation Q_i denotes the capacity flow at cell i which corresponds to the maximum flow that can leave cell i, and Q_{i+1} represents the capacity flow at cell i+1 which corresponds here to the maximum flow that can enter cell i+1. The inclusion of both capacity flows at adjacent cells is due to the consideration of a non homogeneous section where different locations hold different capacities. When there is no congestion, the traffic moves from one cell to the next cell at free flow speed v_i. The quantity ($v_i\rho_i$(k)) represents the total traffic flow travels from cell i to cell i+1. The quantity ($w_{i+1}(\,\overline{\rho}_{i+1} - \rho_{i+1}(k))$ specifies the available space for incoming traffic at the downstream cell i+1 during time step k. The above formulation covers both congested and uncongested regimes. It is known that the minimum operator in Equation (2) can be formulated as Linear Programming (Lo[2]; Ziliaskopoulos[3]; Lo[4]; Gomes and Horowitz[5], and others). This linearity assumption facilitates the development of several efficient algorithms for cell based estimation and optimization (see examples: Daganzo[6]; Gomes and Horowitz[5]; Kurzhanskiy and Varaiya[7]).

3. Optimal Ramp Metering

Ramp metering is one of the most widely used control methods, which aims to improve the throughput on the main road through limiting access of traffic from on-ramps. This section presents a model-based approach, which enhances the efficiency of motorway operations through modelling and regulating the traffic flowing on it. There are some important applications of the first-order macroscopic model (CTM) which were presented in the literature for traffic modelling and management. Lo[2] presents a novel traffic signal control formulation, which developed through a mixed integer programming technique. Ziliaskopoulos[3] presents a system optimum dynamic traffic assignment problem as a linear programming problem based on CTM. Gomes and Horowitz[5] show that the linearity of CTM enables formulating optimal ramp control problems as a linear programming (LP) problem. In the CTM formulation, traffic dynamics are characterized by flow and density in each cell at each time step. The evolution of traffic flow and density is governed by the principles of flow conservation and propagation. This linearity facilitates the development of several efficient algorithms for cell based estimation and optimization (see examples, Ziliaskopoulos[3]; Sun[8]; Gomes and Horowitz[5]; Sumalee[9]).

In original CTM model, the traffic flow is determined by the available space of upstream and downstream. However, the linearization of the original non-linear flow propagation formulation caused the holding back problem[10] in the earliest analytical formulation[3]. Because it does not require the solution to fall on the fundamental diagram. Therefore, the vehicles will be held at some upstream cells even there is enough space in the downstream cells. Gomes and Horowitz[5] introduce total-travel-distance (TTD) to the objective function to eliminate the holding-back problem. Therefore, the delay is adopted as the objective function because TTD is included. Additional constraints may be added for specific applications. For ramp metering, this study adopts total-system-delay (TSD) as the objective function because ramp delay is considered. However, it is noted that the choice of objective function is flexible where different objective functions can be used for different applications. The motorway traffic optimization with ramp metering can be formulated as follows:

$$\min_{c} Z = \sum_{i=1}^{I}\sum_{k=1}^{K}\left(\rho_i(k)\Delta x_i\Delta t - \frac{f_i(k)\Delta x_i\Delta t}{v_i}\right) + \eta\sum_{j=1}^{J}\sum_{k=1}^{K}\left(l_j(k)\Delta t\right)$$

subject to:

$$\begin{aligned}
\rho_i(k+1) &= \rho_i(k) + \frac{\Delta t}{\Delta x_i}\left(f_{i-1}(k) - \frac{f_i(k)}{\overline{\beta}_i} + r_j(k)\right)\\
f_i(k) &\le \rho_i(k)v_i\overline{\beta}\\
f_i(k) &\le Q_i\\
f_i(k) &\le Q_{i+1}\\
f_i(k) &\le w_{i+1}\left(\overline{\rho}_{i+1}(k) - \rho_{i+1}(k)\right)\\
l_j(k+1) &= l_j(k) + \left(\lambda_j(k) - r_j(k)\right)\Delta t\\
l_j(k) &\le \overline{l}_j\\
r_j(k) &\le \overline{r}_j\\
r_j(k) &\ge 0\\
r_j(k) &\le \frac{l_j(k)}{\Delta t} + \lambda_j(k)\\
r_j(k) &\le \left(\rho_j - \rho_j(k)\right)\frac{\Delta x_j}{\Delta t}
\end{aligned} \tag{3}$$

The notation $\overline{l}_j$ refers to the maximum ramp queue length at on-ramp j, and the notation $\overline{r}_j$ refers to ramp capacity at on-ramp j. The optimization problem seeks the optimal control policy **cr** to be implemented over time k =1, 2 ... K and cells i = 1, 2 .... I that minimizes the total system delay Z in the system. The first term is the main road delay in the objective function, and the second term is delays on the boundary links. One will get a trivial solution: $r_j(k) = 0$, $\forall j$; k, which simply prohibits any traffic from on-ramps entering the system if this second term is omitted. The notation η is a parameter that adjusts the balance between the main road delay and ramp delay (boundary queues). The sensitivity analysis is presented in Section 4. In this study, the value of η is set to be 1 indicating all road sections are equally weighted.

The above LP problem can be solved by a number of established algorithms such as SIMPLEX or interior-point methods for global optimal solutions (Vanderbei[12]). Nevertheless, it is known that there is no guarantee that the optimal solution can be found in polynomial time while the complexity of a LP problem basically depends on the number of decision variables and constraints involved.

The first five constraints in Equation 3 equivalents to the CTM as shown by Lo[2]; Gomes and Horowitz[5] and others. The holding back problem[3] is addressed here due to the [$-f_i(k)$] term in the objective function which will maximize the outflow from each cell i over time step k. As a consequent, at least one of the constraints (2 - 5) in Equation 3 must be binding. Constraints (2) and (3) in Equation 3 can be regarded as the (demand) limitations on outflow under free flow condition, while constraints (4) and (5) in Equation 3 can be regarded as the (supply) limitations on outflow under congested condition. Constraint (1) in Equation 3 is the conservation equation to update the density in cell i for next time step k+1.

For the ramp queue length, constraint (6) in Equation 3 is used to capture the evolution of queues $l_j(k)$ on on-ramps $j = 1; 2 \ldots J$, where J is the total number of on-ramps. Moreover, one may add a upper bound $\bar{l}_j$ (constraint 7 in Equation 3) for some on-ramps to specify the maximum queue length of the on-ramps such that an unacceptable long queue on the on-ramp will not be obtained as an optimization result.
For ramp inflow, constraints (8-12) in Equation 3 are additional constraints on the control variable $r_j(k)$ to ensure its upper bound and non-negativity respectively. Constraints (11-12) in Equation 3 are constraints on ramp demand and main road space respectively. Note that $\lambda_j(k)$ does not necessarily equal to $r_j(k)$ due to various reasons including gridlock which traffic cannot be freely flowing into the system but subject to downstream traffic condition. The difference between $r_j(k)$ and $\lambda_j(k)$ can also be due to various control strategies such as ramp metering (Gomes and Horowitz[5]) and variable speed limits (Smulders[11]).

4.Sensitivity Analysis of Balance Parameter

This section explores the sensitivity of the parameter η of the optimal ramp metering. A corridor consisting of 25 cells with one on-ramp at cell 21 and one off-ramp at cell 19. The length of each cell (Δx_i) is 500 meters, and the simulation time step (Δt) is 15 seconds. The on-ramp capacity and maximum on-ramp queue are set to be 1500 veh/hr and 60 vehicles respectively. The split ratio at the off-ramp is set to be 0.15. The same free-flow speed (100 km/hr), capacity (3600 veh/hr) and jam density (240 veh/km) are set for all cells. The simulation horizon is 4 hours with the cool down period of 30 minutes. The optimal ramp metering control strategies are calculated with different value of η, then the ramp metering control strategies are simulated on the same CTM simulation platform, and the total system delay is calculated with the η of one. The balance parameter η of main road delay and ramp delay is sensitive around one. Figure 1 and Table 1 show the delays of ramp metering with the value of η lower than one and higher than one respectively.
Table 1 shows the delays including ramp delay, main road delay and total system delay with η equal and higher than one. The delays are nearly constant for $1 \le \eta \le 1.17$. That means η is not sensitive for the region $\eta \in [1, 1.17]$. However, the ramp delay approach to zero for $\eta > 1.17$. The higher value of η means the system puts more weight on ramp delay than on main road delay. It is important to note that the ramp delay equals to zero under the extreme case ($\eta = 1.19$ in this case study). It is not worthwhile to wait the vehicles on the ramp for the large η. Therefore, the vehicle is served when it arrives the ramp. For η lower than one value (1.19 in this case), the total system delay nearly no difference under different value of η, and the ramp delay reduces as the value of η increases and becomes constant when η lower than one value. However, ramp delay equals to zero for η higher than the particular value, then there is no metering on ramps.

Table 1 Delays of ramp metering with η higher than one (veh-hr)

η	1.0	1.11	1.12	1.13	1.14	1.15	1.16	1.17	1.18	1.19
Ramp	**166.65**	**163.31**	**163.29**	**163.26**	**163.25**	**163.20**	**163.11**	**162.86**	**0.16**	**0.00**
Main	**538.20**	**541.55**	**541.56**	**541.59**	**541.61**	**541.66**	**541.79**	**542.06**	**733.47**	**733.66**
Total	**704.85**	**704.85**	**704.85**	**704.86**	**704.86**	**704.87**	**704.88**	**704.92**	**733.63**	**733.66**

Figure 1 shows the main road delay and ramp delay of ramp metering. The horizontal axis is the value of η between 0.97 and 1.0. The delays of η lower than 0.97 are not shown in the figure due to the delays are constant for $\eta \le 0.97$. That means there is no more space to store more vehicles to relieve the main road congestion even the lower η gives more weight on main road delay. The left and right vertical axes are the main road delay and ramp delay respectively. Figure 1 shows that the main road delay (green solid line with circles) reduces as the increases of η and becomes insignificant when the η approach to one. On the contrary, the ramp delay (blue dotted line with squares) has the opposite trend as shown in the figure.

However the total system delay of ramp metering is constant at the value of 704.85 veh-hr for $\eta \leq 1$. Compared with no control case (733.66 veh-hr), there is 28.8 veh-hr reduction on total system delay.

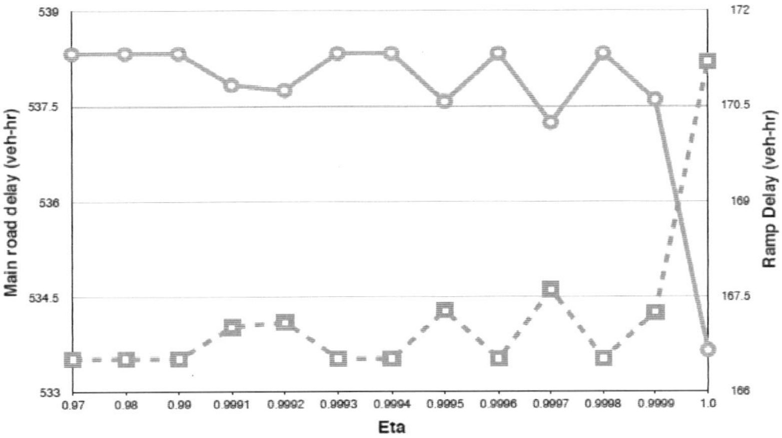

Figure 1: Delays of ramp metering with η lower than one

5.Case Study

The London's orbital M25 Motorway is one of the busiest roads in the United Kingdom, which is used by 250,000 vehicles per day[13]. It is closely monitored and managed by the Highways England. Therefore, the section of the London's orbital M25 Motorway is selected to illustrate overall calibration performance and test the optimal ramp metering control strategy. Moreover, the clockwise direction is selected as it contains data of better quality.

Following Daganzo[6,14], the motorway stretch is divided into a series of cells where the length of all cells Δx_i is 500 meters, which is the standard MIDAS detector spacing. The motorway stretch contains 25 detector stations, which are configured such that the center of upstream and downstream boundaries of each cell will coincide with the location of the associated detector station. The on-ramps and off-ramps are located in the beginning and the end of the cell respectively.

The time step size Δt is set such that $\Delta t \leq \min_i \frac{\Delta x_i}{v_i}$, where $\min_i \frac{\Delta x_i}{v_i}$, refers to the smallest ratio of cell length to the associated free flow speed along the section. The above condition is known as the Courant-Friedrichs-Lewy (CFL) condition[15]. This condition is used to ensure the numerical stability by constraining the traffic flow not to travel further than the length of the cell in one simulation time step. Consequently, the simulation time step Δt is set to 15-sec instead of 1-min as it stored in the dataset.

The data collected on 4 September 2014 (Thursday). The optimization model is applied to manage afternoon peak hour traffic [14:00 - 21:00] at the congested region. The size of the simulation time step is set to be 15-sec, which gives the total number of time steps K = 1680 for a 7 hours [14:00 - 21:00] planning horizon. The optimal control problem is implemented and solved by IBM ILOG CPLEX Optimization Studio V12.5 running on a desktop computer with Intel Core i5-2400 3.1GHz Processor, 4GB RAM, and Windows 7 64-bit operating system. It takes about four minutes to solve.

To illustrate some fundamental features of the optimal solutions, the optimal ramp metering policy that minimizes the total delay along the section of motorway is considered. The problem consists of a total of 6,720 decision variables (ramp inflows, $r_j(k)$, in which 4 (on-ramps) × 1,680 (time-steps) = 6,720). To produce applicable results, the maximum ramp queue constraint needs to be considered. The maximum allowable queue length $\overline{l_j}$ at all on-ramps is set to be 60 (veh), which means the situation where an optimal metering is applied at the on-ramps and the ramp queues are not allowed to exceed 60 vehicles. Figure 2 compares the main road density with and without the optimal metering control, and the maximum queue length is set to be 60 vehicles at all on-ramps. The color scale represents the level of traffic density at the corresponding time and location. In this case, a modest reduction in main road delay is observed due to the consideration of ramp queues. For the 60 maximum ramp queue ramp metering case, the main road delay is 25,802 veh-hr, and the associated ramps' delay is 484.00 veh-hr with metering which gives a total system delay of 27,516 veh-hr, which is smaller than the original 28,345 veh-hr.

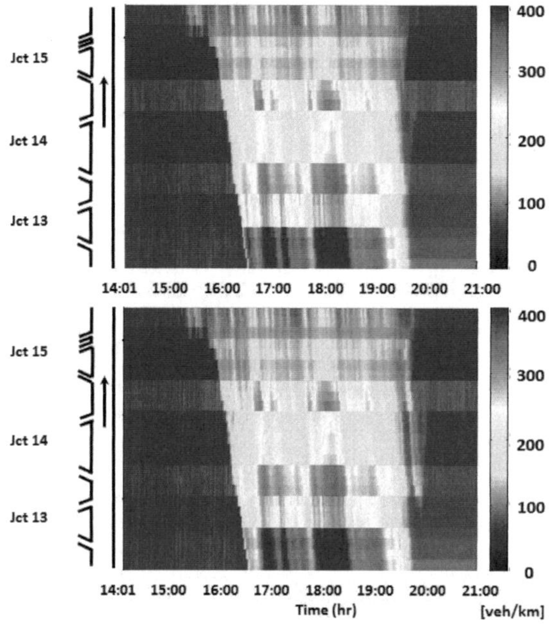

Figure 2: Comparison of main road densities (Three Junctions)
upper - no control; lower - metered (60 veh)

6.Summary

This paper presents a mathematical framework that seeks the optimal ramp metering strategy. Cell transmission model is calibrated with traffic data and implemented to model a section of motorway. The validation result reveals the mean absolute percentage error is 11.5%. With the piecewise linear structure of CTM, the optimal ramp metering problem can be formulated as a LP, which can be solved by a range of established solvers for the global optimal solution. This LP formulation is applied to a scenario of the M25 Motorway where an optimal ramp metering strategy is derived that minimises the total system delay over a fixed space-time horizon. It is shown that optimal solutions are obtainable through CPLEX in a reasonable computational time. We note that the application to motorway traffic is only an illustration and the methodology is indeed generally applicable for other transport networks. We also conduct a sensitivity analysis on the effect of ramp separation on the effectiveness of the ramp metering.

7.Acknowledgments

This work is supported by the Fundamental Research Funds for the Central Universities (310824171001, 300102248201, 300102248401), and National Natural Science Foundation of China (61473047). The authors would like to acknowledge UK Highways Agency for providing the MIDAS traffic data and IBM Academic Initiative for providing the software and license for using the CPLEX Optimization Studio.

References

[1] Daganzo C F 1994 The cell transmission model: A dynamic representation of highway traffic consistent with the hydrodynamic theory. *Transportation Research Part B: Methodological*, 28(4), 269-287.

[2] Lo H K 1999 A novel traffic signal control formulation. *Transportation Research Part A: General*, 33(6), 433-448.

[3] Ziliaskopoulos A K 2000 A linear programming model for the single destination system optimum dynamic traffic assignment problem. *Transportation Science*, 34(1), 37-49.

[4] Lo H K 2001 A cell-based traffic control formulation: strategies and benefits of dynamic timing plans. *Transportation Science*, 35(2), 148-164.

[5] Gomes G and Horowitz R 2006 Optimal freeway ramp metering using the asymmetric cell transmission model. Transportation Research Part C: *Emerging Technologies*, 14(4), 244-262.

[6] Daganzo C F 1994 The cell transmission model: A dynamic representation of highway traffic consistent with the hydrodynamic theory. *Transportation Research Part B: Methodological*, 28(4), 269-287.

[7] Kurzhanskiy A A and Varaiya P 2010 Active traffic management on road networks: a macroscopic approach. Philosophical Transactions of the Royal Society A: Mathematical, *Physical and Engineering Sciences*, 368(1928), 4607-4626.

[8] Sun X T, Munoz L and Horowitz R 2003 Highway traffic state estimation using improved mixture Kalman filters for effective ramp metering control. *In Proceedings of the 42nd IEEE Conference on Decision and Control* (Volume: 6), December 9-12, 6333-6338.

[9] Sumalee A, Zhong R X, Pan T L and Szeto W Y 2011 Stochastic cell transmission model (SCTM): A stochastic dynamic traffic model for traffic state surveillance and assignment. *Transportation Research Part B: Methodological*, 45(3), 507-533.

[10] Doan K and Ukkusuri S V 2012 On the holding-back problem in the cell transmission based dynamic traffic assignment models. *Transportation Research Part B: Methodological*, 46(9), 1218-1238.

[11] Smulders S 1990 Control of freeway traffic flow by variable speed signs. *Transportation Research Part B: Methodological*, 24(2), 111-132.

[12] Vanderbei R J 2007 Linear Programming: Foundations and Extensions. *International Series in Operations Research & Management Science*. 3rd Edition, Springer, New York.

[13] Seaman E 2010 Modelling the UK's busiest road-using detailed data to inform design at junction 15 on the M25. *In European Transport Conference*, October 11-13, Glasgow, Scotland, United Kingdom.

[14] Daganzo C F 1995 The cell transmission model, part II: network traffic. *Transportation Research Part B: Methodological*, 29(2), 79-93.

[15] Lax P and Wendro B 1960 Systems of conservation laws. *Communications on Pure and Applied Mathematics*, 13(2), 217-237.

IOP Conf. Series: Materials Science and Engineering **359** (2018) 012008 doi:10.1088/1757-899X/359/1/012008

Layout Study and Application of Mobile App Recommendation Approach Based On Spark Streaming Framework

H.T. Wang[1], T.T. Chen[2], C Yan[2] and H Pan[2]
[1] Network Information Center, Wuhan University of Technology , Hubei Wuhan 430070, China
[2] School of Computer Science and Technology, Wuhan University of Technology , Hubei Wuhan 430070,China

Abstract. For App recommended areas of mobile phone software, made while using conduct App application recommended combined weighted Slope One algorithm collaborative filtering algorithm items based on further improvement of the traditional collaborative filtering algorithm in cold start, data matrix sparseness and other issues, will recommend Spark stasis parallel algorithm platform, the introduction of real-time streaming streaming real-time computing framework to improve real-time software applications recommended.

1. Introduction

Nowadays, data mining in big data has been applied to all walks of life and become a new subject of internet. Based on the mobile terminal equipment, this paper puts forward a discussion on the recommended method for mobile application software. In the field of information recommendation, due to the huge data, the problem of overload of information against big data becomes more and more obvious, and it is difficult to select recommendation information suitable for individual users from hundreds of millions of massive data. To solve this problem, we propose the introduction of big data computing platform Spark, to handle large amounts of data is recommended, as well as the introduction of Spark Streaming real-time computing framework to improve inefficiencies recommend Recommended present in the system. This paper collaborative filtering recommendation algorithm and the Slope One algorithm, and for those existing algorithm problems are also discussed: the introduction of implicit information feedback, to make the non-score behavior of the user quantified as score values, by using predict initial scoring matrix by Slope One recommendation algorithm Grading, solve the data sparseness problem of the recommended algorithm. Meanwhile make the improved recommendation algorithm parallelization running on the Spark platform, realized mobile phone App application under the big data real-time recommendations. This article grabs Google application store and Baidu mobile application market of mobile phone App information and application rating data as the experimental sample data sets, has done the tests comparing the improved recommendation algorithm and traditional collaborative filtering algorithm, then calculate the mean absolute error of prediction score, and recommended Accuracy and recall rate. Finally, the experimental results obtained by the introduction of weighting Slope One algorithm based on the pre-filled article collaborative filtering algorithm can achieve better recommendations.

2. Item-based collaborative filtering and Slope One algorithms

2.1 Principle of item-based collaborative filtering recommendation

Item-based collaborative filtering (Item-based CF) recommendation is to calculate the degree of similarity of items and then select the most similar item in accordance with information of items for its recommendation to users, which users are interested in; and it is schematically shown in Figure 1.

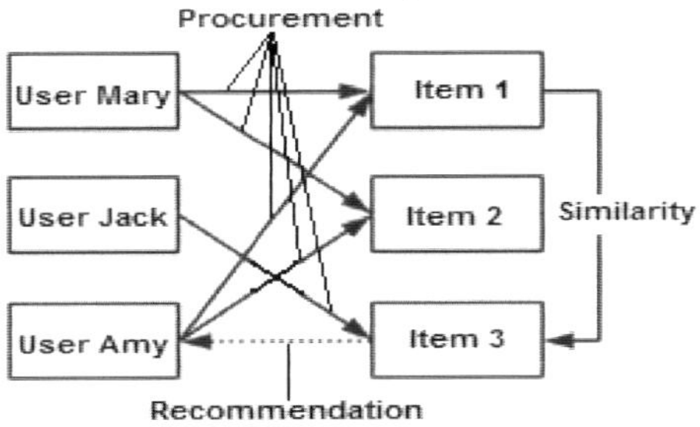

Figure 1. Schematic diagram of item-based recommendation

Users A and B simultaneously downloaded Softwares X and Z and Users C and B downloaded Softwares X and Y, respectively. It is assumed that downloading and non-downloading are scored as 5 and 0 and eigenvectors for Softwares X, Y and Z are defined as (5, 5, 5), (0, 5, 0) and (5, 5, 0), respectively; and it is assumed that the calculated Pearson Correlation Coefficients between Softwares X and Y and X and Z are $\sqrt{(1/3)}$ and $\sqrt{(2/3)}$, respectively, Item X is more similar to Item Z rather than Item Y. Thus, the system prefers to recommend Item Z to User C while he or she scores Item X or has other positive behaviors to Item X.

2.2 Limitations of recommendation of mobile Apps for traditional collaborative filtering algorithms

2.2.1 Data sparsity of the rating matrix: in accordance with the software applications with recommendation modules in the current market, the number of applications is far more than the number of scoring applications; and some users do not present any comment due to their own habits. As for more and more items, the matrix shall be very sparse while a user-item rating matrix is necessarily built; subsequently, the similarity calculation is more complicated and the results are less accurate.

2.2.2 Cold start: as for an App recommendation application, cold start means nay system recommendation may not be performed for a new user without any historical behavior; similarly, while a new mobile App is never downloaded, it may also be recommended t users.

2.2.3 Real-time capability: Due to the rapid updating of mobile Apps, new data is constantly increasing and users may only download an App for a certain period of time so that more timely and efficient real-time recommendation shall be necessary. Because calculation of a lot of data needs more time for a traditional collaborative filtering algorithm, regular updating of user data is usually necessary to update recommendations; however, this recommendation method is inefficient and may not meet the needs of users' getting the latest recommendations.

2.3 Principle and procedures of Slope One: The core concept of Slope One (or ascent algorithm) is also collaborative filtering, which is to predict scores of non-scoring Apps through calculation of deviations between mobile Apps and then in accordance with the similar Apps scored by users, whose deviations are smaller. It is assumed that there's sort of a straight-line relationship ($y = ax + b$) for the preference between two items. For modification of over-fitting while calculating the relevance, it is simplified as:

$$y = x + b \tag{1}$$

Determination of Parameter b is a preprocessing procedure of Slope One. It is assumed that each user's scoring vector set for Items A and B as $S = \{< A_1, B_1 >, < A_2, B_2 >, \dots\dots, < A_n, B_n >\}$, where: $A_1, A_2, \dots\dots,$ and A_n represent scores of Users 1, 2, ……, and n for Item A, respectively. Similarly, $B_1, B_2, \dots\dots,$ and B_n represent scores of Users 1, 2, ……, and n for Item B, respectively. Thus, the score difference (b) between Items A and B is estimated as:

$$b_{AB} = \frac{\sum_{i=1}^{n} B_i - A_i}{n} \quad (2)$$

Where: n represents the number of those users who evaluated Items A and B at the same time. b_{AB} represents the mean preference deviation of Items A and B.
Then, Users may predict scores of Items A and B.
If User u wants to predict scores of Item i, his or her scoring set ({User x, User y, User z,...}) shall be first selected and then his or her scored item set ({Item x, Item y, Item z,...}) shall be selected so as to form his or her scoring matrix in smaller dimensions; thus, his or her predicted scores (P(uA)) for Item A is expressed as:

$$P(uA) = \frac{\sum_{i \in P(u)} (b_{A,i} + r_{u,i})}{k} \quad (3)$$

Where: P(u) represents existing scores of User u; k represents the number of elements of Set P(u); and $r_{u,i}$ represents the User u's scores for Items i.
Because each App is independently scored, the weighted-calculation method (namely Weighted Slope One) is applied to balance effects of scores of various items, where the number (w_{AB}) of users who evaluated two items together with User u is regarded as the weight. Eq. (3) is modified as:

$$P(uA) = \frac{\sum_{i \in P(u)} (b_{A,i} - r_{u,i}) \cdot w_{AB}}{\sum_{i \in P(u)} w_{AB}} \quad (4)$$

In accordance with Slope One, scoring records of users who scored it are only involved into calculation while a certain item is predicted; thus, the results of calculating may be affected greatly for a smaller number of users; on the other hand, the collaborative filtering recommendation cold start may be relieved to some extent. Its workflow processes are shown in Figure 2.

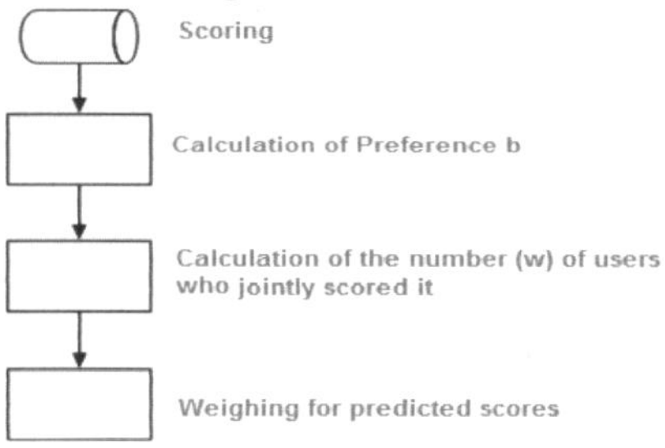

Figure 2. Slope One recommendation workflow processes

3. Improved scheme

Although the collaborative filtering recommendation algorithm is one most popular method, it still has a number of limitations. Our improvements are as follows:
Collecting users' implicit feedback data: the so-called implicit data are preference information which may be gained by means of observation of users' operation records rather than scores users intuitively present. While any user clicks, browses, stores, search or download any information in the recommendation system, these behaviors are quantified as the corresponding scores, which are filled into the user-item matrix to make up the defect (too sparse rating data).

Table 1. Quantization of users' implicit feedback scores

Behavior	Corresponding scores
Click	2
search	3
View details	3
Share	4
Download	5

In accordance with analysis of actual situations, while an App is scored below 2 scores by a user, this suggests that the App is not satisfied; on the other hand, while he or she presents any behavior in Table 1, this suggests he or she is interested in this App; thus, the scoring threshold is 2 scores. While a user performs various behaviors over the same time period, the corresponding maximum score shall be selected as the final score.
For improving those issues such as cold start due to being short of data, his or her downloaded mobile Apps as the initial data for any new user, which are scored as 5 scores for further recommendation. While information of any user's downloaded Apps may not be gained, recommendation may only be performed by means of those by means of Apps in Top Charts. In view of any new App, while scoring data of these Apps are not recorded, recommendation may be performed by searching keywords of Apps.
Next, the simple Slope One may be utilized to gain better results in case of relatively sparse behavior records and scoring data of users; thus, the user-item scoring matrix may be filled first with scores predicted by means of Slope One and then degree of similarity of items may be calculated; finally, the most similar item for recommendation may be determined in accordance with the scoring records of the recommended users.
For improvement of recommendation delay, the Spark Streaming real-time calculating framework was applied here, online and offline calculation is performed first to data; and Spark Streaming real-time receives scoring and implicit data from users. The recommendation model may be trained by using a lot of offline data in the recommendation engine so that these data may be processed for recommendation; moreover, the modified recommendation algorithm is implemented in parallel in the Spark platform; and recommendation results appear finally in the recommendation system. The process decomposition of offline and real-time data is schematically shown in Figure 3.

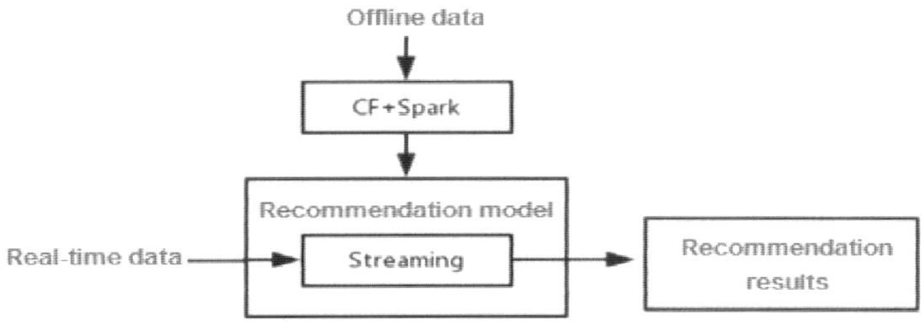

Figure 3. Schematic diagram of processing offline and online data

The entire modified recommendation flow chart is shown as follows:

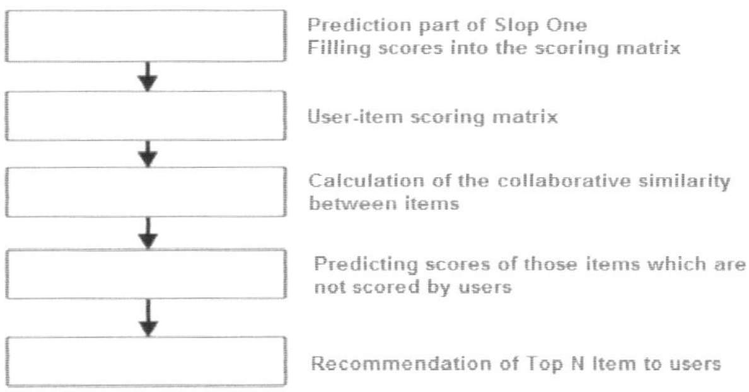

Figure 4. Recommendation flow chart modified by integrating Slope One

4. Experimental analysis of performances of the algorithm

4.1 Preparation

Recommendation of mobile Apps is faced with continuous and dramatic growth of user behavior data and calculation may be more intensive under the standalone mode. For optimization and future expansion of recommendation performances, recommendation calculation is arranged in the distributed Spark cluster to minimize calculation of a lot of data.

Data here were scraped from websites by means of the crawler developed with the python language tools, which come from the Baidu mobile application market and Google Play, respectively. More than 400,000 scoring data were used for the recommendation algorithm in Spark, which were uploaded to the HDFS file system in advance; and some data in the data set are shown in Table 2.

Table 2. App rating data set

User_id	App_id	Rating
70	417	5
305	15	1
201	370	1
57	409	4
125	173	5

The experimental PC is shown in Table 3.

Table 3. PC configurations

Operation system	CentOS 7
Cluster environment	Hadoop 2.4
Spark platform	Spark 2.0.0
IDE	IntelliJ IDEA2016
Single Node Memory	4G
Node number	2 (1 Master and 1 Worker)
Total host memory	32G

4.2 Experimental evaluation criteria

There are no uniform evaluation criteria for the recommendation algorithms and the recommendation effects of each system may be measured by means of various methods, some of which are measured by means of the user feedback mode or calculated recall and so on. Generally, the mean error between the scores predicted based on the test sample data set and the actual scores is regarded as the basis for determining the recommendation precision. The real-time capability is also required for some recommendation systems; thus, real-time recommendation performances are also a part of assessment of recommendation effects.

In general, the evaluation indexes for a recommendation system include Mean Absolute and Root Mean Square Errors (MAE and RSME), which are as follows:

$$\mathrm{MAE} = \frac{\sum_{i=1}^{m} |p_i - r_i|}{k} \tag{5}$$

$$\mathrm{RMSE} = \sqrt{\frac{\sum_{i=1}^{m} (p_i - r_i)^2}{k}} \tag{6}$$

Where: p_i represents the actual scores of Item i from users; r_i represents the predicted scores of Item i from users; and k represents the number of items scored by users.

Smaller MAE or RMSE indicates that the predicted scores more approach the actual scores. In accordance with our experiments, the difference between the predicted and actual RMSEs is larger; thus, a higher recommendation precision is required.

In view of implementation of the entire recommendation, two concepts (namely precision and recall) are put forward, which are expressed as follows:

$$\mathrm{Precision} = \frac{\sum_u |R(u) \cap P(u)|}{\sum_u |P(u)|} \tag{7}$$

$$\text{Recall} = \frac{\sum_u |R(u) \cap P(u)|}{\sum_u |R(u)|} \tag{8}$$

Where: R(u) represents the set of items actually operated by User u; and P(u) represents the set of items recommended to users. In view of each recommendation scene, its precision and recall are different. As for recommendation of the mobile APPs, the precision represents the users' actual download rate of recommended Apps and Recall represents the percentage of recommended Apps in the software actually downloaded by users.

4.3 Experimental design and results analysis

*4.3.1 Absolutely flat All wrong MAE poor.*Indicates that a smaller MAE represents a lower difference between the actual and predicted scores. The data set was divided into two parts (namely the training and measurement sets) in accordance with the ratio (8:2) during our experiments. The final value shall be finally stabilized to a certain value after a few trainings.

The crawled App scoring data set was only utilized for comparison of Slope One and Weighted Slope One. The calculated MAEs are given in Table 4.

Table 4. Comparison of Slope One and Weighted Slope One

	Weighted Slope One	Slope One
MAE	0.845	0.877

The above MAEs indicate that the weighted Slope One has a lower predicted score error so that it shall be selected to fill some predicted scores into the original scoring matrix; and the recommendation algorithm shall be modified based on the weighted Slope One.

Experimental results for the modified recommendation algorithm are as follows:

Table 5. Comparison of MAEs for the recommendation algorithm under various data sets

Data Set	**Item-based CF**	**Item-based CF+slope one**
Movielens 100k	0.8325	0.812
Movielens 1M	0.8075	0.799
App rating	0.852	0.84

The comparison results are shown in Figures 5 and Figures 6, which indicate that:

- The larger the data set or the number of the training set is, the smaller MAE is and the better the recommendation effects are.
- 2) The minimum MAEs for the optimization and item-based CF algorithms are 0.792 and 0.8075, respectively. While data are fewer and the matrix is sparse, the precision of the modified algorithm is higher.
- 3) Due to our data acquired from 27,775 users and scoring information got from only 1691 Apps, the scoring matrix is too sparse so that the effects for prediction of App scores shall be the weakest.

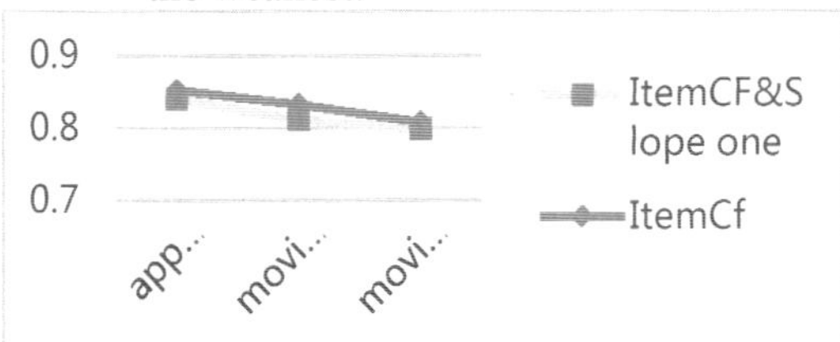

Figure 5. Comparison line chart for various algorithms under various data sets

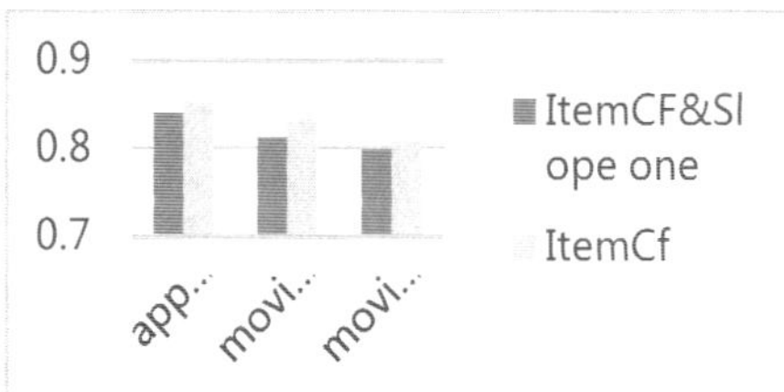

Figure 6. Comparison bar chart for various algorithms under various data sets

4.3.2 Recall and Precision. As for recommendation of mobile Apps, the precision is represented by means of the users' download rate in the recommendation results, which may be regarded as the number of Apps scored as 5 scores by users in the recommendation table based on the implicit feedback quantization scores; on the other hand, the recall may be represented by the percentage of the Apps scored as 5 scores by all users in the mobile App software scoring table. Thus, higher precision and recall necessarily depend on setting a rational recommendation number (N), which is set as 5, 10, 15, 20 and 25 for 5 experimental groups, respectively; and the rate of the training and measurement sets is 8:2. The data set is movielens 1M scoring.

During the experimental processes, the integer (5 scores) seldom occurs because calculation results are not integers. For ensuring the experimental effects, the predicted scores (above 4.8 scores) may regarded as the standards for calculation of the precision and recall; and the corresponding precision and recall results are shown in Table 6 and the line chart is presented in Figure 7.

Table 6. Precision and recall results under various N values

Evaluation value \N	5	10	15	20	25
ItemCF precision	24.32%	19.38%	18.77%	17.65%	17.01%
ItemCF recall	12.08%	13.31%	16.41%	18.59%	22.7%
Precision of the modified algorithm	24.97%	21.87%	19.36%	18.03%	17.55%
Recall of the modified algorithm	13.51%	14.22%	17.86%	19.57%	23.01%

Figure 7 reveals the precision and recall trends for the general item-based CF algorithm and the modified CF algorithm integrated with Slope One under various N values, where the precision and recall are slightly higher for the latter. Concepts of precision and recall indicate that: the recall may fall while the precision rises and they tends to the balance within a certain interval; the precision is among 16-25% and tends to decline along with the growth of the recommendation number (N) while the precision is only taken into account; namely, a larger recommendation number leads to falling of the percentage of Apps liked by users; on the other hand, the recall rises among 12-23%;and such case indicates the users download these favorable Apps in all likelihood; thus, recommendation Apps account for more Apps downloaded by users. The optimal N values for curve intersections are among 15-20.

The modified recommendation algorithm is parallelized in the Spark engine and the scoring data are updated by means of the Streaming calculation framework to perform real-time calculation of big data. Analysis was carried out to those aspects such as recommendation results change, evaluation of recommendation effects and real-time processing time by means of simulation of streaming data for recommendation results, respectively.

By real-time inserting 1,000 new scoring data in batches in the netcat terminal, the batch duration for the Streaming context receiving streaming data and the Top N are set as 20s and 15, respectively. The operation period is set as 1 hour; and it is predicted that real-time data will be received in 180 batches and one calculation period is 20s.

The data calculation time curve is shown in Figure 8.

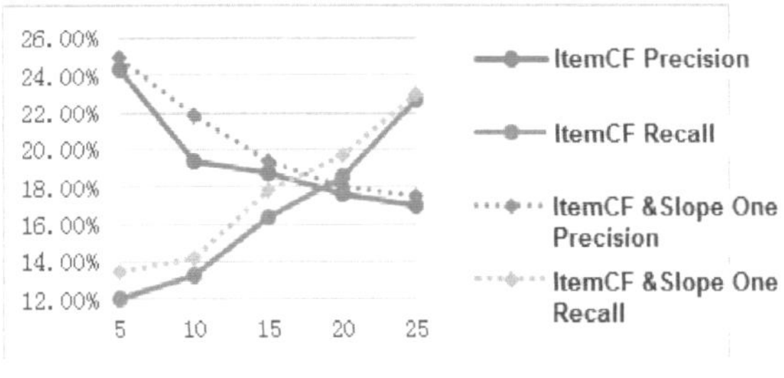

Figure 7. Recall and precision

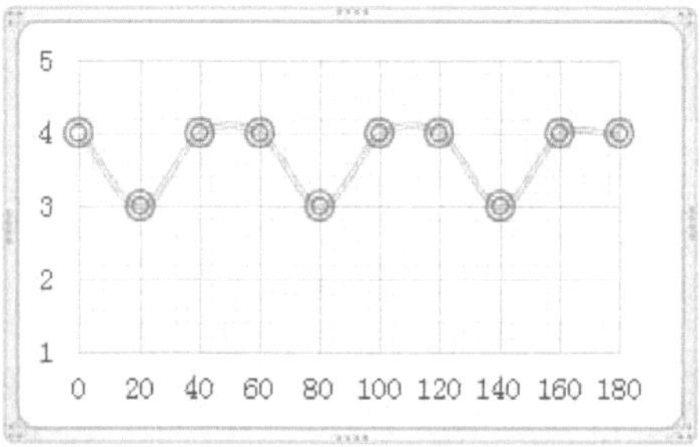

Figure 8. Training time consumption curve

The horizontal and vertical coordinate axes (Figure 8) are batches and training time consumption, respectively. The period for each batch is 20s and the mean training time consumption is 3.7s.
The receiving data streaming bars are shown in Figure 9.

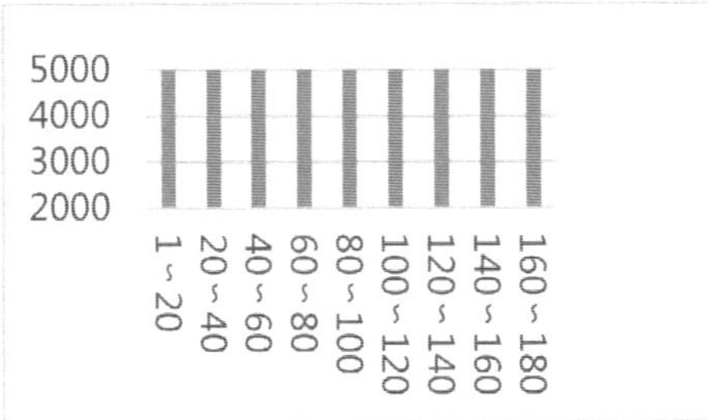

Figure 9. receiving data streaming bars

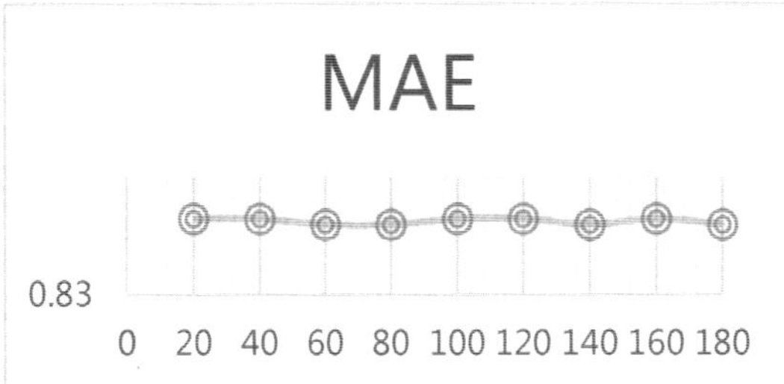

Figure 10. MAE changes

Streaming contexts receive new scoring data real-time input per batch, to which calculation is carried out by means of streaming. Figure 9 indicates that the statistical unit is 20 batches and 5,000 new data were received in total; moreover, while about 1,000 data input in 5 batches are selected stochastically and those data input in other batches are 0 from the above data, the entire process may not be broken down or restarted for calculation.
MAE changes are shown in Figure 10 after inputting the real-time data. Calculated MAEs indicate that they do not change greatly, which are among 0.84-0.845; thus, calculation of new input data may not greatly change recommendation errors.

5. Conclusions

Based on study and application of recommendation of mobile Apps in the Spark Streaming framework, discussion was carried out here by focusing three aspects (namely those recommendation algorithms (such as Item-based CF and Slope One algorithms), Spark real-time streaming calculation in the distributed big data platform and mobile App Store clients based on the mobile devices.
As for the recommendation algorithm, evaluation experiments were carried out for verification of the effectiveness of the modified CF recommendation algorithm; and MAEs for Item-based CF and Slope One algorithms were calculated under various data sets, respectively; and the CF algorithm was then performed in the distributed big data platform Spark where real-time processing may be quickened and whose fault tolerance and expandability are more intensive than those of Hadoop (as a similar big data processing platform).

6. References

[1] Hu Yuxiang 2015 *Design and implementation of recommendation system based on Spark [D]* (Zhejiang University)
[2] Chen Yuzhao 2014 *Study and implementation of Personalized Recommendation based on Big Data [D]* (Xidian University)
[3] Guo Xiaobo, Zhao Shuliang and Niu Dongpan 2015 *A novel combination recommendation method for solving sparse and cold start problems [J.* (Journal of University of Science and Technology of China) 45(10):804-812
[4] Li Xue, Zuo Wanli, Hao Fengling, et al 2009 *An Improved Item-Based Collaborative Filtering Recommendation Algorithm [C]*
[5] Zhang Xiande 2016 *Research and Implementation of Real Time Stream Computing Recommendation System Based on Spark Platform [D]* (Jiangsu University)
[6] Li J, Feng and P, Lv J 2013 *An improved slope one algorithm for collaborative filtering [C]* (Ninth International Conference on Natural Computation IEEE) pp 1118-1123.
[7] Panigrahi S, Lenka R K and Stitipragyan A 2016 *A Hybrid Distributed Collaborative Filtering Recom-mender Engine Using Apache Spark [J]* (Procedia Computer Science) 83:1000-1006.

IOP Conf. Series: Materials Science and Engineering **359** (2018) 012009 doi:10.1088/1757-899X/359/1/012009

Open-closed-loop iterative learning control for a class of nonlinear systems with random data dropouts

X Y Cheng*, H B Wang, Y L Jia and YH Dong

The Key Lab of Industrial Computer Control Engineering of Hebei Province, Yanshan University, Qinhuangdao 066004, China

E-mail: 94180102@qq.com

Abstract: In this paper, an open-closed-loop iterative learning control (ILC) algorithm is constructed for a class of nonlinear systems subjecting to random data dropouts. The ILC algorithm is implemented by a networked control system (NCS), where only the off-line data is transmitted by network while the real-time data is delivered in the point-to-point way. Thus, there are two controllers rather than one in the control system, which makes better use of the saved and current information and thereby improves the performance achieved by open-loop control alone. During the transfer of off-line data between the nonlinear plant and the remote controller data dropout occurs randomly and the data dropout rate is modeled as a binary Bernoulli random variable. Both measurement and control data dropouts are taken into consideration simultaneously. The convergence criterion is derived based on rigorous analysis. Finally, the simulation results verify the effectiveness of the proposed method.

1. INTRODUCTION

Iterative learning control (ILC) is a control method applied to systems where the same task is executed repetitively and periodically in a finite time interval, such as a robotic manipulator in a manufacturing environment or a chemical reactor in a batch processing application. By ILC, there is no need to know the system model completely, but the previous and/or current information is used to update the control law in order to achieve a better performance as the iteration process repeats. Because of its simplicity and effectiveness, ILC has been studied extensively by experts and scholars with significant progress in both theory and application since it was proposed by Arimoto in 1984 [1]. However, among all the achievements that have been made so far, the most centralized topics are the convergence analysis [2-3] and the robustness [4-6] of the proposed ILC methods.

On the other hand, owing to the rapid development of sensing, information processing, and communication technologies, the research in the emerging area of networked control systems (NCSs) has attracted considerable attention in the research community [7]. As a result, the introduction of the communication network to control systems has made great changes on the conventional control structure gradually. Compared with the traditional point-to-point control systems, such NCSs bring some new merits such as low power consumption and cost, easy installation and maintenance, reduced system wiring, and increased system flexibility. However, during data packets transfer between the remote ILC controller and the controlled plant, data dropout is inevitable and unpredictable due to restrictions such as network failure and the limited bandwidth in network communication. It is well known that ILC is a data-based method and data dropouts are potential sources of instability and poor

performance for the system. So, it is rewarding to pay attention to the study of ILC with data dropouts in NCSs.
Some pioneer work about iterative learning control with data dropouts has been made by scholars. Ahn and his co-workers adopt an existing Kalman filtering approach to design robust iterative learning controller in [8]. Each component in the multivariable output vector of the plant must have the same data dropout rate, which is a limitation in applications. To get rid off the limitation in [8], the situation is considered in [9] where each component in the multivariable output vector of the plant subjects to an independent data dropout rate. For simplicity, both [8] and [9] just take the measurement data dropout into account. In [10], both measurement and control data dropouts are considered by Ahn and his co-workers. The framework established in [10] is able to cover various uncertainties, which makes it possible to deal with more general engineering problems. Bu and his co-workers concentrate on ILC algorithms for discrete linear systems subjecting to measurement dropouts in their early works [11-14]. Super-vector approach, which is based on a lifting technique, is widely used in their studies on account of the advantage that the two-dimensional problem of ILC can be changed into the one-dimensional multi-input multi-output problem. Notably, the stability of high order ILC schemes with data dropouts is discussed in [12-13]. In [14], H_∞ iterative learning controller design has been discussed. Then they extend their studies to systems where data dropouts are modeled as stochastic variables satisfying the Bernoulli binary random distribution [15-17]. As super-vector approach can't be used in the nonlinear systems, a contraction mapping approach is proposed in [15-16]. In [17], the ILC design has been investigated based on 2-D stochastic Roesser system and the condition for mean-square asymptotic stability is given by a linear matrix inequality (LMI) technique. In paper [18], the bad effect of data dropouts on convergence speed and stability of the system is discussed. Shen and his co-workers focus on the study of stochastic iterative learning control (SILC) which is defined as ILC for systems containing system noises, measurement noises, random packet losses and other stochastic signals [19-22]. The data dropout is modeled by an arbitrary stochastic sequence with bounded length requirement and the almost sure convergence analysis is given, which are the remarkable differences from the above. In addition, Liu et al. apply an averaging ILC scheme in [23] to overcome the random factors in nonlinear non-affine processes with random data dropouts. Wang et al. give the convergence conditions for the robust ILC method based on contraction mapping approach and a forgetting factor is introduced to obtain a better performance in [24]. Except for the achievements presented above, there are still many unmentioned scholars and researchers who have made prominent contributions to the study of the ILC with data dropouts.
However, all the aforementioned papers are restricted to open-loop ILC update laws. It is very possible that the tracking error can become quite large before finally converging to zero even when the applied ILC algorithm has been proved to be convergent, because the learning control structure is basically open-loop. Inspired by [25], a closed-loop is added to the existing open-loop ILC method for a class of nonlinear systems suffering data dropouts in this paper. In addition to eliminating the huge overshoot, faster convergence speed and better performance in stability and robustness of the system are obtained because of the introduction of the closed-loop. The main contributions of this paper different from [25] are stated as follows: i) Linear systems are considered in [25] while a class of nonlinear systems, which is more complex, are studied in this paper. ii) The condition of convergence is derived with the analytical methods of 2–D system in [25] while contraction mapping approach is used in this paper. iii) When data dropout occurs, the current data will be zero in [25], while the previous data will be used instead in this paper, which can weaken the bad effect on the system caused by data dropout in some degree.
The remainder of this paper is organized as follows. The problem formulation for a class of single-input-single-output (SISO) nonlinear systems and design of controllers are described in section 2. In section 3, the convergence condition for the SISO nonlinear system is given and the convergence analysis is provided. In section 4, the result of the SISO nonlinear system is extended to a class of multi-input-multi-output (MIMO) nonlinear systems. In section 5, a numerical example is given to verify the effectiveness of the proposed method. Finally, the conclusion is given in section 6.

2. PROBLEM FORMULATION AND DESIGN OF CONTROLLERS

2.1. Problem formulation

Consider a SISO nonlinear discrete-time system [16] described in the following form:

$$\begin{cases} x_i(k+1) = f(x_i(k)) + b(x_i(k))u_i(k), \\ y_i(k) = g(x_i(k)) + d(x_i(k))u_i(k), \end{cases} \quad (1)$$

IOP Conf. Series: Materials Science and Engineering **359** (2018) 012009 doi:10.1088/1757-899X/359/1/012009

where the subscript i denotes the iteration number and k denotes the time instance in an iteration process, $i = 0,1,2,L$, $k \in [0,1,2,L\ N]$; $x_i(k) \in R^1$, $u_i(k) \in R^1$ and $y_i(k) \in R^1$ are the state, control input and output of the system, respectively.

Some relevant assumptions of the system (1) are given as follows.

Assumption 1: The desired trajectory $y_d(k)$ is iteration invariant.

Assumption 2: For $y_d(k)$, there exist unique $u_d(k)$ and $x_d(k)$ satisfying the following equations:

$$\begin{cases} x_d(k+1) = f(x_d(k)) + b(x_d(k))u_d(k), \\ y_d(k) = g(x_d(k)) + d(x_d(k))u_d(k), \end{cases} \tag{2}$$

where $u_d(k)$ is the desired control input and $x_d(k)$ is the desired state.

Assumption 3: The resetting condition below is satisfied for every iteration.

$$x_i(0) = x_d(0), \tag{3}$$

where $x_d(0)$ is the initial value of the desired state and $x_i(0)$ is the initial value of the ith iteration.

Assumption 4: The nonlinear function $f(\cdot)$, $b(\cdot)$, $g(\cdot)$, and $d(\cdot)$ are global Lipschitz in x, that is, $\forall k \in [0, N]$, $i = 0,1,2,3L$, $\exists$ positive bounded constants k_f, k_b, k_g and k_d, satisfying the following inequations:

$$|f(x_1(k)) - f(x_2(k))| \le k_f |x_1(k) - x_2(k)|, \tag{4}$$
$$|b(x_1(k)) - b(x_2(k))| \le k_b |x_1(k) - x_2(k)|, \tag{5}$$
$$|g(x_1(k)) - g(x_2(k))| \le k_g |x_1(k) - x_2(k)|, \tag{6}$$
$$|d(x_1(k)) - d(x_2(k))| \le k_d |x_1(k) - x_2(k)|, \tag{7}$$

for any pair $(x_1(k), x_2(k))$ in $R \times R$, where $x_1, x_2 \in x_i(k), i = 0,1,2,3L\ , k = 0,1,2,L\ , N$.

The structure of the NCS in this paper is illustrated in Fig. 1 and the main difference from the general NCS is that there are two controllers in the system rather than one. Thus, the control input $u_i(k)$ consists of two parts, one is $u_{ff,i}(k)$ produced by the feed-forward controller and the other is $u_{fb,i}(k)$ produced by the feed-back controller. This can be described as follows:

$$u_i(k) = u_{ff,i}(k) + u_{fb,i}(k), \tag{8}$$

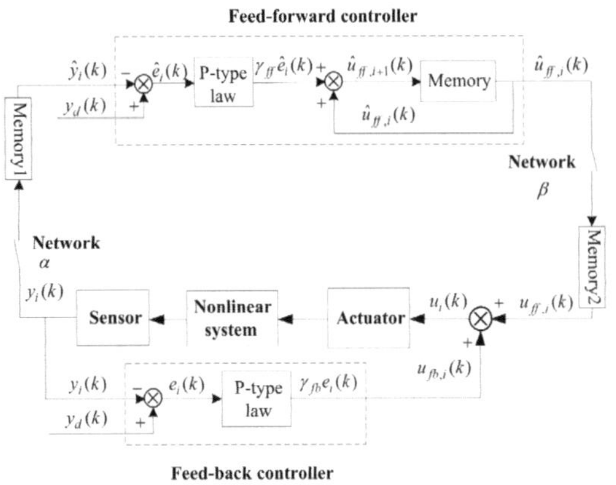

Fig. 1. The block diagram of the NCS

It's obvious that only the data of the feed-forward controller is interchanged by the network. However, as the local controller, the data of the feed-back controller is delivered in the point-to-point way, which makes it no sense to consider data dropouts in feed-back loop. Then, the following discussion will concentrate on the feed-forward loop. As shown in Fig. 1, there are two different types of data dropouts. The first one occurs when the measurement output $y_i(k)$ is transmitted between the sensor and the remote controller, and the other one occurs when the control input $\hat{u}_{ff,i}(k)$ generated by the remote controller is delivered to the actuator. In order to describe this situation vividly, the communication network can be regarded as a switch that opens and closes in a random way. Suppose data successful transfer rate α and β are stochastic parameters, which consist of independent and identically distributed Bernoulli random variables taking the value of 0 or 1 with

$$\text{prob}\{\alpha=1\}=\varepsilon\{\alpha\}=\bar{\alpha},\ 0\le\bar{\alpha}\le 1,$$
$$\text{prob}\{\beta=1\}=\varepsilon\{\beta\}=\bar{\beta},\ 0\le\bar{\beta}\le 1,$$

where $\bar{\alpha}$ and $\bar{\beta}$ are known constants and $\varepsilon\{\cdot\}$ stands for the mathematical expectation of $\{\cdot\}$. It is assumed that α and β are independent of each other. When the switch is open, the current data packet is lost and the previous data stored in the Memory1/Memory2 will be used instead, otherwise, the current data will be delivered successfully and the data stored in the Memory1/ Memory2 will be updated synchronously. That is, the actual value $\hat{y}_i(k)$ received by the feed-forward controller is the current output $y_i(k)$ with the probability of $\bar{\alpha}$ and the previous value $y_i(k-1)$ with the probability of $1-\bar{\alpha}$. Similarly, the actual input $u_{ff,i}(k)$ transmitted to the actuator is the current feed-forward controller output $\hat{u}_{ff,i}(k)$ with the probability of $\bar{\beta}$ and the previous value $\hat{u}_{ff,i}(k-1)$ with the probability of $1-\bar{\beta}$. These can be represented as follows:

$$\hat{y}_i(k)=\alpha y_i(k)+(1-\alpha)y_i(k-1), \tag{9}$$

$$u_{ff,i}(k)=\beta\hat{u}_{ff,i}(k)+(1-\beta)\hat{u}_{ff,i}(k-1). \tag{10}$$

2.2. Design of controllers

An open-loop P-type ILC law is adopted for the remote controller while a closed-loop P-type law is used for the local controller. The ILC laws are shown below:

$$\hat{u}_{ff,i}(k)=\hat{u}_{ff,i-1}(k)+\gamma_{ff}\hat{e}_{i-1}(k), \tag{11}$$

$$u_{fb,i}(k)=\gamma_{fb}e_i(k), \tag{12}$$

where γ_{ff} and γ_{fb} are the learning gain of the remote controller and the local controller, respectively. $\hat{e}_{i-1}(k)=y_d(k)-\hat{y}_{i-1}(k)$ and $e_i(k)=y_d(k)-y_i(k)$.

3. CONVERGENCE ANALYSIS

In this section, the convergence condition of the controllers for the SISO nonlinear system is given and proved.

Theorem 1: For system (1), assume Assumptions 1-4 hold, applying the update laws (11) and (12) to the feed-forward controller and the feed-back controller, respectively, if $\rho_{i,k}=\left|\dfrac{1-(\bar{\beta}\gamma_{ff}\bar{\alpha}-\gamma_{fb})d(x_i(k))}{1+\gamma_{fb}d(x_{i+1}(k))}\right|<1$ is satisfied for all *i* and *k*, then $\lim_{i\to\infty}\varepsilon\{|y_i(k)|\}=y_d(k)$ is guaranteed for all *k*.

Proof: By introducing equations (10), (11) and (12) into equation (8) and making subtraction of the adjacent iterations, we have

IOP Conf. Series: Materials Science and Engineering **359** (2018) 012009 doi:10.1088/1757-899X/359/1/012009

$$\begin{aligned}
&u_{i+1}(k)-u_i(k)\\
&=u_{ff,i+1}(k)+u_{fb,i+1}(k)-u_{ff,i}(k)-u_{fb,i}(k)\\
&=\beta(\hat{u}_{ff,i+1}(k)-\hat{u}_{ff,i}(k))+(1-\beta)(\hat{u}_{ff,i+1}(k-1)\\
&-\hat{u}_{ff,i}(k-1))+\gamma_{fb}(e_{i+1}(k)-e_i(k))\\
&=\beta\gamma_{ff}\hat{e}_i(k)+(1-\beta)\gamma_{ff}\hat{e}_i(k-1)+\gamma_{fb}(e_{i+1}(k)\\
&-e_i(k)),
\end{aligned} \tag{13}$$

where $\hat{e}_i(k)$ and $e_i(k)$ conform to the following relationship

$$\begin{aligned}
&\hat{e}_i(k)\\
&=y_d(k)-\alpha y_i(k)-(1-\alpha)y_i(k-1)\\
&=\alpha e_i(k)+(1-\alpha)e_i(k-1).
\end{aligned} \tag{14}$$

Substituting (14) for $\hat{e}_i(k)$ in (13) leads to

$$\begin{aligned}
&u_{i+1}(k)\\
&=u_i(k)+\gamma_{fb}e_{i+1}(k)+(\beta\gamma_{ff}\alpha-\gamma_{fb})e_i(k)\\
&+[\beta\gamma_{ff}(1-\alpha)+(1-\beta)\gamma_{ff}\alpha]e_i(k-1)\\
&+(1-\beta)\gamma_{ff}(1-\alpha)e_i(k-2).
\end{aligned} \tag{15}$$

Note that

$$\begin{cases}
y_i(k)=g(x_i(k))+d(x_i(k))u_i(k),\\
y_d(k)=g(x_d(k))+d(x_d(k))u_d(k),
\end{cases} \tag{16}$$

thus

$$\begin{aligned}
&e_i(k)\\
&=y_d(k)-y_i(k)\\
&=g(x_d(k))+d(x_d(k))u_d(k)-g(x_i(k))\\
&-d(x_i(k))u_i(k)\\
&=g(x_d(k))-g(x_i(k))+\left[d(x_d(k))\right.\\
&\left.-d(x_i(k))\right]u_d(k)+d(x_i(k))\delta u_i(k),
\end{aligned} \tag{17}$$

where $\delta u_i(k)=u_d(k)-u_i(k)$.

For simplicity in writing, define

$$\begin{aligned}
&h(x_i(k))\\
&=g(x_d(k))-g(x_i(k))\\
&+\left[d(x_d(k))-d(x_i(k))\right]u_d(k),
\end{aligned} \tag{18}$$

then (17) can be rewritten as

$$e_i(k)=h(x_i(k))+d(x_i(k))\delta u_i(k). \tag{19}$$

Replacing $e_i(k)$ for (19) in (15) leads to

$$\begin{aligned}&u_{i+1}(k)\\&=u_i(k)+l_1\left[h(x_{i+1}(k))+d(x_{i+1}(k))\delta u_{i+1}(k)\right]\\&+l_2\left[h(x_i(k))+d(x_i(k))\delta u_i(k)\right]+l_3\left[h(x_i(k-1))\right.\\&\left.+d(x_i(k-1))\delta u_i(k-1)\right]+l_4\left[h(x_i(k-2))+\right.\\&\left.d(x_i(k-2))\delta u_i(k-2)\right],\end{aligned}\tag{20}$$

where

$$l_1=\gamma_{fb},\ l_2=\beta\gamma_{ff}\alpha-\gamma_{fb},$$

$$l_3=\beta\gamma_{ff}(1-\alpha)+(1-\beta)\gamma_{ff}\alpha,$$

$$l_4=(1-\beta)\gamma_{ff}(1-\alpha).$$

Subtracting both sides of the equation (20) from $u_d(k)$, we have

$$\begin{aligned}&\delta u_{i+1}(k)\\&=\delta u_i(k)-l_1\left[h(x_{i+1}(k))+d(x_{i+1}(k))\delta u_{i+1}(k)\right]\\&-l_2\left[h(x_i(k))+d(x_i(k))\delta u_i(k)\right]-l_3\left[h(x_i(k-1))\right.\\&\left.+d(x_i(k-1))\delta u_i(k-1)\right]-l_4\left[h(x_i(k-2))+\right.\\&\left.d(x_i(k-2))\delta u_i(k-2)\right],\end{aligned}\tag{21}$$

where $\delta u_{i+1}(k)=u_d(k)-u_{i+1}(k)$.
After arranging and combining like terms, (21) can be rewritten as

$$\begin{aligned}&\left[1+l_1d(x_{i+1}(k))\right]\delta u_{i+1}(k)\\&=\left[1-l_2d(x_i(k))\right]\delta u_i(k)-l_3d(x_i(k-1))\cdot\delta u_i(k-1)\\&-l_4d(x_i(k-2))\delta u_i(k-2)-l_1h(x_{i+1}(k))\\&-l_2h(x_i(k))-l_3h(x_i(k-1))-l_4h(x_i(k-2)).\end{aligned}\tag{22}$$

Considering the Lipschitz condition in Assumption 4, (18) gives

$$\begin{aligned}&\left|h(x_i(k))\right|\\&\le\left|g(x_d(k))-g(x_i(k))\right|+\left|u_d(k)\right|\cdot\\&\left|d(x_d(k))-d(x_i(k))\right|\\&\le\left|k_g+k_db_{ud}\right|\left|\delta x_i(k)\right|,\end{aligned}\tag{23}$$

where $b_{ud}=\sup\limits_{k\in[0,N]}\left|u_d(k)\right|$, $\delta x_i(k)=x_d(k)-x_i(k)$.

Similarly, considering the Lipschitz condition in Assumption 4 for (22) and substituting (23) for $\left|h(x_i(k))\right|$ result in

$$\begin{aligned}&\left|1+l_1d(x_{i+1}(k))\right|\left|\delta u_{i+1}(k)\right|\\&\le\left|1-l_2d(x_i(k))\right|\left|\delta u_i(k)\right|+\left|l_3d(x_i(k-1))\right|\cdot\\&\left|\delta u_i(k-1)\right|+\left|l_4d(x_i(k-2))\right|\left|\delta u_i(k-2)\right|\\&+\left|k_g+k_db_{ud}\right|\left(\left|l_1\right|\left|\delta x_{i+1}(k)\right|+\left|l_2\right|\left|\delta x_i(k)\right|\right.\\&\left.+\left|l_3\right|\left|\delta x_i(k-1)\right|+\left|l_4\right|\left|\delta x_i(k-2)\right|\right).\end{aligned}\tag{24}$$

Taking expectation in both sides of (24), we have

$$\begin{aligned}
&\left|1+\bar{l}_1 d(x_{i+1}(k))\right| \varepsilon\left\{\left|\delta u_{i+1}(k)\right|\right\} \\
&\leq\left|1-\bar{l}_2 d(x_i(k))\right| \varepsilon\left\{\left|\delta u_i(k)\right|\right\}+\left|\bar{l}_3 d(x_i(k-1))\right| \cdot \\
&\varepsilon\left\{\left|\delta u_i(k-1)\right|\right\}+\left|\bar{l}_4 d(x_i(k-2))\right| \varepsilon\left\{\left|\delta u_i(k-2)\right|\right\} \\
&+\left|k_g+k_d b_{ud}\right|\left(\left|\bar{l}_1\right| \varepsilon\left\{\left|\delta x_{i+1}(k)\right|\right\}+\left|\bar{l}_2\right| \varepsilon\left\{\left|\delta x_i(k)\right|\right\}\right. \\
&\left.+\left|\bar{l}_3\right| \varepsilon\left\{\left|\delta x_i(k-1)\right|\right\}+\left|\bar{l}_4\right| \varepsilon\left\{\left|\delta x_i(k-2)\right|\right\}\right),
\end{aligned} \tag{25}$$

where

$$\begin{aligned}
&\bar{l}_1=\gamma_{fb},\ \bar{l}_2=\bar{\beta}\gamma_{ff}\bar{\alpha}-\gamma_{fb}, \\
&\bar{l}_3=\bar{\beta}\gamma_{ff}(1-\bar{\alpha})+(1-\bar{\beta})\gamma_{ff}\bar{\alpha}, \\
&\bar{l}_4=(1-\bar{\beta})\gamma_{ff}(1-\bar{\alpha}).
\end{aligned}$$

Both sides of (25) are divided by $\left|1+\bar{l}_1 d(x_{i+1}(k))\right|$, and then we can obtain that

$$\begin{aligned}
&\varepsilon\left\{\left|\delta u_{i+1}(k)\right|\right\} \\
&\leq \rho_{i,k}\varepsilon\left\{\left|\delta u_i(k)\right|\right\}+\rho_{i1,k}\varepsilon\left\{\left|\delta u_i(k-1)\right|\right\}+ \\
&\rho_{i2,k}\cdot\varepsilon\left\{\left|\delta u_i(k-2)\right|\right\}+\nu_{i+1,k}\varepsilon\left\{\left|\delta x_{i+1}(k)\right|\right\} \\
&+\nu_{i,k}\varepsilon\left\{\left|\delta x_i(k)\right|\right\}+\nu_{i1,k}\varepsilon\left\{\left|\delta x_i(k-1)\right|\right\} \\
&+\nu_{i2,k}\varepsilon\left\{\left|\delta x_i(k-2)\right|\right\}
\end{aligned} \tag{26}$$

where

$$\begin{aligned}
&\left|1+\bar{l}_1 d(x_{i+1}(k))\right| \neq 0,\ \rho_{i,k}=\frac{\left|1-\bar{l}_2 d(x_i(k))\right|}{\left|1+\bar{l}_1 d(x_{i+1}(k))\right|}, \\
&\rho_{i1,k}=\frac{\left|\bar{l}_3 d(x_i(k-1))\right|}{\left|1+\bar{l}_1 d(x_{i+1}(k))\right|},\ \rho_{i2,k}=\frac{\left|\bar{l}_4 d(x_i(k-2))\right|}{\left|1+\bar{l}_1 d(x_{i+1}(k))\right|}, \\
&\nu_{i+1,k}=\frac{\left|k_g+k_d b_{ud}\right|\left|\bar{l}_1\right|}{\left|1+\bar{l}_1 d(x_{i+1}(k))\right|},\ \nu_{i,k}=\frac{\left|k_g+k_d b_{ud}\right|\left|\bar{l}_2\right|}{\left|1+\bar{l}_1 d(x_{i+1}(k))\right|}, \\
&\nu_{i1,k}=\frac{\left|k_g+k_d b_{ud}\right|\left|\bar{l}_3\right|}{\left|1+\bar{l}_1 d(x_{i+1}(k))\right|},\ \nu_{i2,k}=\frac{\left|k_g+k_d b_{ud}\right|\left|\bar{l}_4\right|}{\left|1+\bar{l}_1 d(x_{i+1}(k))\right|}.
\end{aligned}$$

It is clear that

$$\begin{aligned}
&\delta x_i(k) \\
&=x_d(k)-x_i(k) \\
&=f(x_d(k-1))-f(x_i(k-1)) \\
&+\left[b(x_d(k-1))-b(x_i(k-1))\right]u_d(k-1) \\
&+b(x_i(k-1))\delta u_i(k-1),
\end{aligned} \tag{27}$$

taking absolute value of both sides of the equation (27) and applying Lipschitz condition lead to

$$\begin{aligned}
&|\delta x_i(k)| \\
&\le |f(x_d(k-1)) - f(x_i(k-1))| \\
&+|b(x_d(k-1)) - b(x_i(k-1))||u_d(k-1)| \\
&+|b(x_i(k-1))||\delta u_i(k-1)| \\
&\le (k_f + k_b b_{ud})|\delta x_i(k-1)| + |b(x_i(k-1))| \\
&\cdot|\delta u_i(k-1)| \\
&= k_1 |\delta x_i(k-1)| + k_2 |\delta u_i(k-1)|,
\end{aligned} \tag{28}$$

where $k_1 = k_f + k_b b_{ud}$, $k_2 = \sup_{k\in[0,N]} b(x_i(k))$.

Since $|\delta x_i(0)| = |x_d(0) - x_i(0)| = 0$, by substituting the relation recursively for $|\delta x_i(j)|$, $j = 0,1,2,3L\ k$, the following expression can be obtained

$$|\delta x_i(k)| \le \sum_{p=1}^{k} k_1^{p-1} k_2 |\delta u_i(k-p)|, \tag{29}$$

thus, we have

$$\begin{aligned}
&\varepsilon\{|\delta u_{i+1}(k)|\} \\
&\le \rho_{i,k}\varepsilon\{|\delta u_i(k)|\} + \rho_{i1,k}\varepsilon\{|\delta u_i(k-1)|\} + \rho_{i2,k} \\
&\cdot\varepsilon\{|\delta u_i(k-2)|\} + v_{i+1,k}\sum_{p=1}^{k} K_{0,p}\varepsilon\{|\delta u_{i+1}(k-p)|\} \\
&+v_{i,k}\sum_{p=1}^{k} K_{1,p}\varepsilon\{|\delta u_i(k-p)|\} \\
&+v_{i1,k}\sum_{p=2}^{k} K_{2,p}\varepsilon\{|\delta u_i(k-p)|\} \\
&+v_{i2,k}\sum_{p=3}^{k} K_{3,p}\varepsilon\{|\delta u_i(k-p)|\},
\end{aligned} \tag{30}$$

where

$$K_{0,p} = K_{1,p} = k_1^{p-1}k_2,$$

$$K_{2,p} = k_1^{p-2}k_{2,}\quad K_{3,p} = k_1^{p-3}k_2.$$

Expanding expression (30) from $k = 0$ to $k = N$, we have

$$\varepsilon\{|\delta u_{i+1}(0)|\} \le \rho_{i,0}\varepsilon\{|\delta u_i(0)|\},$$

$$\begin{aligned}
&\varepsilon\{|\delta u_{i+1}(1)|\} \\
&\le \rho_{i,1}\varepsilon\{|\delta u_i(1)|\} + \rho_{i1,1}\varepsilon\{|\delta u_i(0)|\} \\
&+v_{i+1,1}K_{0,1}\varepsilon\{|\delta u_{i+1}(0)|\} + v_{i,1}K_{1,1}\varepsilon\{|\delta u_i(0)|\} \\
&\le \rho_{i,1}\varepsilon\{|\delta u_i(1)|\} \\
&+(\rho_{i1,1} + v_{i,1}K_{1,1} + \rho_{i,1}v_{i+1,1}K_{0,1})\varepsilon\{|\delta u_i(0)|\} \\
&= \rho_{i,1}\varepsilon\{|\delta u_i(1)|\} + L_{1,0}\varepsilon\{|\delta u_i(0)|\},
\end{aligned}$$

$$
\begin{aligned}
&\varepsilon\left\{\left|\delta u_{i+1}(2)\right|\right\} \\
&\le \rho_{i,2}\varepsilon\left\{\left|\delta u_i(2)\right|\right\} + \rho_{i1,2}\left\{\left|\delta u_i(1)\right|\right\} + \rho_{i2,2}\varepsilon\left\{\left|\delta u_i(0)\right|\right\} \\
&+\nu_{i+1,2}K_{0,1}\varepsilon\left\{\left|\delta u_{i+1}(1)\right|\right\} + \nu_{i+1,2}K_{0,2}\varepsilon\left\{\left|\delta u_{i+1}(0)\right|\right\} \\
&+\nu_{i,2}K_{1,1}\varepsilon\left\{\left|\delta u_i(1)\right|\right\} + \nu_{i,2}K_{1,2}\varepsilon\left\{\left|\delta u_i(0)\right|\right\} \\
&+\nu_{i1,2}K_{2,2}\varepsilon\left\{\left|\delta u_i(0)\right|\right\} \\
&\le \rho_{i,2}\varepsilon\left\{\left|\delta u_i(2)\right|\right\} + \left(\rho_{i1,2} + \rho_{i,2}\nu_{i+1,2}K_{0,1} + \nu_{i,2}K_{1,1}\right) \\
&\cdot\left\{\left|\delta u_i(1)\right|\right\} + \left(\rho_{i2,2} + L_{1,0}\nu_{i+1,2}K_{0,1} + \rho_{i,2}\nu_{i+1,2}K_{0,2}\right. \\
&\left.+\nu_{i,2}K_{1,2} + \nu_{i1,2}K_{2,2}\right)\left\{\left|\delta u_i(0)\right|\right\} \\
&= \rho_{i,2}\varepsilon\left\{\left|\delta u_i(2)\right|\right\} + L_{2,1}\left\{\left|\delta u_i(1)\right|\right\} + L_{2,0}\left\{\left|\delta u_i(0)\right|\right\}
\end{aligned}
$$

$$
\begin{aligned}
&\varepsilon\left\{\left|\delta u_{i+1}(N)\right|\right\} \\
&\le \rho_{i,N}\varepsilon\left\{\left|\delta u_i(N)\right|\right\} + \rho_{i1,N}\varepsilon\left\{\left|\delta u_i(N-1)\right|\right\} + \rho_{i2,N}\cdot \\
&\varepsilon\left\{\left|\delta u_i(N-2)\right|\right\} + \nu_{i+1,N}\sum_{p=1}^{N}K_{0,p}\varepsilon\left\{\left|\delta u_{i+1}(N-p)\right|\right\} \\
&+\nu_{i,N}\cdot\sum_{p=1}^{N}K_{1,p}\varepsilon\left\{\left|\delta u_i(N-p)\right|\right\} \\
&+\nu_{i1,N}\sum_{p=2}^{N}K_{2,p}\varepsilon\left\{\left|\delta u_i(N-p)\right|\right\} \\
&+\nu_{i2,N}\sum_{p=3}^{N}K_{3,p}\varepsilon\left\{\left|\delta u_i(N-p)\right|\right\} \\
&\le \rho_{i,N}\varepsilon\left\{\left|\delta u_i(N)\right|\right\} + \left(\rho_{i1,N} + \nu_{i+1,N}K_{0,1}\rho_{i,N} + \nu_{i,N}K_{1,1}\right) \\
&\cdot\varepsilon\left\{\left|\delta u_i(N-1)\right|\right\} + \left(\rho_{i2,N} + \nu_{i+1,N}K_{0,1}L_{N-1,N-2} + \nu_{i+1,N}\right. \\
&\left.\cdot K_{0,2}\rho_{i,N} + \nu_{i,N}K_{1,2} + \nu_{i1,N}K_{2,p}\right)\varepsilon\left\{\left|\delta u_i(N-2)\right|\right\} + \cdots \\
&+\left(\nu_{i+1,N}\sum_{p=1}^{N}K_{0,p}L_{N-p,0} + \nu_{i,N}K_{1,N} + \nu_{i1,N}K_{2,N}\right. \\
&\left.+\nu_{i2,N}K_{3,N}\right)\cdot\varepsilon\left\{\left|\delta u_i(0)\right|\right\} \\
&= \rho_{i,N}\varepsilon\left\{\left|\delta u_i(N)\right|\right\} + L_{N,N-1}\varepsilon\left\{\left|\delta u_i(N-1)\right|\right\} \\
&+L_{N,N-2}\varepsilon\left\{\left|\delta u_i(N-2)\right|\right\} + \cdots + L_{N,0}\varepsilon\left\{\left|\delta u_i(0)\right|\right\},
\end{aligned}
$$

where $L_{m,n}(m,n \in [0,N])$ is a known constant.

The above expressions can be arranged in a matrix form as below

$$\Delta U_{i+1} \le \Phi_i \Delta U_i,$$

where

$$
\Delta U_i = \begin{bmatrix}
\varepsilon\left\{\left|\delta u_i(0)\right|\right\} \\
\varepsilon\left\{\left|\delta u_i(1)\right|\right\} \\
\varepsilon\left\{\left|\delta u_i(2)\right|\right\} \\
\vdots \\
\varepsilon\left\{\left|\delta u_i(N)\right|\right\}
\end{bmatrix},
$$

$$\Phi_i = \begin{bmatrix} \rho_{i,0} & 0 & 0 & \cdots & 0 \\ L_{1,0} & \rho_{i,1} & 0 & \cdots & 0 \\ L_{2,0} & L_{2,1} & \rho_{i,2} & \cdots & \vdots \\ \vdots & \cdots & \cdots & \cdots & 0 \\ L_{N,0} & L_{N,1} & \cdots & L_{N,N-1} & \rho_{i,N} \end{bmatrix}.$$

If all the eigenvalues of Φ_i lie interiorly to the unit circle, then $\varepsilon\{\|\delta u_i(k)\|\} \to 0$ as $i \to \infty$ for all $k \in [0,1,2,\cdots N]$. Since Φ_i is a lower triangular matrix, all its eigenvalues at the ith iteration are $\rho_{i,k}$. Therefore, if $\rho_{i,k} = \dfrac{\left|1-\left(\bar{\beta}\gamma_{ff}\bar{\alpha}-\gamma_{fb}\right)d(x_i(k))\right|}{\left|1+\gamma_{fb}d(x_{i+1}(k))\right|} < 1$ holds for all i and k then $\lim_{i\to\infty} \varepsilon\{\|\delta u_i(k)\|\} = 0.$ That is $\lim_{i\to\infty} \varepsilon\{\|u_i(k)\|\} = u_d(k),$ which, in terms of Assumption 2, implies that $\lim_{i\to\infty} \varepsilon\{\|y_i(k)\|\} = y_d(k)$ is guaranteed for all $k \in [0,1,2,\cdots N]$. This is the end of proof. □

Remark 1. Theorem 1 is the convergence condition for the P-type ILC algorithms as shown in (11) and (12). It is obvious that this proof method is also suitable for the PD-type and PID-type ILC algorithms. The only difference existing in the convergence condition is that the adjustable parameters are different in each algorithm.

4. Extension to a class of MIMO nonlinear systems

The result of Theorem 2 can be extended to a class of MIMO nonlinear systems satisfying the following form:

$$\begin{cases} \mathbf{x}_i(k+1) = \mathbf{f}(\mathbf{x}_i(k)) + \mathbf{b}(\mathbf{x}_i(k))\mathbf{u}_i(k), \\ \mathbf{y}_i(k) = \mathbf{g}(\mathbf{x}_i(k)) + \mathbf{d}(\mathbf{x}_i(k))\mathbf{u}_i(k), \end{cases} \tag{31}$$

where i and k denote the same meaning as in system (1). $\mathbf{x}_i(k) \in R^l$, $\mathbf{u}_i(k) \in R^m$ and $\mathbf{y}_i(k) \in R^n$ are the state vector, control input vector and output vector of the system, respectively.

Some relevant assumptions of the system (31) are given as follows.

Assumption 5: The desired trajectory $\mathbf{y}_d(k)$ is iteration invariant.

Assumption 6: For $\mathbf{y}_d(k)$, there exist unique $\mathbf{u}_d(k)$ and $\mathbf{x}_d(k)$ meeting the following equations:

$$\begin{cases} \mathbf{x}_d(k+1) = \mathbf{f}(\mathbf{x}_d(k)) + \mathbf{b}(\mathbf{x}_d(k))\mathbf{u}_d(k), \\ \mathbf{y}_d(k) = \mathbf{g}(\mathbf{x}_d(k)) + \mathbf{d}(\mathbf{x}_d(k))\mathbf{u}_d(k), \end{cases} \tag{32}$$

where $\mathbf{u}_d(k)$ is the desired control input vector and $\mathbf{x}_d(k)$ is the desired state vector.

Assumption 7: The resetting condition below is satisfied for every iteration.

$$\mathbf{x}_i(0) = \mathbf{x}_d(0), \tag{33}$$

where $\mathbf{x}_d(0)$ is the initial value of the desired state vector and $\mathbf{x}_i(0)$ is the initial value of the ith iteration.

Assumption 8: The nonlinear function vectors and matrices $\mathbf{f}(\cdot) \in R^l$, $\mathbf{b}(\cdot) \in R^{l\times m}$, $\mathbf{g}(\cdot) \in R^n$ and $\mathbf{d}(\cdot) \in R^{n\times m}$ are global Lipschitz in $\mathbf{x}$, that is $\exists$ positive bounded constants k_F, k_B, k_G and k_D, satisfying the following inequations:

$$\|\mathbf{f}(\mathbf{x}_1(k)) - \mathbf{f}(\mathbf{x}_2(k))\| \le k_F \|\mathbf{x}_1(k) - \mathbf{x}_2(k)\|, \tag{34}$$

$$\|\mathbf{b}(\mathbf{x}_1(k)) - \mathbf{b}(\mathbf{x}_2(k))\| \le k_B \|\mathbf{x}_1(k) - \mathbf{x}_2(k)\|, \tag{35}$$

$$\|\mathbf{g}(\mathbf{x}_1(k)) - \mathbf{g}(\mathbf{x}_2(k))\| \le k_G \|\mathbf{x}_1(k) - \mathbf{x}_2(k)\|, \tag{36}$$

$$\|\mathbf{d}(\mathbf{x}_1(k)) - \mathbf{d}(\mathbf{x}_2(k))\| \le k_D \|\mathbf{x}_1(k) - \mathbf{x}_2(k)\|, \tag{37}$$

for any pair of $(\mathbf{x}_1(k), \mathbf{x}_2(k))$ in $R^l \times R^l$, where $\mathbf{x}_1, \mathbf{x}_2 \in \mathbf{x}_i(k), i = 0,1,2,3\cdots, k \in [0, N]$.

$\|\cdot\|_v$ ($v = 1,2,\text{L} ,\infty$) denotes the Euclidean norm and any kind of norm can be taken in (34)-(37) due to the equivalency of different norms. The only difference is the Lipschitz constant because of the selection of different norms. ∞- norm is considered in (34)-(37).

$$\|\mathbf{M}\|_\infty = \max_{1\le j\le n} |m_j|,$$

where $\mathbf{M} \in R^n$, m_j is the *j*th element of the vector $\mathbf{M}$.

$$\|N\|_\infty = \max_{1\le i\le m} \sum_{j=1}^{n} |n_{ij}|,$$

where $\mathbf{N} \in R^{m\times n}$, n_{ij} is the element of the matrix $\mathbf{N}$.

As shown in Fig.1, two types of data dropouts are considered in the MIMO nonlinear system. Suppose each component in the multivariable output or input vector of the nonlinear system suffers the same data dropout rate. Then the relationship can be expressed as follows:

$$\hat{\mathbf{y}}_i(k) = \mathbf{A}\mathbf{y}_i(k) + (\mathbf{I}-\mathbf{A})\mathbf{y}_i(k-1), \qquad (38)$$

$$\mathbf{u}_{ff,i}(k) = \mathbf{B}\hat{\mathbf{u}}_{ff,i}(k) + (\mathbf{I}-\mathbf{B})\hat{\mathbf{u}}_{ff,i}(k-1), \qquad (39)$$

where

$$\mathbf{A} = \begin{pmatrix} \alpha & & \\ & \text{O} & \\ & & \alpha \end{pmatrix}_{n\times n}, \quad \mathbf{B} = \begin{pmatrix} \beta & & \\ & \text{O} & \\ & & \beta \end{pmatrix}_{m\times m},$$

and α , β denote the same meaning as in the SISO system. Then (38) and (39) can be rewritten as:

$$\hat{\mathbf{y}}_i(k) = \alpha\mathbf{y}_i(k) + (1-\alpha)\mathbf{y}_i(k-1), \qquad (40)$$

$$\mathbf{u}_{ff,i}(k) = \beta\hat{\mathbf{u}}_{ff,i}(k) + (1-\beta)\hat{\mathbf{u}}_{ff,i}(k-1), \qquad (41)$$

Then the P-type ILC law for the feed-forward controller and P-type law for the feed-back controller are given as follows:

$$\hat{\mathbf{u}}_{ff,i}(k) = \hat{\mathbf{u}}_{ff,i-1}(k) + \boldsymbol{\theta}\hat{\mathbf{e}}_{i-1}(k), \qquad (42)$$

$$\mathbf{u}_{fb,i}(k) = \boldsymbol{\omega}\mathbf{e}_i(k), \qquad (43)$$

where

$$\boldsymbol{\theta} = \begin{pmatrix} \gamma_{ff1} & & \\ & \text{O} & \\ & & \gamma_{ffm} \end{pmatrix}_{m\times n}, \quad \boldsymbol{\omega} = \begin{pmatrix} \gamma_{fb1} & & \\ & \text{O} & \\ & & \gamma_{fbm} \end{pmatrix}_{m\times n},$$

are the learning gain matrices of the remote controller and the local controller, respectively.

Theorem 2: For system (31), assume Assumptions 5-8 hold, applying the update laws (42) and (43) to the feed-forward controller and the feed-back controller, respectiely, if $\rho_{i,k} = \left\| \dfrac{\mathbf{I}-(\bar{\beta}\boldsymbol{\theta}\bar{\alpha}-\boldsymbol{\omega})\mathbf{d}(\mathbf{x}_i(k))}{\mathbf{I}+\boldsymbol{\omega}\mathbf{d}(\mathbf{x}_{i+1}(k))} \right\|_\infty < 1$ is satisfied for all *i* and *k*, then $\lim_{i\to\infty} \varepsilon\{|\mathbf{y}_i(k)|\} = \mathbf{y}_d(k)$ is guaranteed for all *k*.

The proof process of the Theorem 2 is similar to the Theorem 1, so the proof process is omitted here.

5. NUMERICAL EXAMPLE

Consider the following SISO nonlinear system

$$\begin{cases} x_i(k+1) = \sin(x_i(k)) + (1 - x_i(k))u_i(k) \\ y_i(k) = \cos(x_i(k)) + u_i(k) \end{cases}, \tag{44}$$

where $k \in [0, 200]$. The desired tracking trajectory is $y_d(k) = \sin(\pi k / 300) + \cos(\pi k / 400)$, the initial condition is $x_i(0) = x_d(0) = 0$.

Three different cases of data dropouts are taken into account for the SISO nonlinear system (44) :

Case 1: $\bar{\alpha} = \bar{\beta} = 1$, no data dropout.

Case 2: $\bar{\alpha} = \bar{\beta} = 0.9$, 10% data dropouts.

Case 3: $\bar{\alpha} = \bar{\beta} = 0.8$, 20% data dropouts.

For further verifying the effectiveness of the open-closed-loop ILC algorithm, we also take the open-loop ILC algorithm into consideration. When the learning gain γ_{fb} is chosen as 0, the open-closed-loop ILC algorithm can be regarded as an open-loop ILC algorithm. Given to Theorem 1, the learning gain is chosen as $\gamma_{fb} = 5, \gamma_{ff} = 4.5$ in the open-closed-loop and $\gamma_{fb} = 0, \gamma_{ff} = 0.9$ in the open-loop.

It is assumed that $d(x_i(k)) = 1$ in system (1) and check the convergence condition in Theorem 1, we can obtain

Case 1: $$\rho_{i,k} = \left| \frac{1 - (\bar{\beta}\gamma_{ff}\bar{\alpha} - \gamma_{fb})d(x_i(k))}{1 + \gamma_{fb}d(x_{i+1}(k))} \right| = 0.25 < 1,$$

Case 2: $$\rho_{i,k} = \left| \frac{1 - (\bar{\beta}\gamma_{ff}\bar{\alpha} - \gamma_{fb})d(x_i(k))}{1 + \gamma_{fb}d(x_{i+1}(k))} \right| = 0.39 < 1,$$

Case 3: $$\rho_{i,k} = \left| \frac{1 - (\bar{\beta}\gamma_{ff}\bar{\alpha} - \gamma_{fb})d(x_i(k))}{1 + \gamma_{fb}d(x_{i+1}(k))} \right| = 0.52 < 1,$$

so $\lim_{i\to\infty} \varepsilon\{|y_i(k)|\} = y_d(k)$ is granted for all k.

Through the theoretical analysis in section 3, it can be easily concluded that the smaller $\rho_{i,k}$ is, the faster convergence speed can be obtained. That is why the convergence speed gets slower and slower as shown in Fig. 2, Fig. 4 and Fig. 6. However, the maximum tracking error eventually goes to zero as the iteration increases, which can verify the effectiveness of the proposed method. Besides, by comparing the two figures in Fig. 2, Fig. 4 and Fig. 6, respectively, it could be better understood that, under the same data dropout rate, in addition to a faster convergence speed, the tracking error is also reduced greatly in the open-closed loop compared with the open loop.

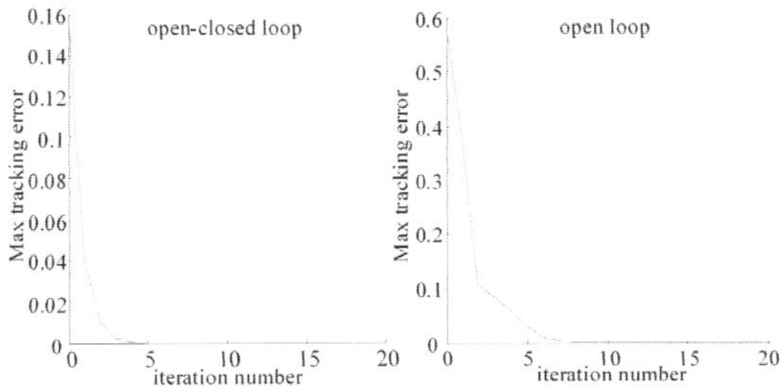

Fig. 2. The maximum tracking error with no data dropout

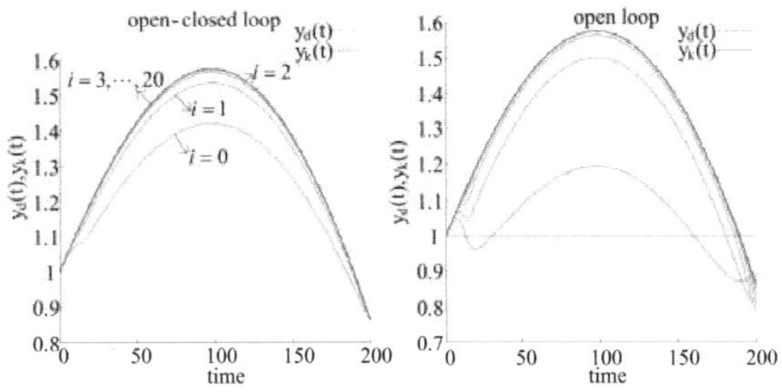

Fig. 3. The output of 20 times iteration with no data dropout

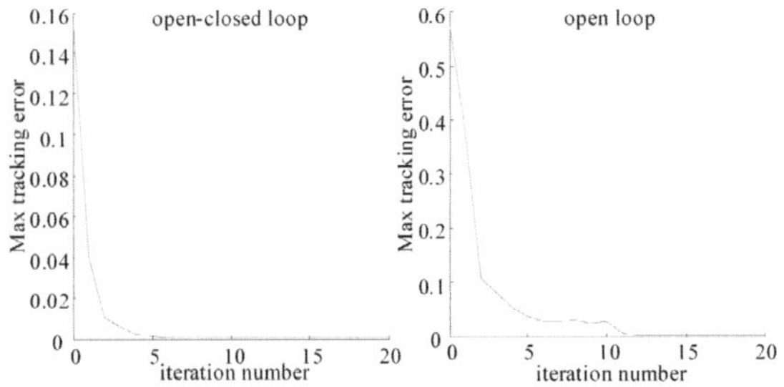

Fig. 4. The maximum tracking error with 10% data dropouts

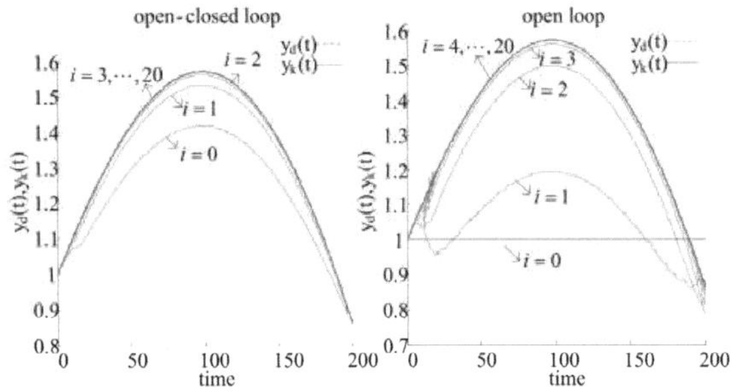

Fig. 5. The output of 20 times iteration with 10% data dropouts

Fig. 3, Fig. 5 and Fig. 7 describe the initialized output of the system ($i = 0$), the output of 20 times iteration ($i = 1, 2, L\ , 20$) and the desired output, that is to say, there are twenty-two lines in Fig. 3, Fig. 5 and Fig. 7, respectively. However, the actual output lines will be coincident with the desired output line after some times of iteration, which is why the lines are less than twenty-two visually. It can be seen that the actual output of the system gets closer and closer to the desired output as the iteration number increases and the desired trajectory can be tracked completely after some iterations.

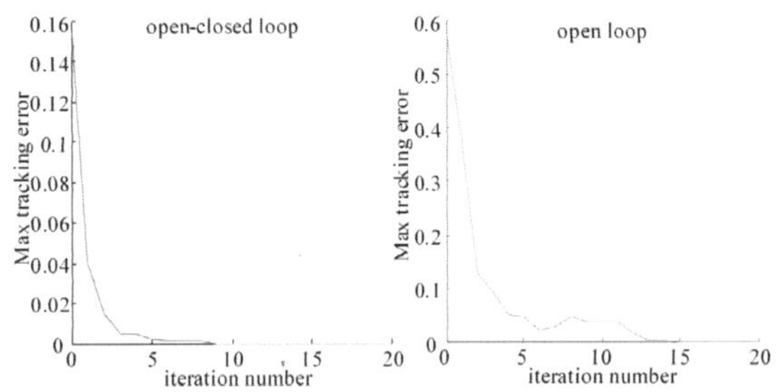

Fig. 6. The maximum tracking error with 20% data dropouts

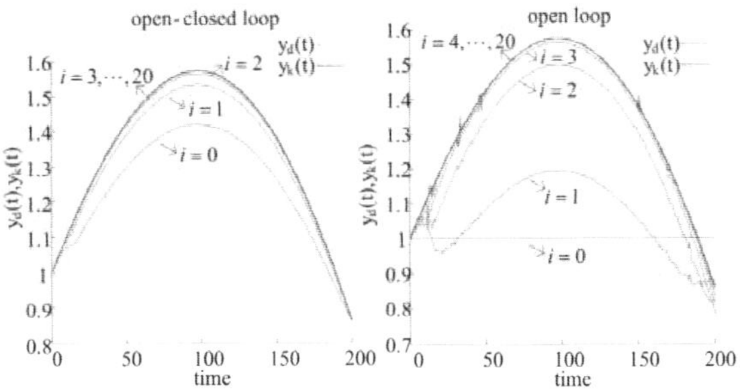

Fig. 7. The output of 20 times iteration with 20% data dropouts

6. CONCLUSION

An open-closed-loop ILC method for a class of nonlinear systems with random data dropouts is introduced in this paper. The main contribution of this paper is adding a closed-loop to the existing ILC method for nonlinear systems. Though the data dropouts have a bad effect on the convergence speed, the ILC algorithm can achieve convergence eventually as long as the packet dropout is incompletable. It has faster convergence speed, greatly reduced tracking error and better performance in stability and robustness due to the open-closed-loop ILC.

When the current data packet loses during the transmission, the previous data stored in the memory will be used instead in this paper. However, some more efficient compensation methods for data dropouts will be considered as one of our further research topics, especially when the data dropout rate is high. The other is the extension to a class of MIMO systems where each component in the multivariable output or input vector of the system suffers different data dropout rate.

7. ACKNOWLEDGEMENT

This paper is supported by National Nature Science Foundation of China: 61473248.

REFERENCES

[1] S. Arimoto, S. Kawamura, and F. Miyazaki, "Bettering operation of robots by learning," *Journal of Robotic Systems*, vol. 1, no. 2, pp. 123-140, 1984.

[2] K. L. Moore, Y.-Q. Chen, and V. Bahl, "Monotonically convergent iterative learning control for linear discrete-time systems," *Automatica*, vol. 41, no. 9, pp. 1529-1537, June 2005.

[3] X.-E. Ruan and Z.-Z. Li, "Convergence characteristics of PD-type iterative learning control in discrete frequency domain," *Journal of Process Control*, vol. 24, no. 12, pp. 86-94, November 2014.

[4] J. Van de Wijdeven, T. Donkers, and O. Bosgra, "Iterative learning control for uncertain systems: robust monotonic convergence analysis," *Automatica*, vol. 45, no. 10, pp. 2383-2391, August 2009.

[5] D. Meng, Y. Jia, J. Du, and F. Yu, "Robust iterative learning control design for uncertain time-delay systems based on a performance index," *Control Theory & Applications, IET*, vol. 4, no. 5, pp. 759-772, 2010.

[6] T. Y. Doh, J. R. Ryoo, and D. E. Chang, "Robust iterative learning controller design using the performance weighting function of feedback control systems," *International Journal of Control, Automation and Systems*, vol. 12, no. 1, pp. 63-70, 2014.

[7] K.-Y. You and L.-H. Xie, "Survey of recent progress in networked control systems," *Acta Automatica Sinica*, vol. 39, vol. 2, pp. 101-117, February 2013.

[8] H. S. Ahn, Y.-Q. Chen, and K. L. Moore, "Intermittent iterative learning control," *Proc. of the 2006 IEEE International Symposium on Intelligent Control*, pp. 832-837, Munich, Germany, 4-6 October, 2006.

[9] H. S. Ahn, K. L. Moore, and Y.-Q. Chen, "Discrete-time intermittent iterative learning controller with independent data dropouts," *Proc. of the 2008 IFAC World Congress*, pp. 12442-12447, Seoul, Korea, 2008.

[10] H. S. Ahn, K. L. Moore, and Y.-Q. Chen, "Stability of Discrete-time iterative learning control with random data dropouts and delayed controlled signals in networked control systems," *In Proceedings of International Conference on Control Automation, Robotics and Vision, IEEE*, pp. 757-762, Hanoi, Vietnam, 17–20 December, 2008.

[11] X.-H. Bu, Z.-S. Hou, and H.-R. Dong, "Stability analysis of iterative learning control with data dropouts,"

Proc. of the 2009 IEEE International Conference on Control and Automation, pp. 1146-1151, Christchurch, New Zealand, 9-11 December, 2009.

[12] X.-H. Bu and Z.-S. Hou, "Stability of iterative learning control with data dropouts via asynchronous dynamical system," *International Journal of Automation and Computing*, vol. 8, no. 1, pp. 29-36, February 2011.

[13] X.-H. Bu, Z.-S. Hou, and F.-S. Yu, "Stability of first and high order iterative learning control with data dropouts," *International Journal of Control, Automation and Systems*, vol. 9, no. 5, pp. 843-849, 2011.

[14] X.-H. Bu, Z.-S. Hou, F.-S. Yu, and F.-Z. Wang, " H_∞ iterative learning controller design for a class of discrete-time systems with data dropouts," *International Journal of Systems Science*, vol. 45, no. 9, pp. 1902-1912, 2014.

[15] X.-H. Bu, F.-S. Yu, Z.-S. Hou, and F.-Z. Wang, "Iterative learning control for a class of nonlinear systems with measurement dropouts," *Control Theory & Applications*, vol. 29, no.11, pp. 1458-1464, November 2012.

[16] X.-H. Bu, F.-S. Yu, Z.-S. Hou, and F.-Z. Wang, "Iterative learning control for a class of nonlinear systems with random packet losses," *Nonlinear Analysis: Real World Applications*, vol. 14, no. 1, pp. 567-580, 2013.

[17] X.-H. Bu, Z.-S. Hou, S.-T. Jin, and R.-H. Chi, "An iterative learning control design approach for networked control systems with data dropouts," *International Journal of Robust and Nonlinear Control*, in press.

[18] X.-H. Bu, Z.-S. Hou, F.-S. Yu, and Z.-Y. Fu, "Effect analysis of data dropout on iterative learning control", *Control and Decision*, vol.29, no. 3, pp. 443-448, March 2013.

[19] D. Shen and Y.-Q. Wang, "Survcy on stochastic iterative learning control," *Journal of Process Control*, vol. 24, no. 12, pp. 64-77, May 2014.

[20] D. Shen and Y.-Q. Wang, "Iterative learning control for networked stochastic systems with random packet losses," *International Journal of Control*, pp. 1-10, 2014.

[21] D. Shen, Y. Zhang, and Y.-Q. Wang, "ILC for networked nonlinear systems with random measurement losses and unknown control direction," *The 11th IEEE International Conference on Control & Automation (ICCA)*, pp. 1438-1443, Taichung, Taiwan, 18-20 June, 2014.

[22] D. Shen and Y.-Q. Wang, "ILC for Networked Discrete Systems with Random Data Dropouts: A Switched System Approach," *Proceedings of the 33rd Chinese Control Conference*, pp. 8670-8677, Nanjing, China, 28-30 July, 2014.

[23] C.-P. Liu, J.-X. Xu, and J. Wu, "Iterative learning control for network systems with communication delay or data dropout," *Joint 48th IEEE Conference on Decision and Control and 28th Chinese Control Conference*, pp. 4858-4863, Shanghai, China, 16-18 December, 2009.

[24] R. Wang, X.-H. Ma, and X.-H. Bu, "Robust Iterative Learning Control for a Class of Linear Systems with Data Dropouts," *Mathematical Problems in Engineering*, vol. 2013, pp. 1-9, 2013.

[25] Z. Zhao, Y. Gao, and Z.-L. Liu, "Novel open-closed-loop iterative learning control in networked control systems," *Control Theory & Applications*, vol. 30, no. 10, pp. 1335-1341, October 2013. (in Chinese)

The comparision and analysis of extracting video key frame

OUYANG ShiZhuang[1],ZHONG Luo[2],LUO RuiQi[3]
1 Master graduate, Wuhan University of Technology, research interests include data mining and image processing,CHN
2 Ph.D., Professor, WuHan University of Technology, research interests include data mining and intelligent system,CHN
3 Ph.D., WuHan University of Technology, research interests include picture processing and computer vision, CHN
Abstract Video key frame extraction is an important part of the large data processing. Based on the previous work in key frame extraction, we summarized four important key frame extraction algorithms, and these methods are largely developed by comparing the differences between each of two frames. If the difference exceeds a threshold value, take the corresponding frame as two different keyframes. After the research, the key frame extraction based on the amount of mutual trust is proposed, the introduction of information entropy, by selecting the appropriate threshold values into the initial class, and finally take a similar mean mutual information as a candidate key frame. On this paper, several algorithms is used to extract the key frame of tunnel traffic videos. Then, with the analysis to the experimental results and comparisons between the pros and cons of these algorithms, the basis of practical applications is well provided.
Key words keyframes; frame difference; threshold; mutual trust

1. Introduction

monitoring video is widely distributed in cities at present. The video monitoring analysis that rely on artificial human eyes is tedious and inefficient considering the large amount of data resources in surveillance video. By extracting the keyframe of the video, the video data will be optimized and compressed, which offer the foundation for the further video semantic analysis.

Rong Pan[1] proposes a keyframe extraction method based on cluster, which finishes the shot boundary detection firstly, and then obtains the child video with smaller data, and finally pick out the nearest frame to the center of the cluster as the keyframe. Zhong adopts the histogram intersection and block weighting, improves the traditional histogram algorithm and solves the defect of shot detection. Thepade[3] comes up with the block encoding idea to extract keyframe, which makes use of the failure vector and similarity degree of two consecutive frames. Huang[4] figures out the mutual information method according to information theory, and confirms that mutual information method contributes to improve the extraction efficiency. Mutual information is a method by calculating the amount of information to determine the threshold and dividing initial class to get keyframe sequence. A. Wang [5] exploits the background difference and light flow method to detect motion object. Wang[6] introduces the adaptive weighted affine propagation algorithm to recognize the keyframe of human body movements among three dimensional video sequence, which has a high precision. K.S.Thakre[7] makes a systematic research on compressing video streaming and presents a way of using self-adaptive threshold method to extract the keyframe of compressed video streaming.

The above words give a summary of domestic and overseas scholars' research on extracting video keyframe within the last five years. The paper presents a mutual information method extracting keyframe and gives an overall analysis combining with other four classical keyframe extraction

IOP Conf. Series: Materials Science and Engineering **359** (2018) 012010 doi:10.1088/1757-899X/359/1/012010

algorithm, shot edge analysis, arithmetic average algorithm, histogram averaging algorithm and video content matches.

2. Shot Segmentation

This paper discusses the key frame extraction is based on a single shot key frame extraction. We will be a video for structured analysis， As shown in Figure 1， First, require shot segmentation, and then the keyframe is extracted based on the shot segmentation, and a number of frame images are obtained. The frame image is grouped according to the image content for obtain scene. When shot segmentation, the first need to calculate the frame difference, and the commonly used frame difference method is as follows： Pixel comparison, Histogram comparison, marginal comparison and block matching[8,9]。In this paper, we propose a sub shot segmentation algorithm based on sliding window. The lens segmentation adopts the traditional double threshold method to segment the video for the first time to get the sub shot[10,11].

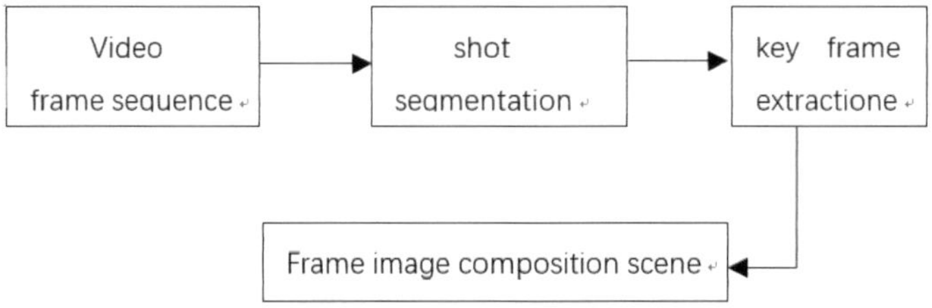

figure1 Video structure analysis flow chart

First, add a sliding window to the video frame sequence and look for possible sub shot boundaries in the small area of the window. The sliding block divides the window into two parts from the middle, and the frame difference between successive frames in the lens is relatively small and can be treated as a class. In this paper, Using fisher linear two kinds of problems. On the two kinds of problems, the idea of Fisher linearity is to project the data samples of multidimensional space into 1-dimensional space, so that the variance between the two classes is as large as possible, and the variance within the class is as small as possible.

Set two classes are C1 and C2， the sample is x, the mean is m1 and m2, the variance is s1 and s2.

$$m_k = \frac{1}{N_k}\sum_{n\in C_k} x_n,(k=0,1) \qquad (1)$$

$$s_k^{\ 2} = \sum_{n\in C_k} (x_n - m_k)^2,(k=0,1) \qquad (2)$$

So that different types of class variance as much as possible, that is, so that the ratio of the two $\frac{(m_2-m_1)^2}{s_1^2+s_2^2}$ as large as possible. When the ratio is maximum, the two classes are optimally divided in 1-dim space.

Ideally, the Fisher criterion inside the sub shot is smaller and approaches zero. In practice, the Fisher criterion at the sub shot boundary due to camera or object motion is larger. Since there is a local maximum due to a small fluctuation inside the sub shot, it should be considered as noise. Thus, a local maxima above the Fisher criterion mean is used as the sub shot boundary.

3. Key Frame Extraction

The key frame extraction[12] in this section is based on the sub shot segmentation.

3.1. Analyze the shot edge to extract keyframes

Analysis of the edge of the lens to extract the key frame is to select the end of these fragments as the key frame.

Take f_1, $f_{N/2}$, f_N as the pre-selected key frame. f_i and f_j between the interframe difference $D(f_i,f_j)$. The calculations $D(f_1,f_{N2})$, $D(f_1,f_N)$, $D(f_{N2},f_N)$ are each compared with the threshold T.

If $D(f_1, f_{N2}), D(f_1, f_N)$, $D(f_{N2}, f_N) < T$. it is considered a similar image, Can take any one of them as a key frame; If $D(f_1, f_{N2})$, $D(f_1, f_N)$, $D(f_{N2}, f_N) > T$,the difference between the three frames is relatively large, and they are all as key frames. In addition, determine their distance, the maximum distance corresponding to the key frame[13].

3.2. Frame average method to extract key frames

Keyframe extraction The most straightforward algorithm is based on the frame image algorithm, including the frame averaging method and the histogram averaging method.
The frame averaging method first takes the mean of all frames of the video clip at i:

$$\overline{R_i} = \frac{(\sum_{k=1}^{n} R_{ki})}{n}, \overline{G_i} = \frac{(\sum_{k=1}^{n} G_{ki})}{n}, \overline{B_i} = \frac{(\sum_{k=1}^{n} B_{ki})}{n} \quad (3)$$

Where n represents the number of frames in the video clip, $\overline{R_i}, \overline{G_i}, \overline{B_i}$ represent the RGB mean, R_{ki}, G_{ki}, B_{ki} representing the RGB value. The difference between each frame and the mean is:

$$D_{ki} = |R_{ki} - \overline{R}| + |G_{ki} - \overline{G}| + |B_{ki} - \overline{B}| \quad (4)$$

Select the maximum and minimum values of D_{ki} as the key frame.

3.3. The Histogram Averaging Method Extracts Key Frames

The histogram averaging is done by selecting each frame in the video clip, dividing it into a histogram and averaging it. Then the histogram of each frame is compared with the average, and the two extremes are obtained and the corresponding Frame to find out as a key frame [14]。
Histogram comparison method is usually based on the image brightness, color, grayscale these three values, respectively, they are divided into N levels, statistics of each level of the number of pixels, drawn into a histogram. The total pixel of the image frame is M, the gray level is N, the pixel with k-level gray scale has f_k , the statistical histogram of the image color feature is a discrete function, and the frequency of k-level gray scale is:

$$h_k = \frac{f_k}{M} (k = 1, 2, \Lambda, N) \quad (5)$$

The histogram shows the gray value distribution of the image .Figure 2 for the video to take a frame image of the gray histogram. The gray scale histogram of all the frames of the lens is averaged to obtain the key frame.

figure2 gray-level histogram Example

3.4. The Visual Content Matching Extracts Key Frames

Through the visual content matching to extract key frames, the algorithm ideas are as follows: Select a video clip, the first frame as the key frame, the same will also be set to the reference frame, and then after each frame and the reference frame for comparison, when the k-th frame and the selected frame is greater than the threshold T, then the first k frame to update the new reference frame, will continue to cycle, to the end of all the video clips are detected[12,13]. The flowsheet shown in Figure 3.

A video stream with a frame count of N,

$I = \{f_1, f_2, \Lambda, f_N\}$;

Select I_1 as the reference frame, define the similarity of the two frames according to the color histogram, and set a threshold T. The correlation coefficients for f_i and f_i are as follows:

$$\rho_{ij} = C_{ij} / (\sigma_i, \sigma_j) \qquad (6)$$

In the above formula, $C_{ij} = (f_i - m)(f_j - m), \sigma_i^2 = C_{ii}$, M is the mean vector. Compare the correlation coefficient with the threshold to arrive at a certain number of key frames.

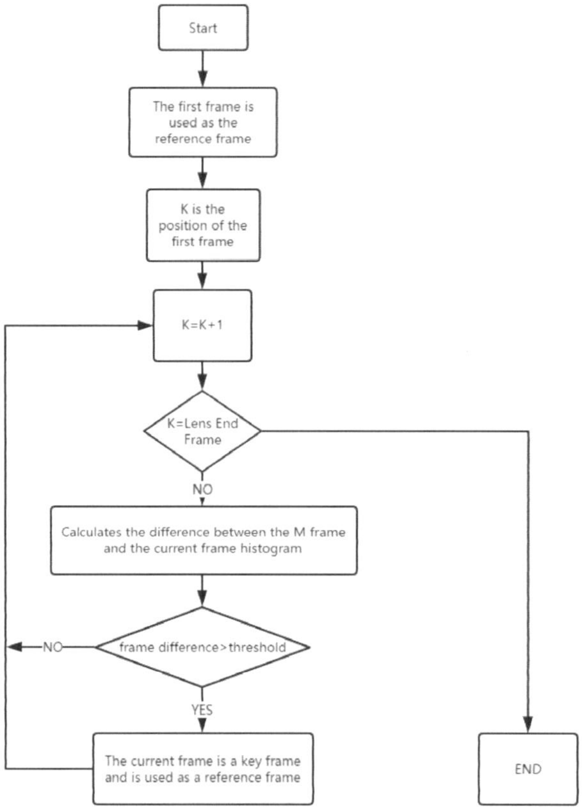

figure3 Extract keyframe flow graphs through visual content matching

4. Key Frame Extraction Base On Mutual Information

The four key frame extraction methods introduced in Section 3 can achieve the extraction of key frames to a certain extent, but there are still some shortcomings. Analysis of the shot edge of the key frame is vulnerable to external influences, when the shot (camera) movement changes, will lead to the shot boundary misjudgment, can not correctly extract the key frame. When the video content is more complex, the frame mean and the histogram method to extract the key frame representation is relatively poor. Through the visual content matching to extract the key frame calculation is moderate, the key frame sequence is reasonable, but does not apply to all the video. Through the above method, this paper proposes a key frame method based on mutual information extraction. The mutual information method introduces the information entropy, can calculate the inter-frame mutual information value, and divide the initial class. At the same time, the method of this paper can calculate the threshold independently [14,15]. In this paper, based on mutual information extraction key frame algorithm steps are as follows.

Step 1: Calculate the higher the ratio of the mutual information I (X, Y) to H (X, Y) for two consecutive frames, the higher the similarity;

$$I(X,Y) = H(X) + H(Y) - H(X,Y) \qquad (7)$$

$$H(X,Y) = -\sum_{i,j} P_{XY}(i,j)\log P_{XY}(i,j) \qquad (8)$$

Where X, Y represent two random variables, H (X) is the information entropy, and I (X, Y) is called the mutual information of X and Y[16]. P_{XY} is the joint probability distribution of X and Y.

The amount of mutual information for the RGB component is calculated from the mutual information formula given above.

$$I^R_{X,Y} = -\sum_{i=0}^{N-1}\sum_{j=0}^{N-1} P^R_{X,Y}(i,j)\log\frac{P^R_{X,Y}(i,j)}{P^R_X(i)P^R_Y(j)} \qquad (9)$$

the amount of information between the G component and the B component can be obtained.

The mutual information of two consecutive frames is:

$$I_{XY} = I^R_{X,Y} + I^G_{X,Y} + I^B_{X,Y} \qquad (10)$$

Step 2: Select the threshold T [17-20].

A video stream with a frame count of N， $I = \{I_{1,2}, I_{2,3}, \Lambda, I_{N-1,N}\}$；

Sort the elements in I to get the new A. $I_{sort} = \{I_1, I_2, \Lambda, I_{n-1}\}$， $I_1 \ge I_2 \ge \Lambda \ge I_{n-1}$

The following five equations（3.11）~（3.15）， Find the value of T.

$$T = \arg\min \delta^2_w \qquad (11)$$

$$\delta^2_w = q_H\delta^2_H + q_L\delta_H, q_H = T, q_L = N - T - 1 \qquad (12)$$

$$\mu_H = \frac{1}{q_H}\sum_{i=1}^{T} I_i, \mu_L = \frac{1}{q_L}\sum_{T+1}^{N-1} I_i \qquad (13)$$

$$\delta^2_H = \frac{1}{q_H}\sum_{i=1}^{T}[I_i - \mu_H]^2 \qquad (14)$$

$$\delta^2_L = \frac{1}{q_L}\sum_{T+1}^{N-1}[I_i - \mu_L]^2 \qquad (15)$$

If the interval between two consecutive frames is $D(k,k+1) \le I_T$， Then from here to start another division.

Finally, the video stream is divided into k segments, such as $part_1, part_2, \Lambda, part_k$.

Step 3: Keyframe extraction

The above steps mainly complete the division of the initial class, and then complete the core algorithm based on the mutual information extraction key frame.

(1) $part_1$ as the first initial class D_1， the mutual information is equal to the value of $\overline{M_1}$。

(2) the calculation of the mutual information

First determine whether A> B is established， if k=k+1， the new initial class is divided； Otherwise, $part_t$ is merged into D_k , Then calculate $\overline{M_k}$.

(3) when t<K ,repeat step(2)。

(4) to determine all the D_k ,take the mutual information similarity to the mean $\overline{M_k}$ of the corresponding frame as a candidate key frame.

Step 4: Fix the results

The method of extracting the key frame there is redundancy, in order to make the shot content is more concise, need to modify the results, the following are two amendments to the principle:

(1) because in the same shot, the image content is similar, so for some of the shorter key frame, can only take one of the frames.

If the amount of mutual information between adjacent frames is less than A, the difference between these two frames is very large, and these two frames are taken as key frames. Otherwise one should be discarded[21].

5. Comparison and Analysis

Experimental environment: Visual Studio2010 and MFC develop kit.

Data resources: the presented data come from the tunnel intelligent monitoring system in Wuhan, Hubei including Fruit Lake, Deposited, Hankou Railway Station and Bayi Road.

The video 1 records 291 frames from 2 to 2:30 p.m. in monitor video. We divide them into three child shot through Sliding window segmentation algorithm and then extract the keyframe using mentioned 5 algorithm. Picture 5 shows the results.

a Frame average method to extract key frames

b The Histogram Averaging Method Extracts Key Frames

c Analyze the shot edge to extract keyframes d The Visual Content Matching Extracts Key Frames

e Key Frame Extraction Base On Mutual Information

Figure 4 Five key frame extraction algorithms

a threshold T=0.23 Key Frame Extraction Base On Mutual Information

b threshold T=0.6 Key Frame Extraction Base On Mutual Information

Figure 5 Threshold Selection Based on Mutual Information Method

There are a red car and two black cars in Video 1. Picture 4.a and 4.d both fail to extract the important information, picture 4.b and 4.c present one car respectively and picture 4.e shows the three cars. When video has less contents, more keyframes have to be extracted for semantic analysis and the keyframes extracted by the first three algorithms is not enough to meet the requirements. The video content matches method only gets one keyframe and misses some important contents. The mutual information method presents the correct result with three cars captured. The experiment results indicate that the keyframe extracted by mutual information method is reliable and reasonable, and reflects the core contents in the video despite some redundant contents are attached. It is noted that the threshold has a significant impact on the final result as the picture 5 shows. The experiments suggest that roughly 0.23 is relatively reasonable[22].

More experiments are performed and analyzed in order to be more persuasive. Table 1 shows the experiment results.

table1 Monitor the video data sheet

video	Test the number of key frames	Actual number of key frames	Mutual information method to extract the number of key frames	duration /s
Shuiguo lake tunnel video 1	6451	160	206	258

Shuiguo lake tunnel video 2	5751	127	156	230
Shuiguo lake tunnel video 3	2441	19	24	81
Shuiguo lake tunnel video 4	15598	390	489	634
Shuiguo lake tunnel video 5	8931	234	302	357
Shuiguo lake tunnel video 6	3543	74	97	142
Shuiguo lake tunnel video 7	5542	148	182	222

The target video of this paper adopts random sampling method to select the video data set from Wuhan tunnel network, and uses social labeling method to determine the truth value.

$$\text{recall} = \frac{\text{Detect the actual number of key frames}}{\text{The actual number of key frames}} \times 100\%$$

$$\text{precision} = \frac{\text{Detect the actual number of key frames}}{\text{The total number key frames}} \times 100\%$$

The results of several key frame extraction methods for video lake monitoring video 1 are compared, as shown in Table 2 and Figure 6:

Table 2 Monitor video key frame extraction method

Key frame extraction method	recall ratio	precision ratio
Frame averaging method	78.6%	48%
Histogram averaging method	84.1%	68.6%
shot boundary method	85.4%	66.7%
Through video content matching	89.4%	70.3%
Mutual information method	89.7%	69.9%

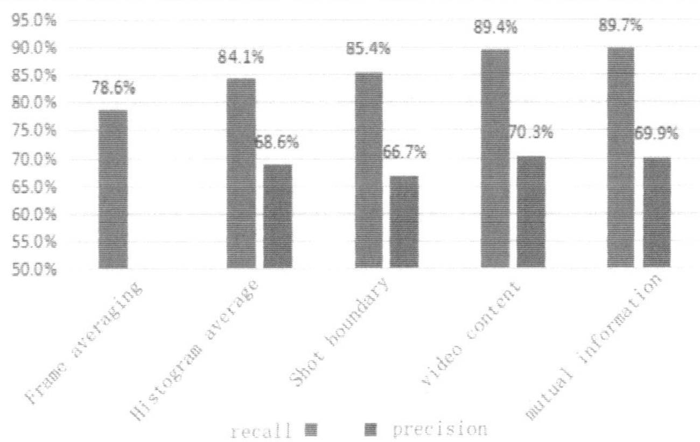

Figure 6 Monitor video histogram

The above analysis proves that the video content matches method and the mutual information method basically fulfill the requirements with a validity and accuracy[23-25]. The two method successfully extract the key contents and remove unnecessary fragments. Meanwhile, the other two method based on frame images give a lower precision ratio with some useless keyframes presented. They are suitable for extracting rough keyframes from a large amount of data since they are easy to understand and operate. The rest shot edge analysis is easily affected by the external factors and lacks of effective features especially with flash video, moved objects or cameras, and so on. Compared with the five method, they should be thoughtfully adopted according to video type and scenario with respective advantages and disadvantages in order to achieve the best efficiency and performance.

6. Conclusion

The paper compares five mainstream keyframe extraction methods and the mutual information method that takes advantage in terms of validity and accuracy is primarily focused. It shows that the keyframe extraction has been mature and the basic semantic analysis approach needs to be enhanced, such as how to recognize the contents of the video. Higher level semantics can be recognized by human like object, behavior, scenarios and events, while artificial intelligence helps to deal with lower level semantics like shape, texture, color. Artificial intelligence has the ability to handle some specific object and scenarios on the basis of pattern recognition and machine learning. However, it still faces challenges in identifying behavior and events, which are the main research fields in the future.

References

[1] Rong Pan, Yumin Tian; Zhong Wang . Key-frame Extraction Based on Clustering [C]. The 2010 IEEE International Conference on Progress in Informatics and Computing, 2010.
[2] Qu Z, Lin L, Gao T, et al. An Improved Keyframe Extraction Method Based on HSV Colour Space[J]. Journal of Software, 2013, 8(7).
[3] Sudeep D. Thepade, Pritam H. Patil. Novel visual content summarization in videos using keyframe extraction with Thepade's Sorted Ternary Block truncation Coding and Assorted similarity measures[C]. Communication, Information & Computing Technology (ICCICT), 2015 International Conference on. IEEE, 2015:1-5.
[4] Huang L, Ye C H. The research and implementation of keyframe extracion methods in content-based teaching video retrieval[C]. Consumer Electronics, Communications and Networks (CECNet), 2012 2nd International Conference on. IEEE, 2012:2179-2182.
[5] A. Wang. Vehicle Detection and Tracking using the Optical Flow and Background Subtraction [J]. china Railway Science . 2014,35 (6) :131-137
[6] Wang X, Sun S, Ding X. A self-adaptive weighted affinity propagation clustering for key frames extraction on human action recognition, J. Vis. Commun. Image R. 2015: 193 - 202.
[7] Thakre K S, Rajurkar A M, Manthalkar R R. Video Partitioning and Secured Keyframe Extraction of MPEG Video[J]. Procedia Computer Science, 2016,78:790-798.
[8] Deng-yin ZHANG,Min YANG.Shot-Segmentation-Based Video Watermarking Algorithm Using Motion Vector. Proceedings of 2014 International Conference on Computer,Network Security and Communication Engineering(CNSCE 2014)
[9] A Sengupta,DM Thounaojam,KM Singh,S Roy. Video shot boundary detection: A review. IEEE International Conference on Electrical , 2015 :1-6
[10] Zhong xian. A Dissertation Submitted in Partial Fulfillment of Requirements for the Degree of Doctor Philosophy in Engineering [D]. Huazhong University of Science and Technology, 2013.
[11] Zhong xian, Yang guang, Lu yansheng. Method Of Key Frames Extraction Base on Double-threshold Values Sliding Window Sub-shot Segmentation and Fully Connected Graph [J].Computer S, 2016, 43(6):289-293.
[12] Pal G, Rudrapaul D, Acharjee S, et al. Video Shot Boundary Detection: A Review[M]. Emerging ICT for Bridging the Future - Proceedings of the 49th Annual Convention of the Computer Society of India CSI Volume 2. Springer International Publishing, 2015:119-127.
[13] BH Shekar , KP Uma. Kirsch Directional Derivatives Based Shot Boundary Detection: An Efficient and Accurate Method. Procedia Computer Science ,2015 , 58 :565-571
[14] Dhagdi S T, Deshmukh P R. New Technique for Keyframe Extraction Using Block Based Histogram[J]. International Journal of Advanced Research in Computer Science, 2012,3(3):794-801.
[15] Zhang Q, Zhang S, Zhou D. Keyframe Extraction from Human Motion Capture Data Based on a Multiple Population Genetic Algorithm[J]. Symmetry, 2014, 6(4):926-937.
[16] Sainui J, Sugiyama M. Minimum dependency key frames selection via quadratic mutual information[C]Tenth International Conference on Digital Information Management. 2015.
[17] SheenaC V. Author links open the author workspace.N.K.Narayanan. Key-frame Extraction by

Analysis of Histograms of Video Frames Using Statistical Methods. Procedia Computer Science.Volume 70, 2015, 36-40

[18] JL Lai， Y Yi. Key frame extraction based on visual attention model. Journal of Visual Communication and Image Representation. 2012, 114-125

[19] Chen B W, Bharanitharan K, Wang J C, et al. Novel Mutual Information Analysis of Attentive Motion Entropy Algorithm for Sports Video Summarization[J]. Lecture Notes in Electrical Engineering, 2014, 260(2):1031-1042.

[20] Zhang Jun,Honghua Tan. Key-Frame Extraction Method from Single-Lens Undersea Video Sequences. Applied Mechanics and Materials, 2013, 514-519

[21] M Chatzigiorgaki,AN Skodras, Real-time keyframe extraction towards video content identificationInternational Conference on Digital Signal Processing. 2009 :934-939

[22] G Liu,X Wen,W Zheng,P He. Shot Boundary Detection and Keyframe Extraction Based on Scale Invariant Feature Transform. Eigth IEEE/ACIS International Conference on Computer & Information Science, 2009 :1126-1130

[23] C Sujatha,U Mudenagudi. A Study on Keyframe Extraction Methods for Video Summary. International Conference on Computational Intelligence & Communication Networks. 2011 :73-77.

[24] N Lv， Z Feng， J Peng. Mutual information based video shot boundary detection. International Conference on Image Analysis & Signal Processing, 2013, 20: 1-5

[25] L Krulikovská,M Mardiak,J Pavlovic,J Polec. Video Analysis Based on Mutual Information. Computer Vision & Graphics-international Conference, 201:6375:73-80.

IOP Conf. Series: Materials Science and Engineering **359** (2018) 012011 doi:10.1088/1757-899X/359/1/012011

Influencing Factors on Shear Strength of Asphalt Mortar with High Viscosity

M L Xing[1,2], H X Chen[2], Z Z Li[2], X Y Ma[2], Y Sun[2], and Y Zhou[3]

[1]Guangxi Key Laboratory of Road Structure and Materials, Guangxi Transportation Research & consulting Co., LTD, Nanning, 530007,China
[2]Engineering Research Centre of Transportation Materials, Ministry of Education, Chang'an University, Xi'an710061,China
[3]Department of Civil Engineering, Tianjin Chengjian University, Key Laboratory of Soft Soil Characteristics and Engineering Environment of Tianjin, Tianjin 300384,China

Abstract. In orderto study the influencing factors on the shear strength of asphalt mortar with high viscosity, many different compositions are considered to analyse theshear strength of asphalt mortar, such as asphalt from different resource, mineral powder, wimble weight, temperature and ratio of filler to bitumen. The results indicate that among these factors, the influence of wimble strength is the smallest, which can be ignored. The shear strength of asphalt mortar decreases with increasing temperature from 25℃~50℃.Meanwhile it changes with the ratio of filler to bitumen by 1.2 under the temperature changing from 25℃~35℃.Also with the rising ratio between filler and bitumen, the shear strength of mortar increases. However, the temperature sensitivity changes differently with its peak. Compared to the type of mineral powder, the influence of asphalt resource on shear strength and temperature sensitivity are more significant.

1. Introduction

Since performance of drainage requires large air void in the mixture of pavement, asphalt and mortar play the important roles in the mixture. Asphalt mortar combined with asphalt and mineral powder plays an important role in asphalt mixture. In dense-graded asphalt mixture, asphalt mortar can fulfil the interspace among the aggregates. And it also forms glue sand with fine aggregate which can coheres with coarse aggregate particles and ensures the overall strength of the mixture[1].Porous asphalt pavement is known as its good performance, such as drainage quickly, good performance in skid resistance. Since porous asphalt mixture is composed with large amount of coarse aggregate, less fine aggregate and high void ratio,it has the frame structure on stone-stone. Asphalt and mortar play the important roles in the mixture. And asphalt mortar is the most important factor in maintaining the stability of the structure [2, 3]. Since porous asphalt mixture lays on the surface of pavement structure, it is affected by the traffic directly. Therefor the shear resistance becomes the most important performance [4-9]. As so far, many studies have done on shear behaviour of mixture, but few on asphalt mortar.

In this paper, influences of asphalt, mineral powder and ratio of filler to bitumen to shear performance are studied by testing the feature of whole performance. And the influences of asphalt, mineral powder, and asphalt mortar are analysed basing on the testing results.

2. Experimental

2.1. Bitumen

Two kinds of modified bitumen are made from matrix bitumen,ESSO70#. One(named B1) is modifiedwith TPSbeing 12% of bitumen in weight, and the other(named B2) is 5%SBS.

2.2. Mineral powder

Two types of limestone powder from different resourceare used in the study. And their specific surface areas and particle sizes are tested by different methods. The table 2 shows the results.

2.3.Asphalt mortar

Asphalt mortar is the mixture of bitumen and mineral powder. Firstly, mineral powder is heated and dried to a constant weight at 105±5℃. Then, a certain quality mineral powder is added into modified asphalt under 160℃. And the mixture is stirred until it isup to the best. So three kinds of asphalt mortar are chosen for studied from 2 types of bitumen and 2 kinds of mineral powder.

Table 1. Indexes of bitumen.

Properties	Test values		
	ESSO70#	B1	B2
Softening point (℃)	49.9	90.2	79.2
Penetration(25℃,0.1mm)	64.1	50.3	56.4
Ductility(cm)	>150(15℃)	44.1(5℃)	37.9(5℃)
Toughness（N·m）	/	29.6	30.2
Tenacity（N·m）	/	16.9	20.9
Dynamic viscosity(60℃,Pa.s)	/	41743	16713

Table 2. Indexes of mineral powder.

Properties		1#	2#
Apparent density（t/m3）		2.64	2.68
Size range	<0.6mm（%）	100	100
	<0.15mm（%）	99.4	98.4
	<0.075mm（%）	86.4	79.4
Specific surface area（<0.075mm）（m^2/kg）		2392.5	3224.6
Average particle size（<0.075mm）（μm）		5.476	4.879

Table 3. Asphalt mortar.

Asphalt mortar	Modified asphalt	Mineral powder	Ratio of filler to bitumen
B1-1	B1	1#	
B1-2	B1	2#	0.4,0.6,0.8,1.0,1.2,1.4,1.6
B2-1	B2	1#	

2.4. Shear strength

The shear strength is calculated by measuring the cone depth of a certain mass at fixed time which is similar to Soil Test. And the shear strength can be calculated by Equation(1).

IOP Conf. Series: Materials Science and Engineering **359** (2018) 012011 doi:10.1088/1757-899X/359/1/012011

$$\tau = \frac{981M\cos^2(\frac{\alpha}{2})}{\pi h^2 \tan(\frac{\alpha}{2})} \tag{1}$$

where, τ is shear strength (kPa), M stands for total weight (g), h is cone depth(0.1mm), and α is the angel between tips.

3. Results and discussion

3.1. Weight of cone

Two cones with different weight are used to investigate their influence on shear strength. And the results in Figure.1 for B1-1 show there is no remarkable influence of cone weight on it.

3.2Temperature

The cone penetration tests forthree kinds of asphalt mortar are carried out under different temperatureranging from25℃~50℃. Then the shear strength is calculated by equation (1) under each condition.

As can be seen in Fig.1, the shear strength increases with the increasing proportion of mineral powder in bitumen. Since the increasing number of filler can raise the viscosity of asphalt mortar which is beneficial to shear behaviour. And this result indicates that ratio of filler to bitumen in mixture should be added to improve shear behaviour as well as to road performance in some circumstances.

Fig.1 shows that the shear strength drops down quickly with increasing of temperature. And it drops obviously with temperature changing from 25℃~35℃. But the drop rate is differ from types of mortar.As to mortar of B2-1, the reduction of shear strength is up to 80% with temperature changing from 25℃~35℃. By contract, the shear strength of mortar B1-1and B1-2 reduces separately around 75% and 60%. When temperature changes from 35℃~50℃, the drop rate of shear strength becomes considerable slow down with range of 14%~20%. Therefore, the whole range of testing temperature can be divided into two parts with 35℃. And 35℃ can be suggested to distinguish the temperature of different asphalt mortar.

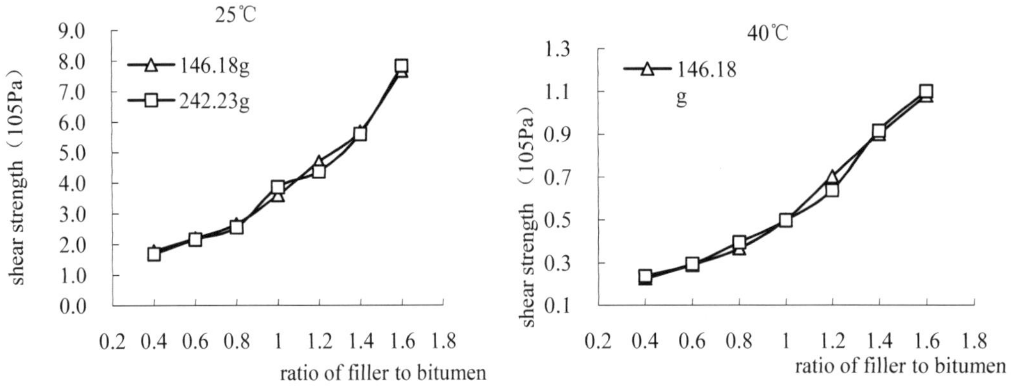

Fig.1. Influence of cone weight (A1-1)

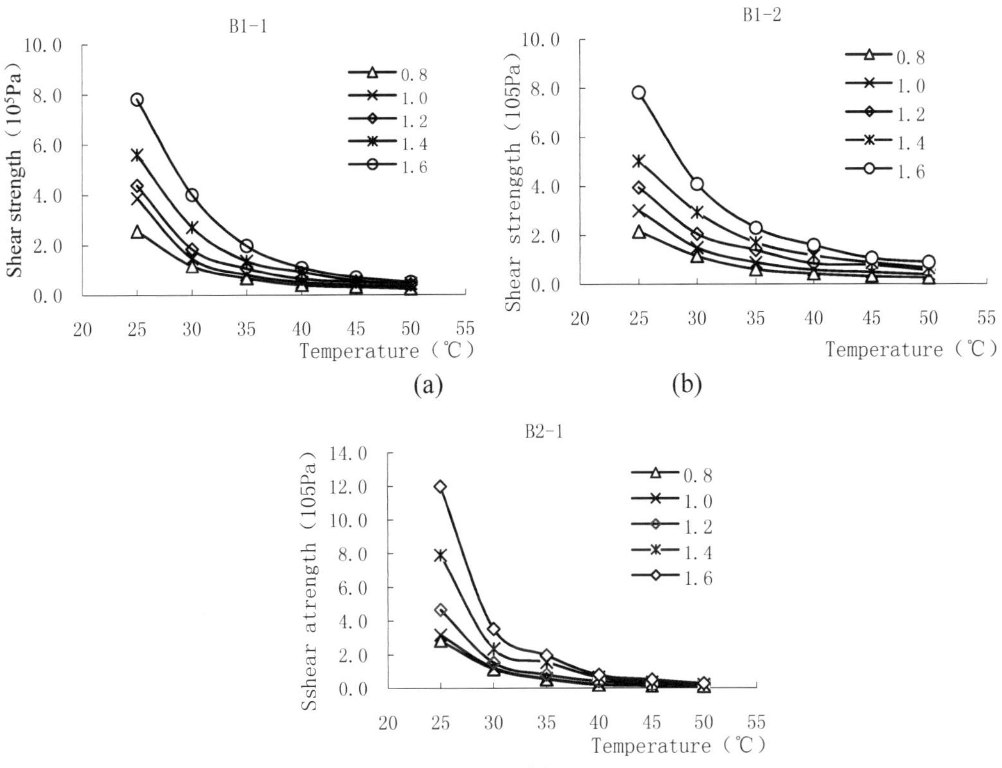

Fig.2.Curves for effect of temperatuire on asphalt mortar

3.3 Asphalt and mineral powder

From Figure2 (a, b, c),the relationship between shear strength and temperature can be represented by the following formula-equation (2):

$$\tau = Kt^{E} \quad (2)$$

Where,τ is shear strength (10^5Pa), t is temperature (℃), and K and E are regression coefficients. Meanwhile, the shear strength increases with the increasing of K value. And|E|value represents the sensitivity of shear strength to the temperature [10]. The larger the|E| value is, the greater sensitivity the shear strength is.

Table 4.K and E for different asphalt mortar

B1-1			B1-2			B2-1		
Ration of filler to bitumen	K	E	Ration of filler to bitumen	K	E	Ration of filler to bitumen	K	E
0.8	14680	-3.4384	0.8	44326	-3.1056	0.8	6E+07	-5.2154
1.0	36958	-3.6239	1.0	33124	-2.9287	1.0	7E+07	-5.2346
1.2	23471	-3.4317	1.2	22426	-2.7407	1.2	3E+07	-4.9305
1.4	85537	-3.7251	1.4	63375	-2.9408	1.4	1E+08	-5.1860
1.6	4E+06	-4.0737	1.6	21082	-3.188	1.6	3E+08	-5.3525

Table 4 shows that there is a minimumfor |E| value with the ration of filler to bitumen beingat 1.2. So it indicates that the sensitivity of shear strength to the temperature is minimal and the mortar has the best temperature stability. Therefore this ratio of filler to bitumen can be suggested to use in the porous asphalt mixture which can also guarantee its characters. At the same ratio of filler to bitumen, mortar of B2-1 is more sensitive than the other two. And the |E| value of B1-2 is the smallest which reveals that the asphalt mortar of B1-2 has the best temperature stability among the three studied. The results suggest that combining with the same asphalt, the larger specific area of mineral powder is, the smaller sensitivity of asphalt mortar has. Because there are sufficient areas on the surface of mineral powder which can adhere to asphalt. Then the structural asphalt are formed which is very important to characters of mixture. Similarly, an asphalt with higher viscosity is beneficial to form the structural asphalt. However comparing with mineral powder, the type of asphalt is more obviousto the influence of temperature sensitivity. So that bitumen with high viscosity is usually chosen in order to improve the shear resistance of porous asphalt mixture, especially used on the crossing.

3.4. Ratio of filler to bitumen

In order to study the influence of ratio of filler to bitumen, shear strength under different temperature with different ratio are tested. Fig.3.shows the results of different mortar under different temperature and ratio of filler to bitumen.

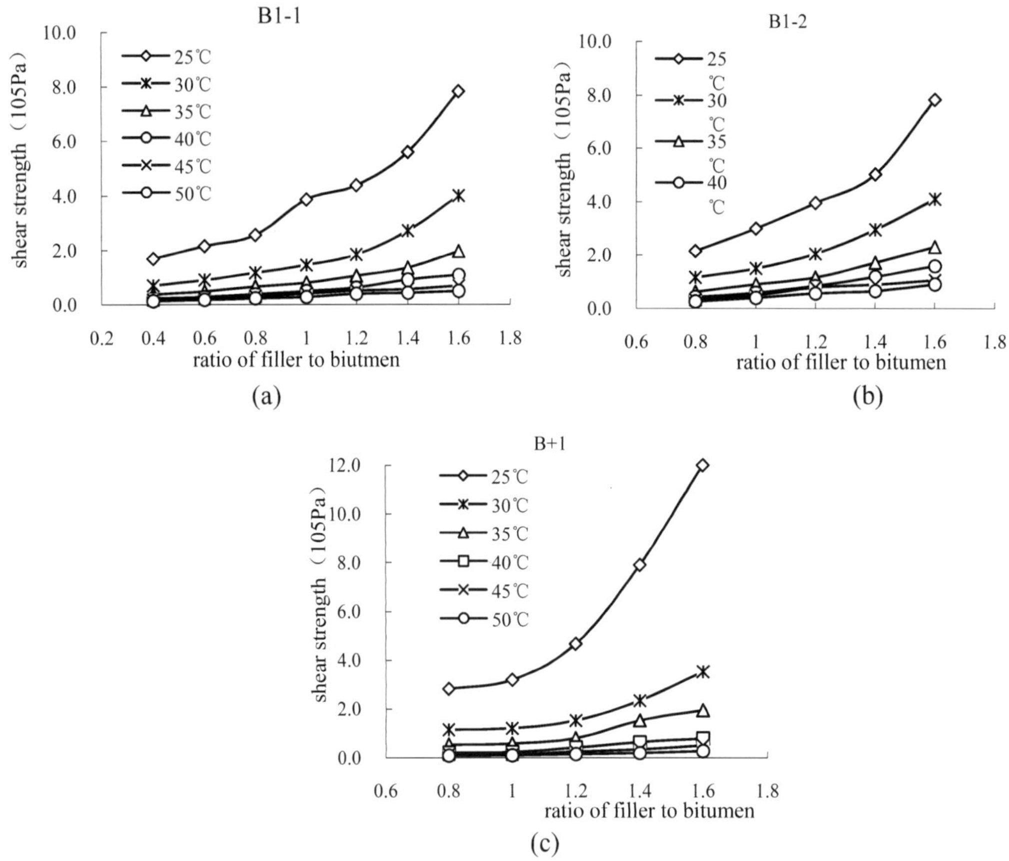

Fig.3. Curves for shear strength with ratio of filler to bitumen

The curves in Fig.3 show that the shear strength rise with the increasing of ratio of filler to bitumen and reduce with the temperature rising. And it is obviously influenced by the ratio of filler to bitumen. With the larger amount of mineral powder in asphalt mortar, the shear strength is higher. Because large amount of filler can provide more surface area for asphalt to adhere, the more structural asphalt can be formed in the mortar which is beneficial for the character in its asphalt mixture. With the same ratio of filler to bitumen, the shear strength of B1-2 is the highest because of the asphalt with high viscosity which can strengthen the adhesion between asphalt andmineral powder. Then the performance of asphalt can be improved effectively.

4. Conclusions

This study investigated the influencing factors of asphalt mortar on shear strength. The influence of cone weight is so small that it can be ignored. The ratio of filler to bitumen influenced the temperature sensitivity and shear strength. When it is at 1.2, the temperature sensitivity of asphalt mortar is the smallest, and it has the best performance at high temperature. Also the shear strength can be improved by adding the asphalt with higher viscosity or finer mineral powder. The drop rate declined with the increasing of temperature. Therefore, 35℃ can be suggested as the division temperature for asphalt mortar.

Acknowledgments

This research work was sponsored by the Key Laboratory of Road Structure and Materials of Guangxi, China Postdoctoral Science Foundation（2015M582592）and Special Fund for Basic Scientific Research of Central Collegesin Chang'an University(310831151080,310831153409)， Tianjin Application Foundation and advanced technique research program(15JCYBJC48800，17JCYBJC22300)， Road Construction and Materials Transportation Industry Key Laboratory Fund Project(310821171114)

Reference

[1] Xiao Y (2008). Fracture Mechanisms of Binder-Aggregate System and Its Effect on Properties of Asphalt Mixtures. Wuhan University of Technology

[2] Xing, M.L(2010).Study on the Mortar and Composition of Pervious Asphalt Mixture. Chang'an University.

[3] Alvarez A.E.,Martin A.E.,Estakhri C. K.(2006). Synthesis of Current Practice on the Design,Construction,and Maintenance of Porous Friction Courses.Report of FHWA,Texas.

[4] SongH, DoJ, SohY. Feasibility study of asphalt-modified mortars using asphalt emulsion. Journal of Construction and Building Materials,**20**(5):332-337

[5] QudaisS.A.,ShweilyH.A.(2007) Effect of aggregate properties on asphalt mixtures stripping and creep behaviour. Journal ofConstruction and Building Materials,**21**(9):1886-1898

[6] Lu W.J.(2010).Influence factors of shear resistance of asphalt mixture,Journal of Southeast University(Natural Science Edition. Journal of Highway.**11**:81-85

[7] Yuan J(2008).Index of Factors Influencing the Shear Strength of Asphalt Mixtures.Journal of Chongqing Jianzhu University.（**30**）06:140-145

[8] He Q.M.(2009). TheResearch on Fatigue Experiment for Shear Properties of Asphalt Concrete. Changsha University of Science and Technology.

[9] Jiang Y., Zhang Y.(2012).Influencing Factors of Shear Strength of Asphalt Mixture.Journal of Highway and Transportation Research and Development**07**:9-14

[10] Chen H.X. (2006) Temperature Susceptibility Analysis of Asphalt Binders. 01:8-11

IOP Conf. Series: Materials Science and Engineering **359** (2018) 012012 doi:10.1088/1757-899X/359/1/012012

End-Cap Modified Unsaturated Polyesters With Low Styrene Emission

S Liu[1], J R Chen[2] and Y Zhang[1,*]

[1]Key Laboratory of Special Functional Polymers and Related Technology of Ministry Education, School of Materials Science and Engineering, East China University of Science and Technology, Meilong Road 130, Shanghai 200237, China
[2] Shanghai Zhongzhi Chemical Technology Co., Ltd, Nanyangwan Road 1068, Shanghai 201612, China

E-mail: [*]yzhang@ecust.edu.cn

Abstract. A benzyl end-cap-unsaturated polyester resin (BP-UPR) with low styrene emission was synthesized. The structure of BP-UPR was characterized by Fourier transform infrared (FT-IR) spectroscopy. According to gel permeation chromatography (GPC) measurement, the BP-UPR has a relative low molecular weight (M_n=1754) and narrow polydispersity of 1.68. The styrene content of BP-UPR is only 25.1%, 8.3% lower than that of a commercial unsaturated polyester resin (C-UPR) with similar viscosity. The styrene emission of BP-UPR and C-UPR was compared by Luong–Walewski method. At the gel point (17 minutes), the styrene emission of BP-UPR is only 38.9g/m^2, 15.9% lower than that of C-UPR. After curing for 2 hours, the total styrene emission of BP-UPR is only 62.5 g/m^2, which is 48.0% lower than that of C-UPR.

1.Introduction

Unsaturated polyester (UP) resinsare widely used in the field of fiber reinforced composites because of their good mechanical strengthand low costs[1-2].And high content of reactive diluentssuch as styreneareusually used for processing requirement[3]. However, styrene, as a hazardous air pollutant (HAP) and volatileorganic compound (VOC),would emit during spray-up processing and curing reaction andresult in severe environment pollution and do great harmful to the human beings. The legislation of limitation VOC has been put forward by the Federal Environmental Protection Agency of the United States[4].And the research on green and low styrene emission (LSE) resins has attracted significant interest.

The volatilization of styrenecan be reduced to some extent byadding styrene suppressant additives or replacing styrene with high boiling point diluents. But the weak interlaminar adhesion or poor dilution effect make them have little commercial values[5-6].

Decreasing the content of styrene dilute inUP resins can reduce styrene emission effectively. Nevertheless, the viscosity of the system may be too high for further processing.The viscosity of the resins mainly depends usually on the molecular weight. By using a monofunctional group monomer to end-cap the growing chains, the precise stoichiometric balance between the reactive acidand alcohol groups would be broken, so the molecular weight of the resins can be controlled effectively.For example, dicyclopentadiene(DCPD) based resin is anend-capped UP resin[7-8]. However, the poor corrosive resistance and high viscosityattributed to side-reactionsof DCPDwould limit its application. Low boiling point monohydric alcoholslikeethyl alcohol have also been used as the end-capped agent[9]. But its high evaporation rate results inlow efficiency of end-cap reaction.

The synthesis of a new unsaturated polyester with low styrene emission is reported in this paper. Benzyl alcohol(BP), a high boiling point monohydric alcohol, is selected as the end-cap agent. It is much easier to control stoichiometric balance between the reactive groups, and also the molecular weight of the resins. At the same time, the solubility of unsaturated polyester in styrenecould also be improved by the replacement of polarcarboxyl acid groups with benzyl groups. The structure and the molecular weight of benzyl alcohol end-capped resin (BP-UPR)were analyzed by FT-IR and GPC, respectively. In addition, the styrene emission of BP-UPRand commercial unsaturated polyester resin (C-UPR)during storage and curing reaction are further followed.

2. Experimental

2.1 Reagents.

Maleic anhydride(AR) was obtained from Sinopharm Chemical Reagent Co., Ltd. Isophthalic acid (AR), and neopentyl glycol (AR), propylene glycol (AR), benzyl alcohol (AR) andstyrene (AR) were purchased from Shanghai Ling Feng Chemical Reagent Co.,Ltd.Commercial unsaturated resin (C-UPR) was provided by Shanghai Zhongzhi Chemical Technology Co., Ltd.

2.2 Measurements.

Fourier transform infrared (FT-IR) spectra was recorded on a Nicolet 5700 spectrometer. The measurement of molecular weight and distribution was performed by a waters 1515 gel permeation chromatography (GPC).Solid content of the resins was determined according to GB/T 7193-2008.Styrene emission test was performed according to Luong–Walewski[10] method at room temperature. 60gUP resinswerecatalyzed by cobaltnaphthenate and methyl ethyl ketone peroxide(MEKP) and mixed for one minute. Then 50g catalyzed sample was put uniformlyinthe watch glass with a diameter of 80mm, then placed in the fume hoodimmediately and weighed every five minutes.The styrene emission of the resins can be calculated according to the following formula:

$$\text{styrene emission}(\text{g/m}^2) = \frac{\text{weight loss (g)}}{\text{area of lid (m}^2\text{)}}$$

2.3 Synthesis.

Propylene glycol,neopentyl glycol and isophthalic acid were placed in a four-neckflask, and reaction lasted for 4 hours with agitation. Then the maleic anhydride and propylene glycol were added into the flask. The reaction temperature was increased to 200~210°C and the acid value (AV) was tested per hour.The benzyl alcohol was added into the flask at an AV of 140~110mg/g. Keep the reaction until the AV of the system reaches to 7~13mg/g.Then the system was cooled down and styrene was added for dilution.

3. Results and Discussion

3.1 The FT-IR analysis.

A comparison of FT-IR spectra of the obtained end-cappedunsaturated polyester resin (BP-UPR) and the unmodified unsaturated polyesteris displayedin Figure 1.The IR spectra of two resins shows a series of typical absorptionpeaks of unsaturated polyester at 1727 cm^{-1} (C=O group), 2965$cm^{-1}$and 2882cm^{-1} (CH_3group), 1646 cm^{-1} (C=Cgroup), respectively. Nevertheless, the intensity of the peak at 1727 cm^{-1} (C=O) for BP-UPR (curve b) is much higher than that of unmodified resin (curve a) by using the peak at 2965cm^{-1} (CH_3) as the internal standard.It demonstrates the esterification reaction between the benzyl alcohol and carboxyl groups at the terminal of UP molecular chains.Furthermore, a new absorption band at 701cm^{-1}corresponding to the C-H flexural of mono-substituted benzene derivatives appears, indicating that the unsaturated polyesters are end-capped by benzyl alcohol.

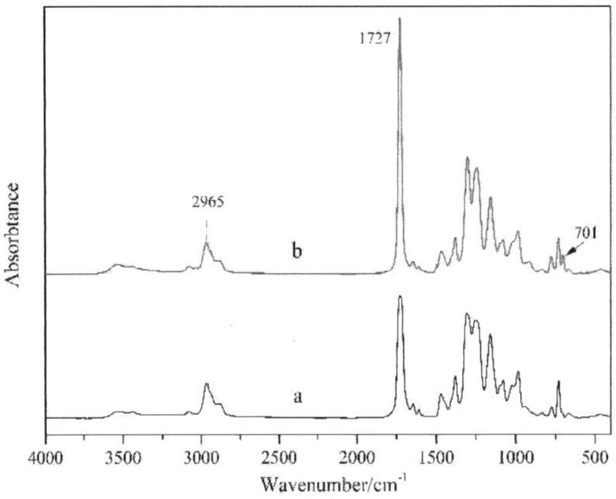

Figure 1.FT-IR spectra of the unmodified unsaturated polyester resin (a) and the BP-UPR (b).

3.2 Molecular weight of BP-UPR.

The GPC spectra of BP-UPR and commercial resin (C-UPR)are compared in Figure 2. The molecular weight as well as its distribution, styrene content of BP-UPR and C-UPR are shown in Table 1.It can be seen that the molecular weight (M_n, and M_w) of BP-UPR is much lower than that of C-UPR because the end-capped reaction of benzyl alcohol has broken the precise stoichiometric balance between reactive acid and alcohol groups. The lower molecular weight of the resins implies lower viscosity of system, so that less diluent (styrene) is needed for processing. The styrene content of BP-UPR is only 25.1%, which is 8.3% lower than that of C-UPR with the same viscosity (1150cp). It means that styrene emission of BP-UPR is much lower than that of C-UPR during storage and shipping.

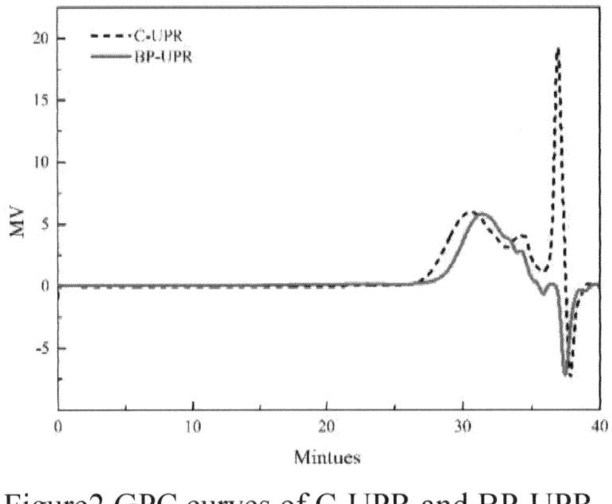

Figure2.GPC curves of C-UPR and BP-UPR.

Table 1.Molecular weight and distribution and styrene content of C-UPR and BP-UPR.

	Mn	Mw	PDI	Styrene content/%
C-UPR	3204	5287	1.65	33.4
BP-UPR	1754	3558	1.68	25.1

3.3 Styrene emission.

A further comparison of styrene emission in curing stage of BP-UPR and C-UPR is shown in Figure 3. At first, the volatilization of styrene in two resins tends to increase with the time linearly because the low viscosity and high styrene diffusivity allow styrene to reach the resin's surface freely[11]. Therefore, the initial emission rate only depends on the styrene content of resins. Due to the low styrene content of BP-UPR (25.1%), its styrene volatilization before gel is only 38.9g/m^2, which is 15.9% lower than that of C-UPR. After curing reaction lasted for 17miutes, the resins began to gel and the three-dimensional net structure was gradually formed. At the same time, the viscosity of the system increased sharply, so the styrene emission rate slowed down. One hour later, there is no styrene emission for BP-UPR while some emission still existed for C-UPR because it has a higher styrene content. More styrene monomers were trapped in the three-dimensional net structure of the cured resins. It would take more time for residual styrene monomers to be completely released. After 2 hours, there was hardly any styrene emission for both C-UPR and BP-UPR. The total styrene emission of BP-UPR is only 62.5 g/m^2, which is 48.0% lower than that of C-UPR.

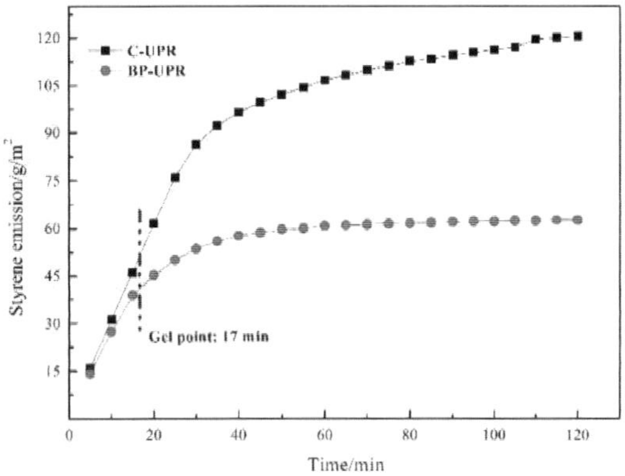

Figure 3. Styrene emission of BP-UPR and C-UPR in curing stage.

4. Conclusions

(1) A benzyl end-capped unsaturated resin (BP-UPR) was synthesized by using benzyl alcohol as the end-capper. FT-IR analysis demonstrates its structure.

(2) Compared with the commercial unsaturated polyester (C-UPR), BP-UPR has a lower molecular weight ((M_n=1754) and a similar polydispersity (PDI=1.68). The styrene content of BP-UPR is only 25.1%, which is 8.3% lower than that of C-UPR with same viscosity (1150cp).

(3) At the first stage of curing reaction, styrene volatilization of two UP resinstends to increase with the time linearly. The volatilization amount of BP-UPR before gel was 38.9g/m^2, 15.9% lower than that of C-UPR. After gel, the styrene emission rate of the two resins began to slow down. After curing reaction lasted for 2 hours, the total styrene emission of BP-UPR was only 62.5 g/m^2, which is 48.0% lower than that of C-UPR.

5. References

[1] Fonseca A C, Costa C S M F, Marques T M P, Coelho J F J and Serra A C 2017 *J. Mater. Sci.***52** 1272-84
[2] Nguyen Q T, Ngo T D, Bai Y and Tran P 2016 *Fire Mater.* **40** 1047-69
[3] Erdem A andKusefoglu S 2011 *242nd National Meeting of the American-Chemical-Society* (Denver: Colorado)
[4] Perrot Y, Davies P, Keboul A and Baley C 2008 *Appl. Compos. Mater.* **15** 87-97
[5] Jaswal S and Gaur B 2015 *Green Proc. and Synth.* **4** 191-202
[6] Youngblood J P, Sellars A B, Clark A J and Coles S R 2015 *Green Mater.* **3** 35-42
[7] Maurin R, Perrot Y, Bourmaud A, Davies P and Baley C 2009 *Composites Part A***40** 1024-32
[8] Chengguo L, Wen L, Zhengchun C, Jianqiang C, Lihong, Yan D and Yonghong Z 2013 *Ind. Crops. Prod.***49** 412-18
[9] Loza R and Hartinger D G 2004 US 6794483 B2[P]
[10] Raul D 2000 *Reinf. Plast.***44** 42-44
[11] Scala J J, Orlicki J A, Jain R, Ulven C A, Palmese G R, Vaidya U K and Sands J M 2009 *Clean Technol. Environ. Policy* **11** 283-92

IOP Conf. Series: Materials Science and Engineering **359** (2018) 012013 doi:10.1088/1757-899X/359/1/012013

Mechanical Properties of Oxide Films on Electrolytic In-process Dressing (ELID) Copper-based Grinding Wheel

J C Kuai[1] , J W Wang[1], C R Jiang[1], H L Zhang[2] and Z B Yang[1]

[1]School of Mechanical and Power Engineering, Henan Polytechnic University, Jiaozuo, Henan, 454003, PR China.

[2]School of Mechanical Engineering, Nantong University, Nantong, Jiangsu, 226019, PR China.

E-mail:hitgjc@163.com

Abstract. The mechanical properties of oxide films on copper based grinding wheel were studied by nanoindentation technique. The analysis of load displacement shows that the creep phenomenon occurs during the loading stage. Results show that the oxide film and the matrix have different characteristics, and the rigidity of the copper based grinding wheel is 0.6-1.3mN/nm, which is weaker than that of the matrix; the hardness of the oxide film is 2000-2300MPa, which is higher than the matrix; and the elastic modulus of the oxide film is 100-120GPa, also higher than the matrix. .

1.Introduction

ELID grinding technology is widely used in the field of ultra-precision grinding of hard and brittle materials, because the grinding wheel surface coated with oxide film, which has auxiliary polishing effect, then can get a high grinding quality[1]. It can be seen that the performance of the oxide film produced by the electrolysis of the grinding wheel surface becomes a key factor affecting the grinding quality, and many researchers have conducted extensive and deep research on the performance of the oxide film[2-5]. Some researchers simplify the oxide film to "spring", one end was connected with the grinding wheel matrix, and the other end was loaded with abrasive grains. In the ultra-precision grinding, the "spring" instead of grinding wheel friction polishing, therefore, ELID grinding can get high precision grinding surface[6]. But researchers doesn't know much about the mechanical properties of the “spring” . Writer have been studied the mechanical properties of the oxide film on the iron based grinding wheel , and the effects of the grinding depth、 feed rate and other process parameters on the hardness, elastic modulus and stiffness of the oxide film on the grinding wheel surface[7]. However, copper based grinding wheel is often used in high efficiency grinding in ELID grinding, and the mechanical properties of the oxide film on the grinding wheel surface are rarely reported.

In this study the mechanical properties of the oxide film on the copper matrix grinding wheel was measured by nanoindentation technique, and tested the hardness, elastic modulus and stiffness of the oxide film on the copper based grinding wheel.

IOP Conf. Series: Materials Science and Engineering **359** (2018) 012013 doi:10.1088/1757-899X/359/1/012013

2.Experimental Instruments and Methods

2.1Experimental instruments. CSM instrument SA UNHT-type ultra nanoindentation instrument; Berkovich diamond indenter; the experiment was carried out in 10000 levels of ultra clean environment, the atmosphere of the environment medium was air, and the ambient temperature was 24±0.5℃; W40 copper based grinding wheel; ELID power supply; ELID electrolyte; MM7140 surface grinder.

2.2Experimental methods. Put the W40 copper based grinding wheel in the MM7140 surface grinder which has the ELID electrolytic device, and start electrolytic polishing and grinding experiment, electrolytic and grinding parameters are shown in Table 1.

Table 1 ELID electrolytic and grinding parameters of copper based grinding wheel

Electrolytic voltage U(V)	Electrolytic current I(A)	Electrolytic pulse width ton(μs)	Electrolytic pulse spacing toff(μs)	grinding speed Vs(m/s)	grinding depth AP(mm)	feed rate F(mm/min)
60-120	20-50	0-99	0-99	15.7	0.001-0.01	20-50

After the end of the ELID electrolytic grinding, use wire cutting machine to cut the surface of the copper based grinding wheel with oxide film part, about 20mm*30mm*10mm, for nanoindentation test; a total of 4 test depth: 0.001mm, 0.002mm, 0.005mm, 0.01mm, each test depth was repeated 4 times and the average was taken as the test result.

3. Results and discussions

3.1Variation of load and displacement with time. As shown in Figure 1a)-1d), it' s nanoindentation loading curve, load displacement are 0.001mm, 0.002mm, 0.005mm, 0.010mm. From Figure 1a), the maximum load is Fmax=59.74mN, the maximum displacement is hmax= 1039.93nm; Figure 1b), the maximum load is Fmax=217.78mN, the maximum displacement is hmax= 2010.02nm; Figure 1c), the maximum load is Fmax=874.23mN, the maximum displacement is hmax= 5024.99 nm; Figure 1d), the maximum load is Fmax=1637.60mN, the maximum displacement is hmax= 10078.94nm. When the load is 0.001 mm, it can be seen from the unloading curve Fn and displacement curve Pd, only a small amount of elastic recovery, large plastic deformation occurred, and elastic recovery is 50 nm; when the load is 0.002 mm, elastic recovery is 300 nm; When the load is 0.005 mm, elastic recovery is 700 nm; When the load is 0.01 mm, elastic recovery is 1700 nm. It is found that the elastic recovery increases with the increase of the load depth, which indicates that the influence of the grinding wheel matrix increases. In addition, from the elastic recovery of the oxide film and the change of the load in figure 1 we can see, the elastic coefficient of the oxide film of the grinding wheel is not constant, but varies. The elastic coefficient increases with the increase of the load depth.

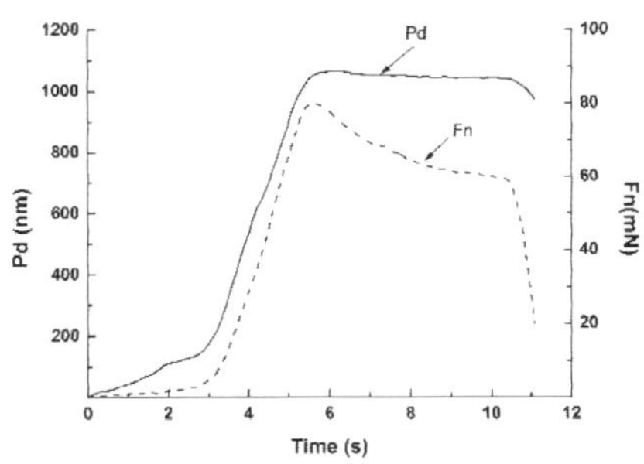

a) load depth 0.001mm

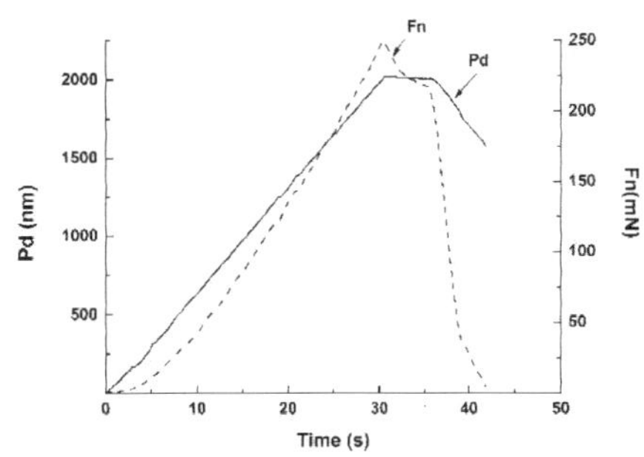

b) load depth 0.002mm

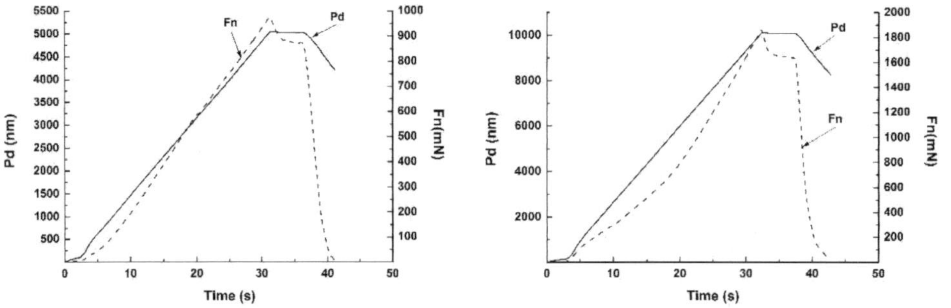

c) load depth 0.005mm d) load depth 0.010mm

Figure 1. Nanoindentation load and displacement variation

3.2Variation of load and displacement. As shown in Figure 2a)-2d), it' s nanoindentation load displacement curve, load displacement are 0.001mm, 0.002mm, 0.005mm, 0.010mm. Fig. 2a) shows that there is creep phenomenon at the displacement 0-100nm; Figure 2b) shows that there is creep phenomenon at the displacement 1250-1500nm; Figure 2c) shows that there is creep phenomenon at the displacement 3000-3500nm.

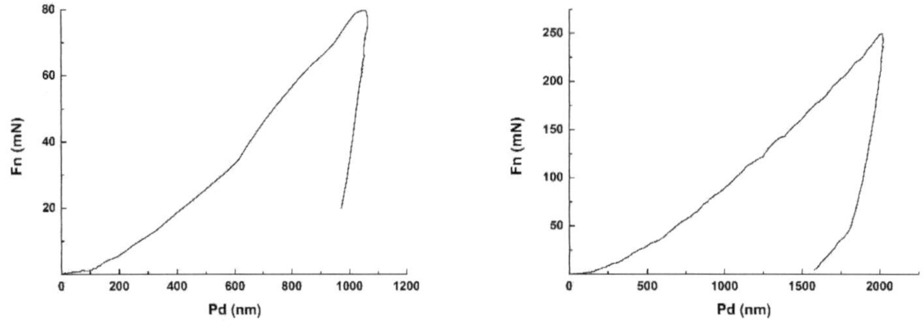

(a) a) load depth 0.001mm (b) load depth 0.002mm

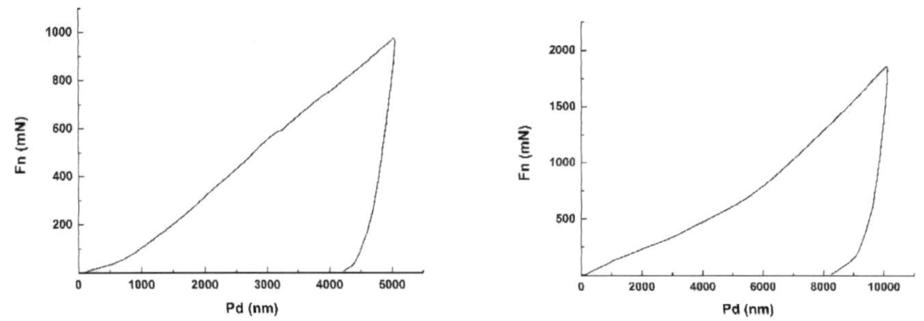

c) load depth 0.005mm d) load depth 0.010mm

Figure 2. Nanoindentation load displacement curve

3.3Variation of stiffness, hardness and modulus of elasticity. Figure 3 shows that the oxide film stiffness increases with the increase of indentation depth. At the initial stage, the slope of the

0.001-0.002mm indentation depth is larger, and then tends to be gentle. This is due to the thin film formed on the copper based grinding wheel, only a few microns. Therefore, the measured indentation depth exceeds the binding site of the oxide film and the matrix, and presents different characteristics in the binding site.

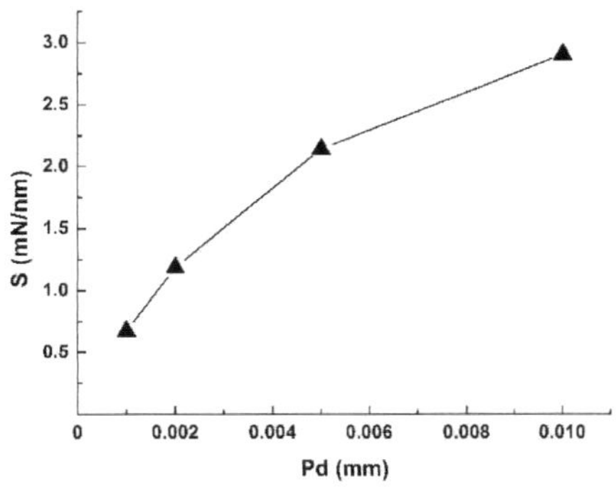

Figure 3. Nanoindentation test of the stiffness of the oxide film

Figure. 4 is the change curve of the hardness of the oxide film and the indentation depth. It can be seen from Fig. 4 that with the increase of the indentation depth, the hardness of the oxide film increases first and then decreases, which shows the two different variation characteristics of the former and the latter. When the indentation depth is 0.001-0.002mm, the hardness of the oxide film is measured at this time, about 2000-2300MPa; as the indentation depth increases to 0.005mm, the indenter is at the binding site of the oxide film and the copper based grinding wheel, and the curve shows that the hardness drops to about 1000 MPa; when the indentation depth is increased to 0.01mm, the indenter has clearly crossed the binding site of oxide film and copper based grinding wheel matrix, and test the composite hardness of the grinding wheel matrix and oxide film, the hardness decreased to 700MPa, shows the different characteristics of oxide film formed after electrolytic and the matrix. The hardness of the oxide film is much higher than that of the copper based grinding wheel matrix. This seems to indicate that the polishing performance of copper based grinding wheel oxide film is poor, and the superhard oxide film seems to be easier to scratch the finished surface.

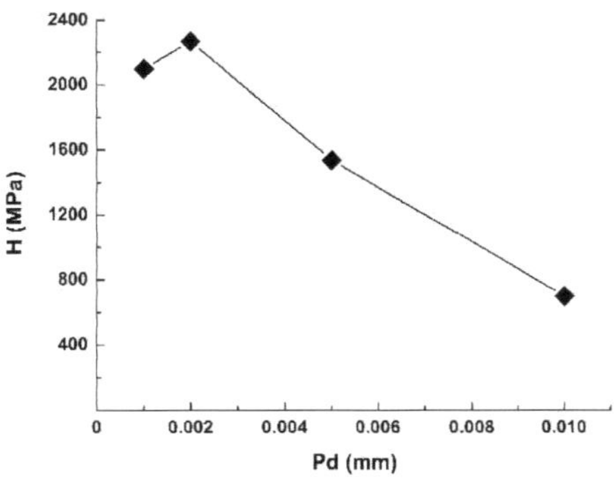

Figure 4. Nanoindentation test of hardness of the oxide film.

Figure. 5 is the change curve of the elastic modulus of the oxide film and the indentation depth. It can be seen from figure 5 that the elastic modulus decreases from 120GPa to 60GPa with the increase of indentation depth. When the indentation depth is 0.001-0.002mm, the elastic modulus of oxide film is measured, about 100-120GPa; when the indentation depth is 0.005-0.010mm, the indenter has been crossed the oxide film, and entered into the copper based grinding wheel matrix at this time. Therefore, the elastic modulus is the composite elastic modulus of oxide film and the matrix, about 60-80GPa. It follows that the elastic modulus of the oxide film is different from that of the matrix, the elastic modulus of the oxide film is higher than that of the matrix.

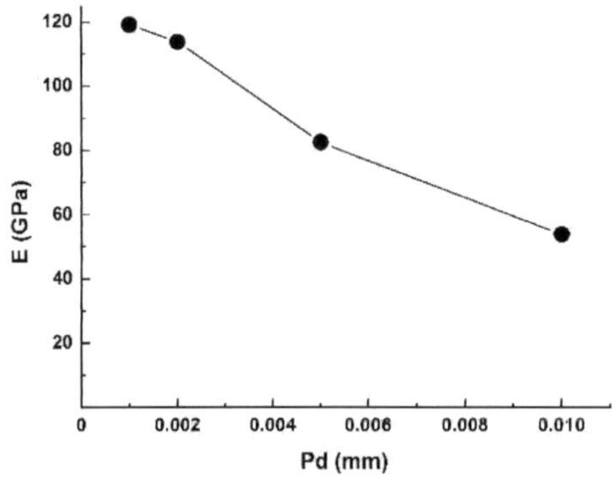

Figure 5. Nanoindentation test of elastic modulus of the oxide film

4.Summary

This paper studied the mechanical properties of the oxide film on the copper based grinding wheel, and obtained some useful results:

1. The elasticity of the oxide film on the copper based grinding wheel is weak, and its elastic coefficient is not constant, should be regarded as a spring damping system with variable elastic coefficient;
2. The hardness of the oxide film on the copper based grinding wheel is as high as 2000-2300MPa, which is much higher than that of the matrix, and that's mean the oxide film seem not to be suitable for the auxiliary polishing in ELID grinding;
3. The oxide film of the copper based grinding wheel has high elastic modulus, as high as 100-120GPa, which is higher than the elastic modulus of the matrix.

5.Acknowledgments

This work was supported by the National Natural Science Foundation, China (Grant No. 51475147), and by the key project of Henan Province Science and Technology Research (Grant No. 13A460341). Also funded by the Natural Science Foundation of the Jiangsu Province, China (Grant No. BK20150406) and the technical project of Nantong, China (Grant No. MS12016015).

References

[1] Gao D X, Liu Q, ZhangB, Xiao C X, Ren C Z. 2016 Research on formation mechanism of grinding wheel oxide film in ELID ultra precision mirror grinding *Materials Science Forum* **861** 90-96.

[2] Tang H, Deng Z H, Guo Y S, Qian J, Reynaerts D. 2015 Depth-of-cut errors in ELID surface grinding of zirconia-based ceramics *International Journal of Machine Tools &Manufacture* **88** 34-41.

[3] Wu M L, Zhang K F, Ren C Z. 2015 Study on the non-uniform contact during ELID groove grinding *Precision Engineering* **39** 116-124.

[4] Zhang H L, Kuai J C. 2017 Forming mechanism of -Fe2O3in the oxide films on iron-bonded diamond wheel surface by ELID grinding *Key Engineering Materials.* **723** 434-438.
[5] Kuai J C. 2013 Friction coefficient of nano-ceramic polished by oxide film on elid grinding wheel *Advanced Materials Research.* **669** 91-94.
[6] Kuai J C, Zhang H L, Zhang F H, Zhang Y. 2010 Research on minimum mechanism of roughness of ELID grinding *Advanced Materials Research* **135** 441-446.
[7] Kuai J C, Zhang F H, Zhang Y. 2010 Mechanical Properties of Oxide Film on ELID Grinding Wheel Surface *Nanotechnology & Precision Engineering.* **8** 447-451.

IOP Conf. Series: Materials Science and Engineering **359** (2018) 012014 doi:10.1088/1757-899X/359/1/012014

Study on Thermal Deformation Behavior of TC4 – ELI Titanium Alloy

Y Song[1],F S Zhang[1], T Huang[2,3] and K X Song[2,3]

[1]Schoolof Mechatronics Engineering, Henan University of Science and Technology, Luoyang 471023,China

[2]School of Materials Science & Engineering, Henan University of Science and Technology, Luoyang 471023,China

[3]Collaborative Innovation Center of Nonferrous Metals of Henan Province, Luoyang 471023,China

huangtao@haust.edu.cn

Abstract. The TC4-ELI titanium alloy was subjected to the hot compression deformation test by the Gleeble-1500D thermal simulation test machine. The thermal deformation behavior of the TC4-ELI titanium alloy was studied under the condition of 850℃-1050℃, $0.001s^{-1}$-$10s^{-1}$ strain rate and 50% deformation; The constitutive equation of TC4-ELI titanium alloy was established based on the hyperbolic sine model of Arrhenius equation. The results show that the flow stress of TC4-ELI titanium alloy decreases with the increase of temperature at high temperature.The calculated heat activation energy of TC4-ELI titanium alloy is 300367.5807J / mol.

1.Introduction

Titanium alloy with high strength, low density, high melting point and good corrosion resistance, is widely used in aviation, aerospace, ship and other fields[1,2].TC4-ELI titanium alloy has good strength, shaping, heat resistance, corrosion resistance, and it is so widely used that amounts for 75% -85% of the total amount of titanium alloy[3].TC4-ELI titanium alloy has large deformation resistance, is difficult to process at room temperature, and is sensitive to thermal processing parameters, so it is difficult to formulate the process in forming process. The oxidation resistance of TC4-ELI titanium alloy is poor at high temperature, which limits its application to some extent [4-8].Therefore, the thermal deformation behavior of TC4-ELI titanium alloy is urgently needed to research.

2.Materials and methods

The chemical composition of TC4-ELI titanium alloy is shown in table 1.The heat treatment process is that the material is heated to 950℃andkept 65minutes, then is air-cooled to room temperature.

The specimens were processed into Φ8 mm × 12 mm cylinders and the two ends of the cylinders were polished with fine sandpaper. The compression test was carried out on a Gleeble-1500D thermal simulator. According to the TC4-ELI titanium alloy transformation point temperature, the hot pressing temperature is selected as 850℃, 900℃, 950℃, 1000℃, 1050℃, the strain rate is$0.001s^{-1}$,$0.01s^{-1}$,$0.1s^{-1}$,$1s^{-1}$, $10s^{-1}$ and total compression deformation is 50%. In order to reduce the friction force, the graphite sheet is added between the test machine indenter and the end of the cylindrical specimen before the test. At the beginning of the test, the specimen was heated to 1100 ° C at a rate of 10℃ / s and held 5 min, then was cooled to each deformation temperature at a rate of 5℃/ s and held 10 seconds to heat evenly. Finally,the specimens were compressed to 50% of the total deformation at specified strain rates. After the completeness of compression process, the specimens were subjected to water quenching to preserve the high temperature deformation. The hot compression test curve is shown in figure 1.

Table.1 The mass fraction of each element of TC4-ELI titanium alloy

element	Al	V	Fe	O	C	N	H
wt%	5.92	3.91	0.188	0.092	0.0081	0.008	0.004

3.Results and analysis

*3.1.*True stress-strain curves of TC4 - ELI titanium alloy

The true stress-strain curves of TC4-ELI titanium alloy at different temperatures are shown in figure1. With the increasing strain rate, the stress of the TC4-ELI titanium alloy increases rapidly to a peak stress and then the stress tends to be stable. The peak stress increases with the increase of strain rate, and decreases with the increase of hot compression temperature. In the initial stage of deformation, a significant work hardening phenomenon that the stress increases sharply is mainly due to the resistance of the dislocation motion increases with the increase of the dislocation density. When the peak stress is reached, the stress enters the steady state or falls a little and then into the steady state (strain rates 0.001 s^{-1} and 10 s^{-1}), showing typical dynamic recovery characteristics on the stress-strain curve. The main reason is that hardening and dynamic softening achieve a relative balance.

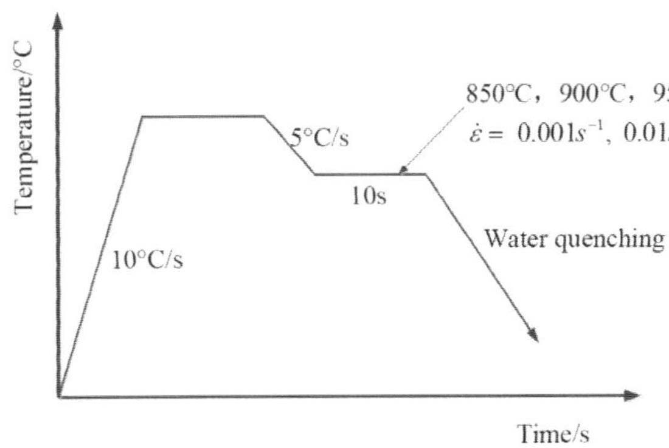

Figure1. The curve of thermal compression

As shown infigure 2,when the hot compression of TC4-ELI titanium alloy is occured, the

IOP Conf. Series: Materials Science and Engineering **359** (2018) 012014 doi:10.1088/1757-899X/359/1/012014

stress peak shiftes to the right and decreases with the increase of the deformation temperature under the same strain rate, which indicates that the decrease of deformation resistance of TC4-ELI titanium alloy with the increase of temperature and the dynamic recrystallization at this time. At the same deformation temperature, the higher the strain rate, the higher the peak value of the stress, the greater the corresponding strain. The greater the strain rate, the shorter the corresponding deformation time. The time of dislocation movement significantly reduces, which is not conducive to dynamic response and dynamic recrystallization.

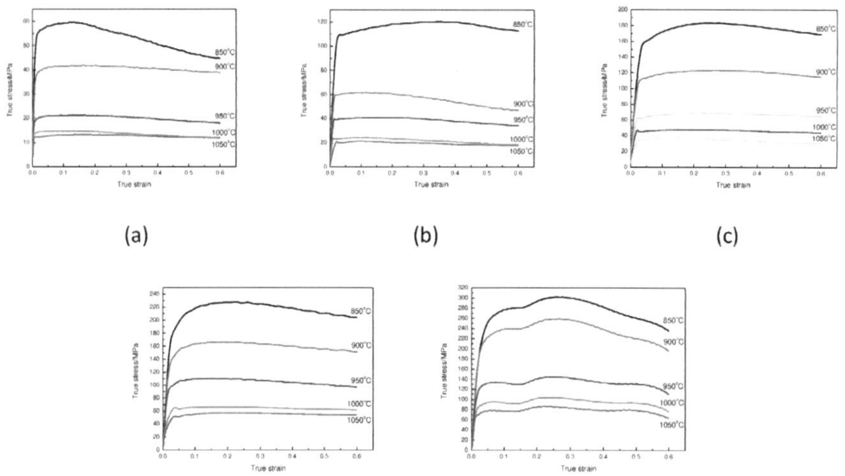

(a) (b) (c)

(d) (e)

(a) $\dot{\varepsilon}$ =0.001s^{-1} (b) $\dot{\varepsilon}$ =0.01 s^{-1}(c) $\dot{\varepsilon}$ =0.1 s^{-1}(d) $\dot{\varepsilon}$ =1 s^{-1}(e) $\dot{\varepsilon}$ =10 s^{-1}

Figure2. True stress-true strain curves of TC4-ELI titanium alloy under different strain rates and temperatures

*3.2.***Constitutive model of TC4-ELI titanium alloy**

The constitutive model of TC4-ELI titanium alloy uses a hyperbolic sine model with the Arrhenius equation, including strain rate, deformation heat activation energy Q, deformation temperature T and molar gas constant R, to describe the relations between strain rate, stress, deformation temperature, etc. As shown in equation (1):

$$\dot{\varepsilon}=AF(\sigma)\exp[-Q/RT] \quad (1)$$

Equation (2) is the formula for the different stress levels in the hyperbolic sine model of the Arrhenius equation.

$$\begin{cases} F(\sigma)=\sigma^{n_1} & (\alpha\sigma<0.8) \quad (a) \\ F(\sigma)=\exp(\beta\sigma) & (\alpha\sigma>1.2) \quad (b) \\ F(\sigma)=[\sinh(\alpha\sigma)]^{n} & (\text{for all stress}) \quad (c) \end{cases} \quad (2)$$

In equation (1) and (2): $\dot{\varepsilon}$ is the strain rate(s^{-1}); σ is flow stress(MPa);Q is the activation of the activation energy(kJ/mol);T is the deformation temperature;R Is the molar gas constant;A,n, α and β are the material constant, $\alpha=\beta/n_1$.

The (a) and(b)of equation (2) are respectively taken into the equation (1),then the logarithm of both sides can be obtained at the same time:

$$\ln\dot{\varepsilon}=\ln A_1+n_1\ln\sigma-Q/RT \quad (3)$$

$$\ln\dot{\varepsilon}=\ln A_2+\beta\sigma-Q/RT \quad (4)$$

By the equation (3) and (4),when the temperature is constant, $\ln\dot{\varepsilon}-\ln\sigma$, $\ln\dot{\varepsilon}-\sigma$ bothhavea linear relationship. $\ln\dot{\varepsilon}-\ln\sigma$ and $\ln\dot{\varepsilon}-\sigma$ were drawn from the experimental data,the linear relationship between the two is fitted by the least squares method,As shown in figure 3 (a), (b).From the equation (3) and (4) , n1and β are slopes of $\ln\dot{\varepsilon}-\ln\sigma$ and $\ln\dot{\varepsilon}-\sigma$ respectively.The average value of n1 is 5.083 and the average of β is 0.085andthe stress level parameter $\alpha=0.017\text{mm}^2\cdot\text{N}^{-1}$ can be obtained by $\alpha=\beta/n_1$ $\text{mm}^2\cdot\text{N}^{-1}$.

The(c)of the equation (2)is taken into equation (1), then the logarithm of both sides can be obtained at the same time:

$$\ln\dot{\varepsilon}=\ln A+n\ln[\sinh(\alpha\sigma)]-Q/(RT) \quad (5)$$

When the deformation temperature is constant, equation (6) is the result of partial derivative of 1/T in equation (5).Drawing the $\ln\dot{\varepsilon}\text{-}\ln\left[\sinh(\alpha\sigma)\right]$ relationship diagram and using the least squares method to fit out the curve, the slope of the average value is the value of n, as shown in figure 3 (c).Finally, the material flow stress index n = 5.028.

$$\frac{\partial\ln\dot{\varepsilon}}{\partial\left(\frac{1}{T}\right)}=n\frac{\partial\ln[\sinh(\alpha\sigma)]}{\partial\left(\frac{1}{T}\right)} \quad (6)$$

When the strain rate is constant, equation (7) is the result of partial derivative of 1/T in equation (6).Drawing the $\ln[\sinh(\alpha\sigma)]\text{-}1/T$ relationship diagram and using the least squares method to fit out the curve, as shown in figure 2 (d), coefficient s = 7188.806.

$$\frac{Q}{Rn}=\left\{\frac{\partial\ln[\sinh(\alpha\sigma)]}{\partial(1/T)}\right\}_{\dot{\varepsilon}} \quad (7)$$

Equation (9)can be obtained by equation (6) and (7), the thermal activation energy of the deformation is Q = 300367.5807J / mol.

$$Q=R\left\{\frac{\partial\ln\dot{\varepsilon}}{\partial\ln[\sinh(\alpha\sigma)]}\right\}_T\cdot\left\{\frac{\partial\ln[\sinh(\alpha\sigma)]}{\partial(1/T)}\right\}_{\dot{\varepsilon}} \quad (8)$$

The strain rate factor Zener-Hollomon parameter Z is used to describe the effect of strain rate and heat distortion temperature T on deformation.

$$Z=\dot{\varepsilon}\exp[Q/RT] \quad (9)$$

Z can be obtained by the equation (1) and (9):

$$Z=A[\sinh(\alpha\sigma)]^n \quad (10)$$

The logarithm of the equation (10) is taken on both sidessimulaneously, the equation (11):

$$\ln Z=\ln A+n\ln[\sinh(\alpha\sigma)] \quad (11)$$

By equation (11), lnA is the intercept of $\ln Z\text{-}\ln\left[\sinh(\alpha\sigma)\right]$.The thermal deformation temperature T, strain rate $\dot{\varepsilon}$ and thermal activation energy Q were taken into equation (11)to calculate Z under various thermal deformation conditions.The relation of $\ln Z\text{-}\ln\left[\sinh(\alpha\sigma)\right]$ is analyzed by the linear regression,as shown in figure 3, $\ln A=55.63916$ is found, and the structure factor of the material is $A=1.458\times10^{24}\text{s}^{-1}$.

From the above analysis and calculation, under the strain rate $0.001s^{-1}-10s^{-1}$ and the thermal deformation temperature of 850℃-1050℃, the constitutive equation of the flow stress can be expressed as follows:

$$\dot{\varepsilon}=1.458\times10^{24}\left[\sinh\left(0.017\sigma\right)\right]^{5.028}\exp\left(-\frac{300367.5807}{RT}\right) \quad (12)$$

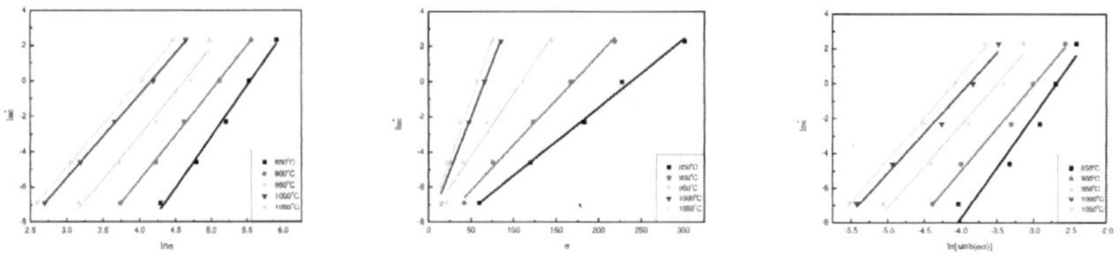

(a) $\ln\dot{\varepsilon}$-$\ln\sigma$ (b) $\ln\dot{\varepsilon}$-σ (c) $\ln\dot{\varepsilon}$ - $\ln\sinh(\alpha\sigma)$

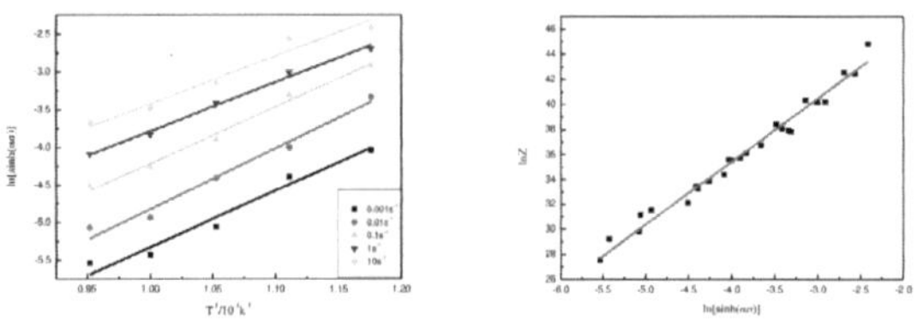

(d) $\ln\left[\sinh\left(\alpha\sigma\right)\right]$-$1/T$ (e) $\ln Z-\ln\left[\sinh\left(\alpha\sigma\right)\right]$

(a) $\ln\dot{\varepsilon}$-$\ln\sigma$ (b) $\ln\dot{\varepsilon}$-σ (c) $\ln\dot{\varepsilon}$ - $\ln\sinh(\alpha\sigma)$ (d) $\ln\left[\sinh\left(\alpha\sigma\right)\right]$-$1/T$ (e)

$\ln Z-\ln\left[\sinh\left(\alpha\sigma\right)\right]$

Figure 3. Theanalysisfigures of constitutive equation

4.Conclusion

1) With the increasing strain rate, the stress of the TC4-ELI titanium alloy increases rapidly to a peak stress and then the stress tends to be stable. The stress peak is shifted to the right and decreases with the increase of the deformation temperature under the same strain rate.The peak stress increases with the increase of strain rate under the same deformation temperature.
2) The thermal deformation activation energy of TC4-ELI titanium alloy was calculated to be 300367.5807 J/mol, and the hyperbolic sine constitutive equation with Z parameters was constructed: It provides reference for future production and finite element simulation analysin.

$$\dot{\varepsilon}=1.458\times10^{24}\left[\sinh\left(0.017\sigma\right)\right]^{5.028}\exp\left(-\frac{300367.5807}{RT}\right)$$

Acknowledgments

This work was financially supported by the National Natural Science Foundation (51741503), National key research and development plan (2016YFB0300400), China Postdoctoral Science Foundation (2016M590677) and Innovative Science and Technology Teamin Henan Province (C20150014).We would like to thank editor and anonymous reviewers for their constructive comments and helpful suggestions for improving the manuscript.

5.References

[1] NieLei,LiFuguo and Fang Yong 2001 J. Aeronaut. Mater. 21 13-18.

[2] Liu Jinhao,LiuJiansheng,XiongYunsen,HeWenwu,ZhangPeng and Liu Xuemin 2009 Rare.Metal. Mat. Eng. 30 1674-77.

[3] Liu Wanying,ZhuYike,LinYuanhua,ShaoDengquan,ShiTaihe and Wu Junjie 2013 Mater. Rev.27 108-111.

[4] Yang Ming,LiangYilong and Zhong Ying 2013 J. Nonferrous Metal Engineering. 322-24.

[5] ZengShangwu,JiangHaitao and Zhao Aimin 2015 Rare. Metal. Mat. Eng. 44 2812-16.

[6] Wang Xiaofang,ChenMinghe,Chen Wei and ShenFalan 2012 J Nanjing U Aeronaut Astronautics. 44 117-120

[7] Liu Jinhao,LiuJiansheng,XiongYunsun,HeWenwu,ZhangPeng and Liu Xuemin 2013 Rare. Metal. Mat. Eng. 42 1675-78.

[8] You Xuelei,HuaJianshe,YaoWei,LiuYingying,LiJie and Wang Kun 2015 T Mater Heat Treat. 36 237-242.

IOP Conf. Series: Materials Science and Engineering **359** (2018) 012015 doi:10.1088/1757-899X/359/1/012015

Strength design of $Zr_{(x)}Ti_{(x)}Hf_{(x)}Nb_{(x)}Mo_{(x)}$ alloys based on empirical electron theory of solids and molecules

Y K Li*, **Y W Chen, X W Cheng, C Wu and B Cheng**

School of Materials Science and Engineering, Beijing Institute of Technology, Beijing 100081, China
National Key Laboratory of Science and Technology on Materials under Shock and Impact

E-mails: 2861209902@qq.com ; liyunkai@bit.edu.cn

Abstract: In this paper, the valence electron structure parameters of $Zr_{(x)}Ti_{(x)}Hf_{(x)}Nb_{(x)}Mo_{(x)}$ alloys were calculated based on the empirical electron theory of solids and molecules (EET), and their performance through these parameters were predicted. Subsequently, the alloys with special valence electron structure parameters were prepared byarc melting.The hardnessand high-temperature mechanical properties wereanalyzedto verify the prediction. Research shows that the influence of shared electron number n_A on the strongest bond determines the strength of these alloysand the experiments are consistent with the theoretical prediction.

1.Introduction

In recent years, high entropy alloys (HEAs) were proposed as a new concept of material design [1-3]. The multi-principal elements provide more composition spaces to explore alloys with excellent performance, which gives rise to a wide range of researches around the world [4-7].The alloys design is an important step in the process of new HEAs discovery, the traditional alloy design of the "cooking" method wastes plenty of manpower and material resources, which does not apply to HEAs research apparently today. In the recent research, the phase prediction from the viewpoint of thermodynamics has been developed. Zhang et al.[5, 6]predicted the single phase forming ability through δ and Ω. Guo et al.[7, 8] further proposed the crystalline structure of single-phaserules through the valence electron concentration (VEC) in HEAs. Through these prediction rules, the brittle phase can be reduced, but the mechanical properties cannot reveal during the alloy design process. Therefore, it is necessarythatthe material design come into the level of the electronic structure, R H Yu et al. proposed an empirical electron theory of solids and molecules (EET) [9, 10] in the last century, which can solve the prediction of properties perfectly. Because the HEAs have simple structure similar to the traditional alloys, the EET has good applicabilityand the experiments results confirm it well.

2.Alloy design process

The bond length difference (BLD) is a new method of computing solid electronic structure, which bases on the experimental data of crystal structure to avoid solving the Schrodinger equation [9]. Before the analysis of BLD, the crystal structure type, the lattice constant and the atomic position need to be confirmed. Consequently, the crystalline constant prediction is the first step in the process of $Zr_{(x)}Ti_{(x)}Hf_{(x)}Nb_{(x)}Mo_{(x)}$ alloy design.Based on the elements constant of these refractory elements, the difference of size parameters (δ), solid solution ability (Ω) and the VEC are calculated, which is in the range of the single phase of body centered cubic.

The average cell was adopted due to the numerous kinds of atoms and complex combination.In this condition, the individual atoms and the position in the cell are roughly equivalent. The average lattice constant a:

$$a=\sum c_i * a_i (i = Zr, Ti, Hf, Nb, Mo) \quad (1)$$

Where$a_i$is the lattice constant of the ith element and c_i is the concentration of the ith element in these HEAs.The distance between any two atoms can be calculated through the lattice geometry, which wasnamed as the test distance of bondD_{n_α} （$\alpha = A, B$） in the EET:

According to the state of atoms through the EET, the bond distance of covalent empirical equation was established:

$$\overline{D_{n_\alpha}}=R_\mu(l)+R_v(l)-\beta \log n_A \quad (2)$$

Where μ and υ are the bonding atoms, $R_\mu(l)$and $R_v(l)$are the half single-bond length of μ andυ, respectively. n_α is the shared electron pair number of the covalent bond, and D_{n_α} is the theoretical bond length, β is equal to 0.06nm in these $Zr_{(x)}Ti_{(x)}Hf_{(x)}Nb_{(x)}Mo_{(x)}$ alloys.The difference between theoretical bond length and test bond lengthis$|\Delta D_{n_\alpha}|$, which is used to justify whether the atomic state matches the actual situation:

$$|\Delta D_{n_\alpha}|<0.05nm \quad (3)$$

If inequation (3) is not satisfied, then the $n_A$value needs to be recalculated and set as other mixed order of atoms. When the $|\Delta D_{n_\alpha}|$ is the minimum value, this stateis regarded as the closest to the reality, the shared electron pair number of the strongest covalent bond is defined asn_A in this condition.This paper computed the results of the calculation process through the MATLAB 2012(b)as they were complex, and displayed them on table 1.

Table1 The electronic structure parameters calculation results: the ability of single phase formation (**Ω**); atomic radius difference (***δ***);the valence electron concentration (VEC); the thelattice constant(***a***); the shared electron pair number of the strongest covalent bond nearest the real value (***n**$_A$*)

Parameters	$ZrTiHf_{2.0}NbMo$	$Zr_{2.0}TiHfNbMo_{2.0}$	$ZrTiHfNb_{0.5}Mo_{0.5}$	$ZrTiHfNb_{2.0}Mo_{2.0}$
Ω	1.5819	1.7447	1.619	2.1289
δ/%	5.32%	5.90%	5.10%	5.33%
VEC	4.5	4.71	4.38	4.86
***a*/nm**	0.3404	0.3371	0.341	0.333
n_A	0.3189	0.3231	0.309	0.3842

3. Experimental

The $Zr_{(x)}Ti_{(x)}Hf_{(x)}Nb_{(x)}Mo_{(x)}$ alloys were prepared through the arc melting under the protective atmosphere, and the cooling system employed the water-cooled copper crucible. In order to ensure the homogeneity of composition, the low-melt bulk materials wereplaced lower in turn and the ingots were flipped and remelted at least six times. The purity of row materials is greater than 99.9%.

The microhardness of the HEAs was measured by 450SVD™ Vickers's hardness tester under 300 g loaded applied for 15s, the ultimate hardness value was averaged by five different positions measurements at the specimen's polished surface.The focused high temperature compressive performances were performed on a Gleeble-3500 thermal simulator. In this study, the thermocouple used Pt-Rh alloy, the specimens adopted cylinder of 6 mm×9 mm, the strain rate was $10^{-3}s^{-1}$, and the amount of compression was set as 40%. Each alloy was heated to 700°C, 800°C, 900°C in turns, the heating rate was set as 300°C and the holding time was 3 minutes.

4.Results and discussion

The relationship between the covalent electron pair number of the strongest covalent bond (n_A) and the strength of $Zr_{(x)}Ti_{(x)}Hf_{(x)}Nb_{(x)}Mo_{(x)}$ alloys is shown in the figure 1. The value of $n_A$reflects the strength of bonding force among the atoms on the micro, and the $n_A$value increases with the bond strength enhance, which can be quantificated through the EET. On the macro, the hardness and high temperature fracture strength have close relationship with the bond strength. Through the EET, the relation between mechanical performance and valence electron parameter is established. The compressive stress at elevated temperature of these alloys is consistent with the hardness. The value of n_A(0.3842) of $ZrTiHfNb_{2.0}Mo_{2.0}$ alloy is far more than others, the hardness (543.3HV) and high temperature fracture strength (1155MPa at 800℃ and 897MPa at 900℃) are the highest among these alloys. The hardness and stress at high temperature are greater when the n_A value increase, which can predict the strength of the alloy with different content in the design process.

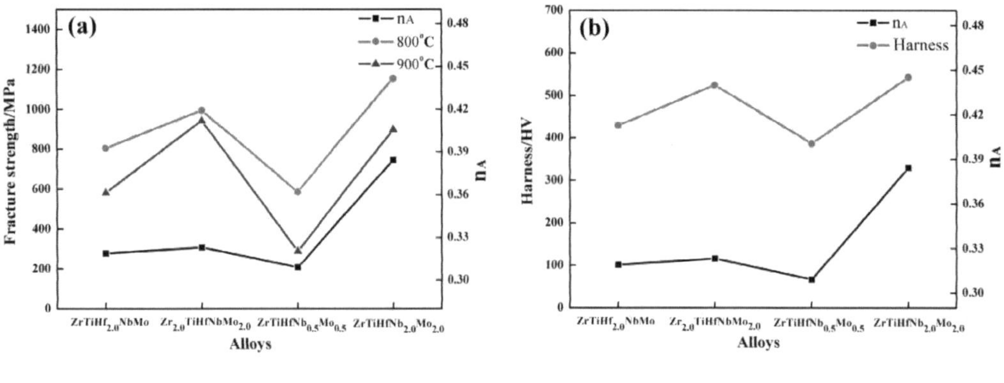

Figure 1. The relationship between n_A value and strength of $Zr_{(x)}Ti_{(x)}Hf_{(x)}Nb_{(x)}Mo_{(x)}$ alloy:(a) is the Harness-n_A relationship, (b) is the high temperature fracture strength-n_A relationship

5. Conclusion

This paper researched the $Zr_{(x)}Ti_{(x)}Hf_{(x)}Nb_{(x)}Mo_{(x)}$ high entropy alloys. Thethermodynamic parameter (Ωandδ) andVECarein the range of the single phase of body centered cubic. Based on the crystal structure,the valence electron structure parameters were calculated to predict the mechanical properties through empirical electron theory of solids and molecules (EET). The covalent electron pair number of the covalent bond (n_A) reflects the bond strength, and the hardness and compressive stress of these alloys increase with the n_A enhance, which can predict the strength of $Zr_{(x)}Ti_{(x)}Hf_{(x)}Nb_{(x)}Mo_{(x)}$ through the EET.

References

[1] Yeh, J.-W., et al., *Anomalous decrease in X-ray diffraction intensities of Cu–Ni–Al–Co–Cr–Fe–Si alloy systems with multi-principal elements.* Materials Chemistry and Physics, 2007. **103**(1): p. 41-46.
[2] Chen, T.K., et al., *Nanostructured nitride films of multi-element high-entropy alloys by reactive DC sputtering.* Surface & Coatings Technology, 2005. **200**(5-6): p. 1361-1365.
[3] Cantor, B., et al., *Microstructural development in equiatomic multicomponent alloys.* Materials Science and Engineering: A, 2004. **375-377**: p. 213-218.
[4] C.-Y. Hsu, J.-W. Yeh, S.-K. Chen, T.-T. Shun, *Wear resistance and hightemperature compression strength of FCC CuCoNiCrAl0.5Fe alloy with boron addition*, Metall. Mater. Trans. A **35A** (2004) 1465-1469.
[5] Stepanov, N.D., et al., *Structure and mechanical properties of a light-weight AlNbTiV high entropy alloy.* Materials Letters, 2015. **142**: p. 153-155.
[6] Maiti, S. and W. Steurer, *Structure and Properties of Refractory High-Entropy Alloys.* Tms 2014 Supplemental Proceedings, 2014: p. 1093-1097.
[7] Zou, Y., H. Ma, and R. Spolenak, *Ultrastrong ductile and stable high-entropy alloys at small scales.* Nature Communications, 2015. **6**.
[8] Yang, X. and Y. Zhang, *Prediction of high-entropy stabilized solid-solution in multi-component alloys.* Materials Chemistry and Physics, 2012. **132**(2-3): p. 233-238.
[9] Zhang, Y., X. Yang, and P.K. Liaw, *Alloy Design and Properties Optimization of High-Entropy Alloys.* Jom, 2012. **64**(7): p. 830-838.
[10] Guo, S. and C.T. Liu, *Phase stability in high entropy alloys: Formation of solid-solution phase or amorphous phase.* Progress in Natural Science: Materials International, 2011. **21**(6): p. 433-446.
[11] Guo, S., et al., *Effect of valence electron concentration on stability of fcc or bcc phase in high entropy alloys.* Journal of Applied Physics, 2011. **109**(10): p. 103505.
[12] R H Yu, *Empirical Electronic theory of solids and molecules,* Chin.Sci.Bull, **23**(1978)217-224.
[13] R L Zhang, *Empirical Electronic theory of solids and molecules, Jiling Science and Technology Pulishing House,* 1993.

IOP Conf. Series: Materials Science and Engineering **359** (2018) 012016 doi:10.1088/1757-899X/359/1/012016

Raindrop Characteristics Analysis by Laser Raindrop Spectrometer

X Q Yang[1,2], N N Liu[3], J Lin[4], W Y Yao[1], P Q Xiao[1] and Z Z Shen[1*]

1. Key Laboratory of Soil and Water Loss Process and Control on the Loess Plateau of the MWR, Yellow River Institute of Hydraulic Research, Zhengzhou, 450003, Henan;

2. Henan Institute of Science and Technology, Xinxiang, 453003, Henan;

3. Ningxia agricultural Institute of survey and Design, 471900, Ningxia;
4. Fuzhou entry-exit inspection & quarantine bureau of P.R.C Fuzhou 350015

Abstract: Raindrop characteristics, including speed and size of raindrops, in Zhengzhou city of Yellow River basin were analyzed through a natural rainfall on the loess slope. Results showed that the process of natural rainfall belonged to a parabola and counts, size and terminal velocity would increase with the rainfall intensity rising. Besides, the size and terminal velocity of natural raindrops were relatively scattered; In the process of individual rainfall, the terminal velocity and its peak value were mainly focused between 1~4.2m/s and 1.4-3.4m/s, respectively. Size of raindrops were mainly consisted of 0.125-0.5mm, The final velocity of 0.125mm is mainly between 1-3.4m/s, and the final velocity of the raindrops with 0.25mm particle size is mainly concentrated on the range of 1-4.2m/s, 0.375mm diameter of the raindrops mainly concentrated on the 1-3.4m/s, the 0.5mm diameter of the raindrops mainly concentrated on the 3.4-4.2m/s.

1. Introduction

Loess Plateau, the region with extremely serious problem of loss of water and soil among the world, has been the main source of sediments in Yellow River and its management is the critical factor to curb sediments[1-2]. However, due to the complexity of soil and water loss and the limitations of observational methods and content [3], some research fields, for example, law of watershed soil and water loss and the mechanism of erosion and sediment yield haven't had a breakthrough, especially the lack of practical soil and water loss model in Loess Plateau[9], even though there being some advances around the world [4-8]. All the limitations restrict the development of soil and water loss management in depth and breadth directly [10]. To have a further study of natural rainfall distribution [11] by combining the field observation and laboratory test were the research focus to explore the mechanism of soil and water loss law, develop optimized model and establish mathematical model of soil and water loss on the Loess Plateau.

In order to establish such models, it's imperative to analyze the scale transformation and rainfall similarity and finally the raindrops characteristics in the soil erosion experiments must be solved [12-15]. This study aims to analyze the terminal velocity and grading of raindrops, combined with the sediment results in different rainfall conditions and offer a technology support for dynamic rainfall simulation.

2. Experiment design and data acquisition

2.1 Experiment design of runoff plots

The experiment was carriedout at the Zhengzhou test base of model Yellow River of Yellow River Institute of Hydraulic Research. The length, width, depth and slope of runoff plots are 5m, 1m, 60cm and 20°, respectively. Structure of plots is brick-concrete and there is a water channel at the outlet in each plots. Soil landfilled in each plots is surface loessal soil from the Mangshan Mountain in Zhengzhou, with the proportion of particle size of 0.05 ~ 0.01 mm, 0.02 ~ 0.05 mm being 43.4% and 35.45% respectively. And soil bulk density is $1.2 g/cm^3$.

2.2 Test method

The 5m long slope was divided into five fractural surface from top to bottom. The LPM laser raindrop spectrometer was used here to collect the information about characteristics of raindrops, including intensity, size, terminal velocity and number under the same terminal velocity. Slope velocity radar gun and steel rule were used to record hydraulics parameters, including speed, width and depth of runoff. Sediments were collected by 1 minute after producing runoff and analyzed the parameters mentioned above. Erosion amounts and sediment concentration were calculated by the method of substitution or oven drying and calculated the parameters like infiltration of runoff according to rainfall and runoff yield.

3. Results and analysis

In this paper, five typical rainfall time points in natural rainfall on August30, 2015 (hereinafter referred to as 20150830) were selected. The length of each time point was 1min. The particle size of raindrops in the min length, the speed of raindrops, and the number of raindrops with the same raindrops were collected by LPM laser raindrop spectrometer. The raindrops characteristics were also analyzed. The results are as follows.

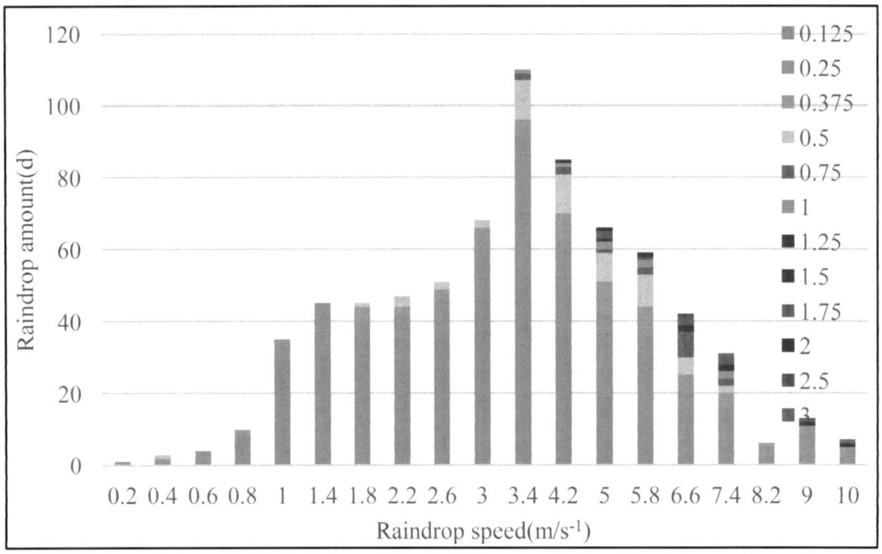

Figure 1: In the first minute during the natural rainfall on 20150830

The distribution of the particle size and the final velocity of the raindrops in the first 1 minute during the natural rainfall on 20150830is shown in Fig.1.It can be seen from the figure that the final velocity of the raindrops in the first 1min of the rainfall is mainly concentrated between 0.8-7.4m/s and the peak velocity is 3.4m/s, and the number of raindrops with the final velocity of the raindrop is about 110, accounting for about 15%;The rainfall is mainly composed of raindrops with particle size of 0.125-0.5mm. The final velocity of 0.125mm is mainly between 0.8-5.8m/s, and the final velocity of the raindrops with 0.25mm particle size is mainly concentrated on the range of 1-9m/s, 0.375mm diameter of the raindrops mainly concentrated on the 1.8-7.4m/s, the 0.5mm diameter of the raindrops mainly concentrated on the 3.4-6.6m/s.

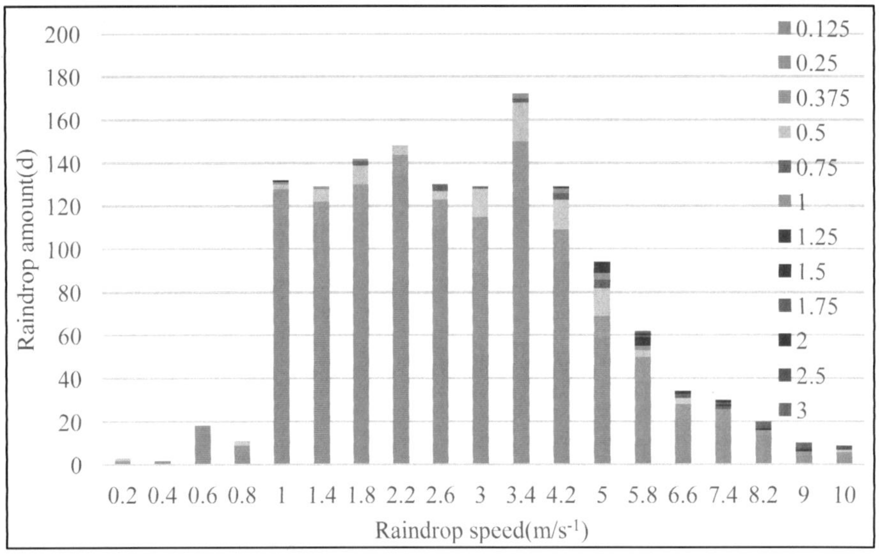

Figure 2: In the second minute during the natural rainfall on 20150830

According to the distribution of the raindrops and the final velocity of the raindrops in the second minute during the natural rainfall process (Figure 2), the final velocity of the raindrops in the second minute of the rainfall is mainly concentrated between 0.6-8.2m/s and the peak velocity is 3.4m/s, with the number of raindrops at the end of the raindrops is about 170, accounting for about 15%; We can also see the rainfall is mainly composed of 0.125-0.5mm diameter raindrops, the final velocity of 0.125mm particle size is mainly concentrated on the 1-5.8m/s, the final velocity of 0.25mm sizeis mainly concentrated in the 1-8.2m/s, the final velocity of 0.375mmparticle size is mainly concentrated in the 1-5.8m/s, thefinal velocity of 0.5mm particle size is mainly concentrated in the 1.4-5m/s.

Fig.3 shows the distribution of the particle size and the final velocity of the raindrops in the third minute during the natural rainfall in 20150830, it can be seen from the figure that the final velocity of the rains in the third minute of the rainfall is mainly concentrated in the range of 0.6-7.4m/s, and the velocity of the raindrops in the third minute of the natural rainfall is analyzed. At the same time, the final velocity and the peak velocity was formed, the value is 1.4m/s and 3.4m/s respectively. That is, most of the raindrop speed is 1.4m/s and 3.4m/s or so, with the number of raindrops about 100 and 85 or so, accounting for about 30% or so; We can also see the rainfall is mainly composed of 0.125-0.5mm diameter raindrops, of which 0.125mm particle size of the end of the rain mainly concentrated in the 0.6-4.2m/s, 0.25mm diameter of the raindrops mainly concentrated in the 1-5.8m/s, the 0.375mm diameter of the raindrops mainly concentrated in the 1.4-4.2m/s,the 0.5mm diameter raindrops is mainly between 1.8-5.8m/s.

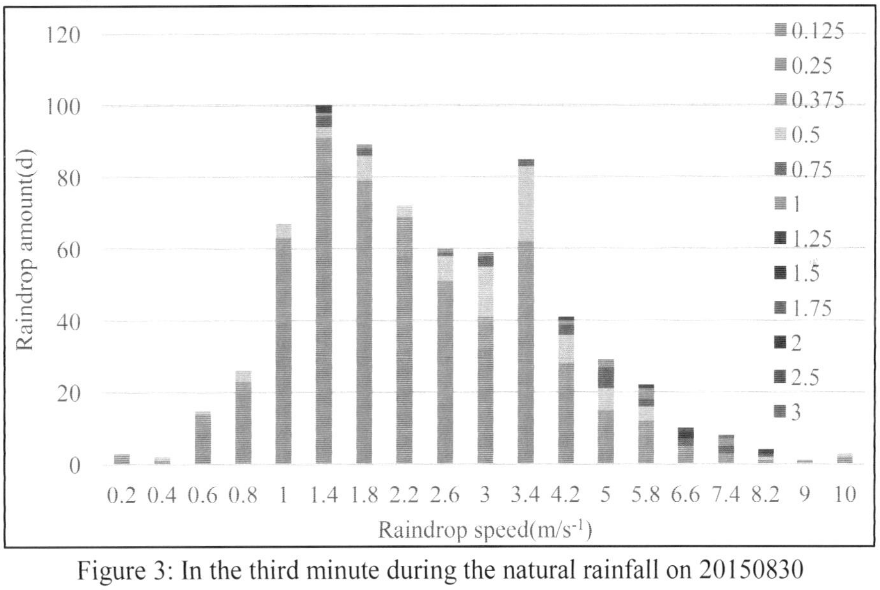

Figure 3: In the third minute during the natural rainfall on 20150830

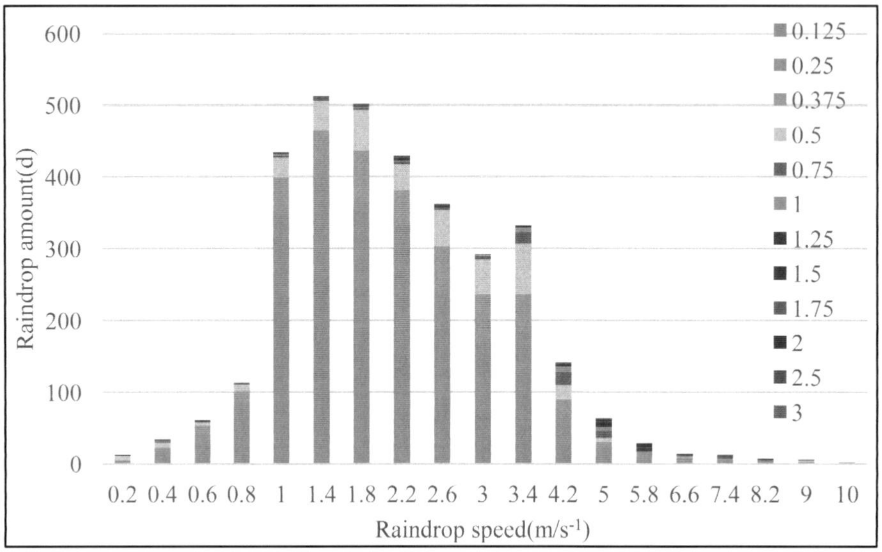

Figure 4: In the fourth minute during the natural rainfall on 20150830

Fig.4 shows the distribution of the particle size and the final velocity of the raindrops in the fourth minute during the natural rainfall in 20150830. It can be seen that final velocity of the raindrops in the fourth minute of the rainfall is mainly concentrated between 0.4-5.8m/s, forming two final velocity peaks, namely 1.4m/s and 3.4m/s respectively, that is, most of the raindrops is mainly 1.4m/s and 3.4m/s or so, with the number of raindrops is 505 and 320 or so, accounting for about 35%; We can also see theparticle size of rainfall is mainly 0.125-0.5 mm raindrops. Of which 0.125, 0.25, 0.375, 0.5 mm particle size of the rain speed is mainly concentrated in 0.6-4.2,1-4.2,1-3.4,1-3.4m/s.

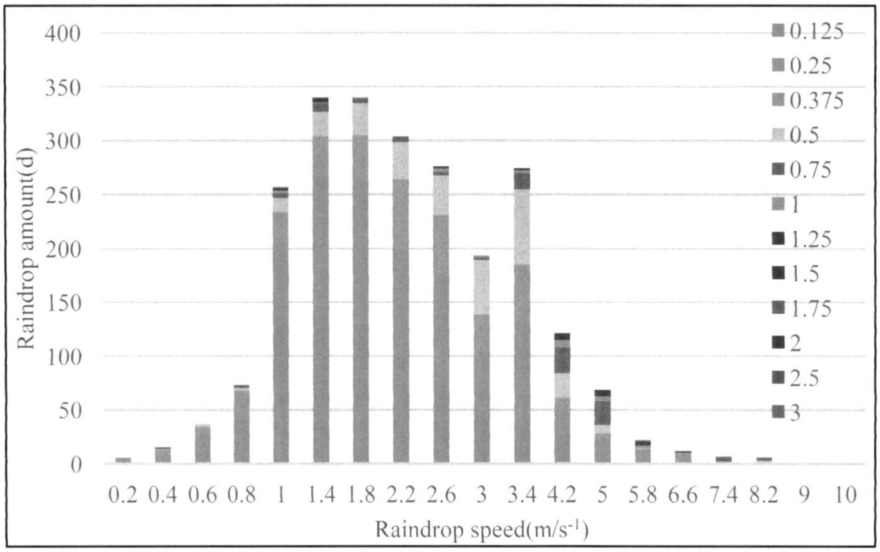

Figure 5: In the fifth minute during the natural rainfall on 20150830

Fig.5 shows the distribution of the particle size and the final velocity of the raindrops in the 5th minute during the natural rainfall, we can see that the final velocity of the raindrops in the first 5min of the rainfall is mainly concentrated between 0.6-5.8m /s, and the velocity of the raindrops in the 5th minute during the natural rainfall is analyzed. Forming a final velocity peak is 1.4m/s, that is, most of the raindrop speed is 1.4m/s, with the raindrop speed of the number of raindrops for about 340, accounting for about 15%; We can also see the rainfall is mainly composed of 0.125-0.5mm diameter raindrops, of which 0.125,0.25,0.375,0.5mm diameter of the raindrops mainly concentrated in 0.6-4.2,0.8-4.2,1-3.4,1.4-3.4 m/s.

4. Conclusions

Through the analysis of natural raindrops, the following aspects have been obtained:
3.1 The natural rainfall process is a process of rainfall intensity parabolic changes, and with the increase of rain intensity, the number of raindrops per unit time, particle size, the final velocity will be a corresponding increase. Natural rainfall raindrops and raindrops are relatively scattered, from small to large are distributed.
3.2 During the process of this rainfall, the final velocity of the rain is mainly concentrated between 1-4.2m/s, the peak of the final velocity is generally 1.4-3.4m/s; rainfall is mainly composed of the particle size of 0.125- 1.25mm raindrops. The velocity of the raindrops of 0.125mm diameter is mainly between 1-3.4m/s, the final velocity of the raindrops of 0.25mm particle size is mainly concentrated between 1-4.2m/s, 0.375mm concentrated in the 1-3.4m/s, the 0.5mm particle size of the raindrops mainly concentrated in the 3.4-4.2m/s.

Acknowledgment
We greatly appreciate all referees for their assistance and laboratory work. This research mainly supported by the National Key R&D Program (2017YFC0504503); National Key Technology R&D Program (2013BAC05B04); Innovation S&T Troop Construction Projects of Henan Province （162101510004）

REFERENCES

[1] Tang Keli, Soil and water conservation in China [M]. Beijing: Science Press, 2004.
[2] Chen Lei, Soil and water conservation in China [J], 2002 (7): 4-6.
[3] Li Zhanbin, Storm erosion test and small watershed sediment yield model of sloping land system in Loess Region [D], Doctoral Dissertation of Shaanxi Institute of Mechanical Engineering, 1991.
[4] Moldenhauer, W C. Procedure for studying soil characteristics using disturbed samples and simulated rainfall [J]. Transactions, American Society of Agricultural Engineer, 1965, 8(1):30-35.
[5] Renard K G, Foster G R, Weesies G A, et al. Predicting soil erosion by water: a guide to conservation planning with the Revised Universal Soil Loss Equation (RUSLE)[J]. Agriculture Handbook, 1997.
[6] Foster G R, Lane L J. User requirements, USDA-Water Erosion Prediction Project (WEPP) [R], NSERLReportNo1, West Lafayette: USDA-ARS National Soil Erosion Research Laboratory, 1987.
[7] Nearing M A, Foster G R, Lane L J, et al. A process-based soil erosion model for USDA-Water Erosion Prediction Project technology [J]. Trans. ASAE, 1989,32(5): 1587-1593.
[8] Laflen J M, Lwonard J L, Foster G R. WEPP a new generation of erosion prediction technology[J]. J of Soil and Water Cons., 1991,46(1):34-38.

[9] YaiWenyi, Several design theories and applications of the Yellow River river channel physical simulation [D], Doctoral Dissertation of Hohai University, 2005.
[10] YaiWenyi, Review and Prospect of mathematical models for erosion and sediment yield in China [J]. Sediment Research, 2011, (2): 65-74.
[11] Shen Zhenzhou, Liu Puling, XieYongsheng et al. Transformation of erosion types on loess slope by REE tracking [J], Journal of Rare Earths, 2007, 25 (4): 67-73.
[12] Shen Zhenzhou, Yao Wenyi, Li Zhanbin, Li Mian, Xiao Peiqing,WU Yan. Transformation of Rainfall Characters on Soil Erosion Similar [J], Journal of Rare Earths, 2014, 32 (S): 130-133.
[13] SHEN Zhenzhou,YAOWenyi, L I Mian,et al. Study on Correlativity Between Erosion and Seeping Under Different Underlying Horizon[J], Journal of Soil and Water Conservation, 2008, 22 (5): 43-46.
[14] SHEN Zhenzhou, YAO Wenyi, LI Mian, et al. Runoff Sediment-Carrying Capacity and Critical Shear Force on Slope [J]. Journal of Tianjin University, 2008, 41(4):149-153.
[15] Shen Zhenzhou, Yao Wenyi, Li Mian et al, Influence factors of runoff energy consumption and denudation amount on loess slope [J]. Science of Soil and Water Conservation, 2009, 7 (6): 9-13.

Processing of Al_2O_3/$SrTiO_3$/PDMS Composites With Low Dielectric Loss

J L Yao[1], M J Guo[1], Y B Qi[2], H X Zhu[1], R Y Yi[1] and L Gao[3]

1. School of Materials Science and Engineering, Wuhan Institute of Technology, Wuhan 430205, China;

2. National Engineering Laboratory for Fiber Optic Sensing Technology, Wuhan University of Technology, 430070, China;

3. School of Chemistry and Environmental Engineering, Jianghan University, Wuhan, 430056, China

Corresponding author e-mail: junlongyao@yahoo.com

Abstract. Polydimethylsiloxane (PDMS) is widely used in the electrical and electronic industries due to its excellent electrical insulation and biocompatible characteristics. However, the dielectric constant of pure PDMS is very low which restricts its applications. Herein, we report a series of PDMS/Al_2O_3/strontium titanate (ST) composites with high dielectric constant and low loss prepared by a simple experimental method. The composites exhibit high dielectric constant (relative dielectric constant is 4) after the composites are coated with insulated Al_2O_3 particles, and the dielectric constant gets further improved for composites with ST particles (dielectric constant reaches 15.5); a lower dielectric loss ($\tan\delta = 0.05$) is also found at the same time which makes co-filler composites suitable for electrical insulation products, and makes the experimental method more interesting in modern teaching.

1 Introduction

Due to the excellent insulation, light-weight and easy transportation and installation, polymeric insulators have attracted more attention, and they are increasingly replacing the traditional ceramic insulators [1]. However, they still have some inherent disadvantages, hindering their application in fields of electrical and electronic industries. To overcome these defects, ceramic fillers have been incorporated into the polymer matrix. Among the polymeric insulators, PDMS has one superior attribution on the electrical and electronic industries because of its superior or comparable physical, electrical and chemical properties as compared to traditional ceramic and glass-based insulators [2~4]. However, the apparent disadvantages of the pure silicon rubber is low dielectric constant [5~6], which limits its application as flexible dielectric materials. Previous studies report that inorganic fillers and ceramics dispersed in the polymers are able to enhance dielectric properties.

Al_2O_3 and ST have remarkable electrical resistivity and high dielectric constants, respectively [7]. The co-filled system of tow kind of inorganic fillers is more useful when some researches display the effect of materials' dielectric properties improved are not obvious. The dielectric properties of PDMS co-filled with Al_2O_3 and ST have not been reported so far. Here present is an investigation of the combined effect from such fillers on dielectric properties of PDMS composites, and an unexpected increase of dielectric constant are obtained by using conventional hot-press method, which be frequently applied in teaching experiment of polymer materials lessons, and make the classic experiment more interesting in modern teaching.

2 Experimental

2.1 Materials

PDMS(parts: A and B) used in present by Shenzhen Hong Ye Jie Technology Co (Shenzhen, China), Aluminum oxide used as a filler was 99.99% metals basis by Aladdin Industrial Co (Shanghai, China), Strontium titanate used as a high dielectric properties filler by Nantong Auxin Elcctronic Technology Co (Jiangsu, China).

2.2 Samples preparations

Part A and Part B of the PDMS were mixed at a volume ratio of 1: 1 for 10min at room temperature. Then, Al_2O_3 and ST mix with PDMS for 10min respectively. The composites were prepared by a simple heating and curing processes at about 70~80℃ for 5h. Fianlly, the samples were silvered by silver-paint.

2.3 Characterizations

The measurement of dielectric constant and dielectric loss were carried out using a LCR Spectrum Analyzer (Frequency range is 10^2~10^6 Hz).

3 Result and Discussion

3.1 The dielectric properties of Al_2O_3/PDMS and ST/ PDMS composites

The dielectric constants and dielectric loss of Al_2O_3/PDMS composites were showed in Fig.1(a), Fig.1(b) respectively. Fig.1(c) and Fig.1 (b) respectively display dielectric constants and dielectric loss of ST/PDMS composites.

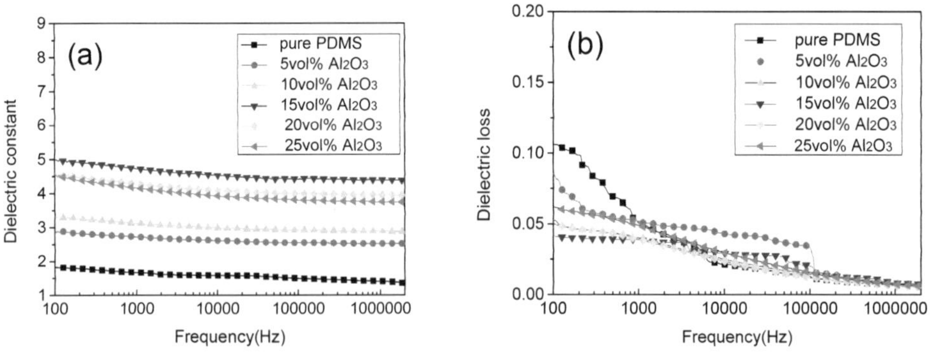

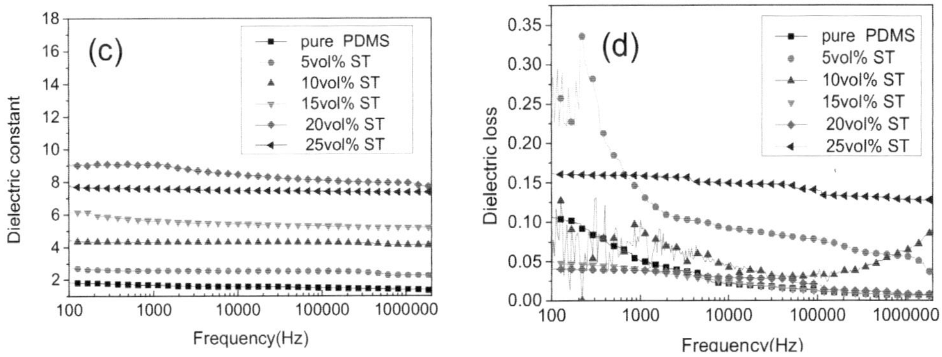

Figure.1 The dielectric constants and dielectric loss of Al_2O_3/PDMS composites (a, b) and ST/PDMS composites(c, d)

Evaluation of dielectric properties of Al_2O_3/PDMS and ST/PDMS composites yielded obvious results (Fig.1).Although the dielectric constant of pure PDMS is as low as 1.8 at 100 Hz, the addition of a series of fillers display excellent dielectric constants and low dielectric loss: the highest value of Al_2O_3/PDMS and ST/PDMS reaches 5 and 9 at 100Hz, respectively, at 15vol% Al_2O_3 and 20vol% ST, at the same time, the dielectric loss of tow composites was below 0.05 from low frequency to high frequency. The results showed the addition of Al_2O_3 or ST both improve material's dielectric constant, and it is obvious that ST/PDMS composites had more higher dielectric constant than Al_2O_3/PDMS composites. At the same times, dielectric loss of materials with numerous ST were unstable when the frequency was blow 1000 Hz . So to overcome tow fillers' disadvantages and improve the dielectric constant further, Al_2O_3 and ST has been add into PDMS together.

3.2 The dielectric properties of Al_2O_3/ST/PDMS composites

The ternary composites' dielectric constants and dielectric loss were showed in Fig.2(a) and 2(b), respectively.

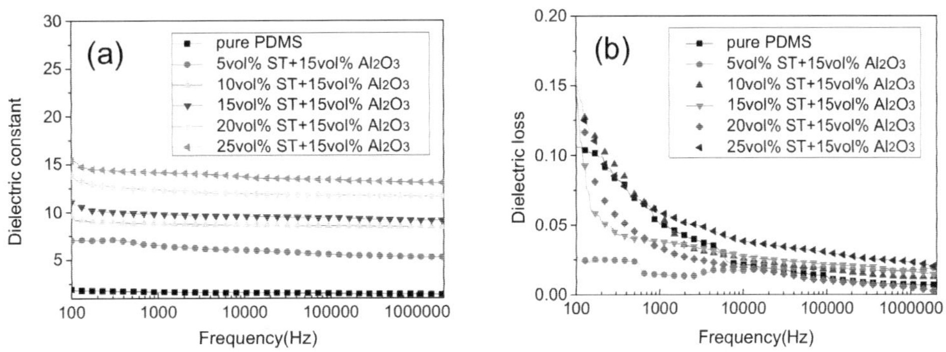

Figure.2 The dielectric constants and dielectric loss of Al_2O_3/ST/PDMS composites (a, b)

The test of dielectric properties of Al_2O_3/ST/PDMS composites yielded remarkable results. Al_2O_3/ ST co-filled system(15vol% Al_2O_3 and 20vol%ST) obtain a the highest value (13)that was seven times than pure PDMS and dielectric loss did not increased and more stable compare with pure PDMS.This result display Al_2O_3/ ST co-filled system is more effective than the only one filler in enhancement of dielectric constant.

4 Conclusion

A series of Al_2O_3/PDMS, ST/PDMS and Al_2O_3/ST/PDMS composites exhibit outstanding dielectric properties and low dielectric loss owing to the Al_2O_3 and ST have remarkable electrical resistivity and high dielectric constants. The highest value of materials' dielectric constant reach 13 at 100Hz and the dielectric loss (0.05) does not obviously increase at the same time. The results display that Al_2O_3/ ST co-filled system obviously enhance dielectric properties of composites.

Acknowledgments

The authors gratefully acknowledge the financial supports from the Project of Education and Teaching Reform of Graduate Student, Jianghan University, Wuhan, China.

References

[1] Zha J, Dong Z and Kang W.2014 Effect of micro-Si_3N_4-nano-Al_2O_3 co-filled particles on thermal conductivity, dielectric and mechanical properties of silicone rubber composites *J. IEEE T DIELECT EL IN*, **12(4)** p 1989~1996

[2] Su Q and Yu Y . 2009 Preparation of High Dielectric Constant Silicone Rubber Composites *J. Word Rubber Industry,* **36(9)** p 27~30

[3] M Tariq Nazir, B T Phung and M Hoffman . 2017 Micro-AlN/nano-SiO_2 co-filled silicone rubber composites with high thermal stability and excellent dielectric properties *J. Materials Letters.* **00** p 1~13

[4] Shao J, Zhao Y M and Yang C M, 2009 Performance of silicone rubber insulator and its application in high voltage *J.High Voltage Apparatus.***45(5)** p 155~158

[5] Sui X, Zhou W and Dong L 2017 A novel fiber-reinforced silicone rubber composite with Al particles for enhanced dielectric and thermal properties *J. Adv Polym Technol.* **00** p 1~10

[6] Guan G, Li H, Zhao S and Guo L, 2017 The surface modification of BaTiO3 and its effects on the microstructure and electrical properties of BaTiO3/silicone rubber composites J. *J VINYL ADDIT TECHN.* p 1~7.

[7] Lu Z P, Wang Y L, Sun Z G and Guan J G. 2010 High dielectric materials, low dielectric loss polymer matrix composites *J. Progress In Chemistry* .**22(8)** p 1619~1625

IOP Conf. Series: Materials Science and Engineering **359** (2018) 012018 doi:10.1088/1757-899X/359/1/012018

Evaluation Method for Low-Temperature Performance of Lithium Battery

H W Wang, Q Ma, Y L Fu, Z Q Tao, H Q Xiao, H Bai and H Bai

Institute of Industrial and Consumer Product Safety, Chinese Academy of Inspection and Quarantine, No. A3, Gaobeidian Bei Lu, Chaoyang District, Beijing, PR China, 100123

Email: wanghongwei-1978@163.com

Abstract. In this paper, the evaluation method for low temperature performance of lithium battery is established. The low temperature performance level was set up to determine the best operating temperature range of the lithium battery using different cathode materials. Results are shared with the consumers for the proper use of lithium battery to make it have a longer service life and avoid the occurrence of early rejection.

1. Introduction

The mobile power, also called as 'Charger', is a portable device integrated the power storage, boost and charge management, can charge or supply power in standby to the mobile phone and other digital devices whenever and wherever possible, and has become a most essential digital device for customers in daily life. Because the majority of mobile powers on the market are composed of 3 major parts combined with lithium battery (including 18650 or polymer), circuit board and shell, but the low temperature performance of lithium battery is poor, the research shows that the capacity and operating voltage of the common-used lithium ion battery will be significantly reduced at -10°C,significantly deteriorated at -20°C, its discharge capacity at -14°C is about 30% of that at room temperature, or individual lithium-ion battery even can't operate at -40°C. Moreover, there is no special standard for mobile power products in China, therefore, the establishment of an evaluation method for low temperature performance of lithium battery is required, and the low temperature performance level should be set up to determine the best operating temperature range of the mobile power of lithium battery using different cathode materials, so that the consumers have a clear perception of the low temperature performance of the mobile power purchased.

2. Method and Procedures

The purpose of this paper is to provide an evaluation method for low temperature performance of lithium batteries, and can be used to evaluate the performance of the mobile power under low temperature condition, thus sorting out the best operating temperature range of the lithium battery using different cathode materials, to provide the consumers with advices for proper use of lithium battery mobile power to make it have a longer service life and avoid the occurrence of early rejection.

To this end, the present paper provides a method for evaluating the low temperature performance of lithium battery (mobile power), the steps are as follows:

IOP Conf. Series: Materials Science and Engineering **359** (2018) 012018 doi:10.1088/1757-899X/359/1/012018

2.1. Discharge Performance Test and Required Results

Discharge the mobile power to termination voltage specified by the manufacturer at $1I_3$current under ambient temperature (20°C±2°C), and then charge the mobile power in the mode specified by the manufacturer under ambient temperature (20°C±2°C). After 10 cycles, put it in a place for 8 hours at 10°C, 20°C and30°C, respectively, and then conduct discharge performance test of the mobile power at $1I_3$ current, the required test results are as shown in table 1.

Table 1 Discharge Performance Test and Required Results

Ambient Temperature	Discharge Capacity(Ah)/Initial Discharge Capacity(Ah)	Discharge Energy (Wh)/Initial Discharge Energy(Wh)	Subentry value	Weight coefficient	Score
-10℃	>90%	>80%	3 points	0.5	S1
	80%≤ratio≤90%	70%≤ratio≤80%	2 points		
	70%≤ratio<80%	60%≤ratio<70%	1 point		
	<70%	<60%	0 point		
-20℃	>80%	>70%	3 points	0.3	
	70%≤ratio≤80%	60%≤ratio≤70%	2 points		
	60%≤ratio<70%	50%≤ratio<60%	1 point		
	<60%	<50%	0 point		
-30℃	>70%	>60%	3 points	0.2	
	60%≤ratio≤70%	50%≤ratio≤60%	2 points		
	50%≤ratio<60%	40%≤ratio<50%	1 point		
	<50%	<40%	0 point		

2.2. Charge Performance Test and Required Results

Charge the mobile power under ambient temperature (20°C±2°C), and discharge to termination voltage specified by the manufacturer at $1I_3$ current, and then charge the mobile power in the mode specified by the manufacturer under ambient temperature (20°C±2°C). After 10 cycles, place it in a place for 8 hours at 10°C,20°C and30°C, respectively, and then conduct charge performance test of the mobile powerat$1I_3$ current, the required test results are as shown in table 2.

Table 2 Charge Performance Test and Required Results

Ambient Temperature	Charge Capacity(Ah)/Initial Charge Capacity(Ah)	Charge Energy (Wh)/Initial Charge Energy(Wh)	Subentry value	Weight coefficient	Score
-10℃	>90%	>80%	3 points	0.5	S2
	80%≤ratio≤90%	70%≤ratio≤80%	2 points		
	70%≤ratio<80%	60%≤ratio<70%	1 point		
	<70%	<60%	0 point		
-20℃	>85%	>75%	3 points	0.3	
	75%≤ratio≤85%	65%≤ratio≤75%	2 points		
	65%≤ratio<75%	55%≤ratio<65%	1 point		
	<65%	<55%	0 point		
-30℃	80%	>70%	3 points	0.2	
	70%≤ratio≤80%	60%≤ratio≤70%	2 points		
	60%≤ratio<70%	50%≤ratio<60%	1 point		
	<60%	<50%	0 point		

2.3. 3) Determination of Low Temperature Performance Rating and Proposal of Operating Temperature

According to table 1 and table 2, obtain the comprehensive score S of the mobile power, and S=S1+S2, the comprehensive score and low temperature performance rating is set as shown in Table 3.

Table 3 Comprehensive Score and Low Temperature Performance Rating

Comprehensive Score S	Low Temperature Performance Rating	Proposal of Operating Temperature
$9<S\le12$	A level	not lower than -20°C
$6<S\le9$	B level	not lower than -15°C
$3<S\le6$	C level	not lower than -10°C
$0<S\le3$	D level	not lower than -5°C

3. Experimental Procedure

Test sample. Li-ion cell: rated voltage 3.7 V, rated capacity 5200 mAh. Dimensions: 96 mm×45 mm×23.9 mm, weight 126g, Input power: 5V DC, output voltage: 5V DC, input current 1000mA, output current 1000mA.

Test Instrument. Test Instrument: Integrated battery tester, MACCOR battery performance test systems, thermostat box, model ZPS-16 and Multimeter (UT33B).

4. Results and discussion

Conduct 10 charge-discharge cycles of 3 mobile powersunder room temperature (20°C±2°C) condition by using constant-current and constant-voltage charge system and constant-current discharge system, i.e. discharge the mobile powers to be tested at 1800 mA under the test temperature condition until the last blue LED light of the mobile power went off or starts flashing, put them in a place for 1 h, and then conduct constant current charge at 1800 mA until the mobile power voltage reaches 3.7V,conduct constant voltage chargeand stop charging as the charging current drops to 180 mA, so repeat 10 charge-discharge cycles, and put the mobile powersin the places at 10°C (one piece), 20°C (one piece), 30°C (one piece), respectively for 8 hours, respectively, conduct discharge performance test of the mobile power at a current of 1800 mA, the test results are as shown in Table 4. Conduct charge performance test of the mobile power at a current of 1800 mA, the test results are as shown in Table 5.

Table 4 the test results

Ambient Temperature	Charge Capacity(Ah)/Initial Charge Capacity(Ah)	Charge Energy (Wh)/Initial Charge Energy(Wh)
-10℃	92.1%	75.6%
-20℃	82.3%	71.6%
-30℃	69.5	57.2

According to Table 1 and Table 4,

$S1=3\times0.5+2\times0.5+3\times0.3+3\times0.3+2\times0.2+2\times0.2=5.1$

Table 5 the test results

Ambient Temperature	Charge Capacity(Ah)/Initial Charge Capacity(Ah)	Charge Energy (Wh)/Initial Charge Energy(Wh)
-10℃	89.4%	74.7%
-20℃	78.3%	64.6%
-30℃	60.5	56.8

According to Table 2 and Table 5,

$S2=2\times0.5+2\times0.5+2\times0.3+1\times0.3+1\times0.2+1\times0.2=3.3$

Finally, Comprehensive Score S is:

$S=S1+S2=5.1+3.3=8.4$

According to table 3, the low temperature performance of the mobile powers is level A, the recommended operating temperature is not less than -20°C.

5. Summary

Based on the foregoing test results, analysis and discussions, the conclusion can be obtained as follows:

1) The evaluation method for low temperature performance of lithium battery mobile power is proposed for the first time.

2) This method uses the commonly used test device with simpleoperating procedures and clear results.

3) This method is easy to popularize.

6. Acknowledgment

The authors would like to acknowledge the financial support provided for this research work by Fund: National key R & D program of China (2017YFF0210703), National key R & D program of China (2017YFF0210002), and Fundamental scientific research projects of Chinese Academy of Inspection and Quarantine (2017JK024).

References

[1] SU Xiao-hui, GAO Jun, YANG Yong. Low temperature electrolytes for Li-MnO_2 batteries. Power Supply Technology, 2009.3 Vol.33 No.3: 221-223.
[2] ZHAO Hui, DUAN Yan-li, GE Gui-lan, CHEN Yi-kui. Study on performance of low temperature type lithium ion battery. Power Supply Technology, 2009.1 Vol.33 No.1: 44-46.
[3] QCT 743-2006 Lithium-ion Batteries for electric vehicles. National Development and Reform Commission, 2006.
[4] IEC 62133: 2002, Secondary cells and batteries containing alkaline or other non-acid electrolytes-Safety requirements for portable sealed secondary cells, and for batteries made from them, for use in portable applications. 2002.
[5] IEC 62660-2-2010, Secondary lithium-ion cells for the propulsion of electric road vehicles - Part 2: Reliability and abuse testing. 2010.
[6] UL 2580-2011, Standards for batteries of use in EV. 2011.
[7] Lei Zhiguo，Zhang Chengning，Li Junqiu, et al. A Study on the Low-temperature Performance of Lithium-ion Battery for Electric Vehicles. Automotive Engineering, 2013(Vol. 35) No. 10: 927-933.
[8] WANG Hong-wei, DU Chun-yu, WANG Chang-bo. Study of low temperature performance of Li-ion battery. Battery bimonthly, Vol. 39, No.4, Aug., 2009: 208-210.

IOP Conf. Series: Materials Science and Engineering **359** (2018) 012019 doi:10.1088/1757-899X/359/1/012019

Effects of Ta_2O_5 Addition on Electrical Properties of ZnO-V_2O_5 Based Varistor Ceramics

J W Fan, H J Zhao and X L Zhang

Zhongyuan University of Technology, Zhengzhou, 450007, China

Email: jwf@zut.edu.cn[1]; zhj@zut.edu.cn[2]; karer2004@163.com[3]

Abstract. ZnO varistors are widely used for the protection of electronic and electrical equipment against transient surges. ZnO–V_2O_5 based varistor system is a potential candidate which can co-fire with Ag, and avoids the use of expensive Pa and Pt as the inner electrode in making multilayer chip varistors. However, the study of ZnO–V_2O_5-based ceramics is still in the initial stage for practical applications. The current work reports the effects of Ta_2O_5 on the electrical properties of ZnO-V_2O_5 based varistor ceramics. It shows that within 850-925°C experimental sintering temperature, the addition of Ta_2O_5 (0.05-0.20 mol%) may not improve the nonlinear coefficient but reduces the breakdown field of ZnO–V_2O_5 varistor ceramics.

1. Introduction

Zinc oxide (ZnO) is non-stoichiometric materials possessing excessive Zn ion and thus it is n-type oxide semiconductor. ZnO is proven as a very promising candidate for application to high performance electrical/electronic devices. Due to their highly nonlinear I-V characteristics and excellent energy handle capabilities, ZnO varistors are widely used for protection of electronic and electrical equipment against transient surge and overvoltage. Since its invention in 1968, ZnO varistor has been intensively studied and developed. The development of ZnO varistors has been one of great successes of functional ceramics [1-7].

Most of commercial ZnO varistors are made based on ZnO-Bi_2O_3 or ZnO-Pr_6O_{11} systems, these two systems exhibit excellent nonlinear I-V characteristics [1-7]. Currently, the research and development in electronic design have tended toward miniaturization and with higher packaging density. Hence many different miniaturized multilayer chip ZnO varistor components are produced. However, due to the low melting temperature of silver (Ag, 961°C), the commercial Bi_2O_3 or Pr_6O_{11} based ZnO varistor ceramics cannot be co-fired with the Ag inner electrode in multilayered chip components because of their relatively high sintering temperature above 1000 °C [8, 9], they can only co-fired with expensive palladium (Pa) or platinum (Pt) metals. Therefore, the new varistor ceramics are required in order to use Ag inner electrodes. Among the various ceramics, one candidate is the ZnO–V_2O_5 based system [10, 11]. This system can be sintered at relatively low temperature of approximately 900°C. This is quite important to miniaturized multilayer chip component applications, because it can be co-sintered with Ag inner-electrode without using expensive Pa or Pt as electrodes.

Since the nonlinearity coefficient α (the parameter of nonlinearity) of binary $ZnO–V_2O_5$ is low [10,11], a study of $ZnO–V_2O_5$-based varistor ceramics is still in the initial stage for practical application. Nahm [12] reported that the ternary $ZnO–V_2O_5–MnO_2$ ceramics could possess improved nonlinear properties with a nonlinear coefficient of 17–19. Some metal oxide are concerned as additives for $ZnO–V_2O_5$ system varistor ceramics. Ta_2O_5 has been used as a dopant in the $ZnO-Bi_2O_3$ based varistor ceramic systems, it can enhance α value of $ZnO-Bi_2O_3$ varistors. Nahm [13] added Ta_2O_5 into $ZnO–V_2O_5–MnO_2$ ceramics and reported the α value increasing with the rising of sintering temperature. However, the effects of Ta_2O_5 doping on the $ZnO–V_2O_5$ ceramics are still unclear. In this study, the investigation of different amount of Ta_2O_5 doped $ZnO–V_2O_5$ ceramics with sintering temperatures are reported and some new results are obtained.

2. Experimental Procedures

The ceramic samples used in this study were based on the compositions: ZnO:(96.5–X)mol%; V_2O_5: 0.5mol%; MnO_2: 2.0mol%; Co_2O_3: 1.0mol%; Ta_2O_5: (X=0, 0.05, 0.10, 0.20mol%). All oxides were analytic grade chemicals. They were mixed in deionized water with agate balls for 20 hours ball milling. After drying, granules were obtained by passing the dried powders through a screen having an aperture mesh of nominally 0.3mm. The small disc shape samples (12 mm in diameter and 2.5 mm in thickness) were pressed by a uniaxial press at 80MPa. The samples were sintered at 850, 875, 900 and 925 °C for 2 hour in air, with both heating and cooling rates at 3°C/min.

The microstructural observation of samples was examined by a scanning electron microscope (JSM-6360LV, Japan). The crystalline phases of samples were identified by X-ray diffractometry (D/MAX2500C, Japan) with CuK radiation.

Before electrical measurements, sliver electrodes were applied to both polished sides of the samples. The nonlinearity coefficients (α) and the breakdown electric field (E_b) were determined by using an instrument for dc parameters of varistors (Type: CJ1001, Chuangjie Ltd. China).

3. Results and Discussion

For crystalline phase characterization of Ta_2O_5 doped $ZnO–V_2O_5$ ceramics, the XRD pattern of the 0.2mol% Ta_2O_5 doped sample sintered at 900°C are shown in Fig. 1. The phese patterns revealed the presence of $Zn_3(VO_4)_2$, and $TaVO_5$ as minor secondary phases in addition to a major phase of hexagonal ZnO, which is agreement to the finding of Nahm.[13] These minor secondary phases may act as liquid-phase sintering aids.[13,14].

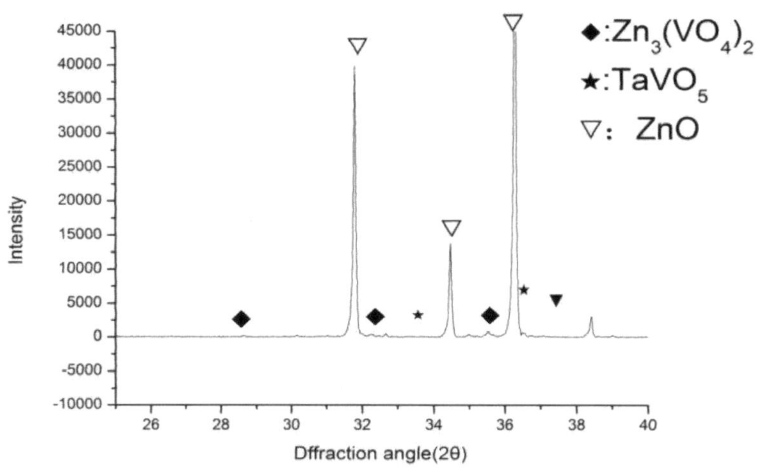

Fig. 1. The XRD pattern of the 0.2mol% Ta_2O_5 doped sample sintered at 900°C

Figure 2 shows the nonlinearity coefficients (α) of Ta_2O_5 doped samples at different sintered temperatures. From Fig. 2 it can be seen that the addition of Ta_2O_5 has a reverse effect on the nonlinearity coefficients of samples sintered at all sintering temperatures. The highest α value, 9.2, was obtained on the samples without Ta_2O_5 doping and sintered at 850°C, and second highest α value, 5.8, was obtained on the samples doped with 0.05mol% Ta_2O_5 doping and also sintered at 850°C. This finding is in the region to that by reported Nahm,[14] which is 8 obtained on the sample doped with 0.05mol%Ta_2O_5 sinetred at 875°C. However, with addition of Ta_2O_5 within the experimental doping levels, the α value of samples droped no matter of at what sintering temperatures, and trended to a certain value between 3 to 6. This result is different with the finding of Nahm [13], who reprted that the α value of 0.05mol% Ta_2O_5 doped sample increased with the increasing of temperature, it could rearch 28 at 950 °C sintered sample, but no results on the α value of samples doped with Ta_2O_5 at higher doping level.

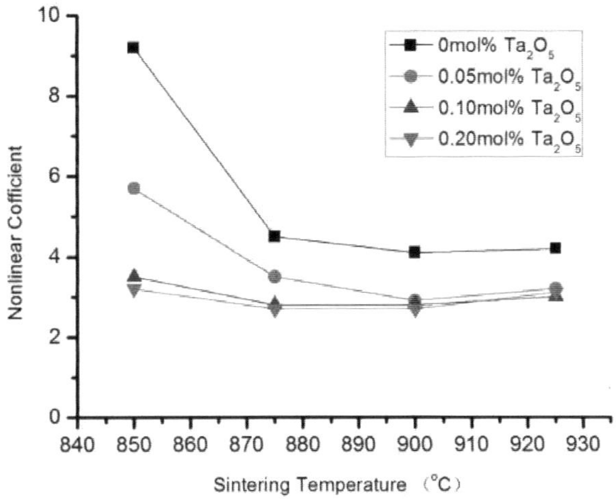

Fig. 2. The non-linearity coefficients (α) of samples as the fuction of sintering temerature

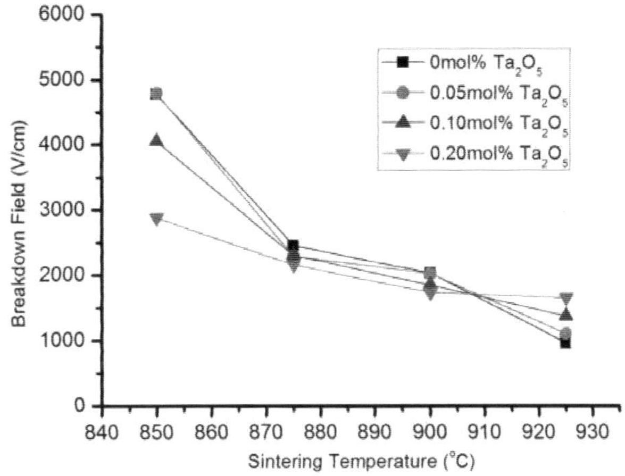

Fig. 3. The breakdown field (E_b) of samples as the fuction of sintering temerature

Figure 3 shows the breakdown field (E_b) as the fuction of sintering temerature. The E_b value obviousy decreased as the content of Ta_2O_5 increases at the 850°C sintering temperature, and this reduction became not obvious at 875°C and 900°C. It can also be seen that the E_b value is quite sensitive to the sintering temperature, the breakdown field (E_b) of samples drop significantly with the increase of sintering temperature. This is because that nonlinear I-V characteristics of ZnO varistors can be ascribed to the double Schottky type electrical barrier at the ZnO grain boundary [1-6]. The parameters controlling the value of E_b are the grain size and barrier voltage (V_{gb}) according to the following equation:

$$E_b = V_{gb}N_g \tag{1}$$

where the N_g is the average number of grains per centimeter [1-6]. From the results of Fig. 3, it suggests that the addition of Ta_2O_5 could enhance the grain size of $ZnO–V_2O_5$ samples. The higher sintering temperature usually results in lower breakdown electric field because of the increasing grain size as a result of increase of sintering temperature.

4. Conclusion

Except main phase ZnO, the $Zn_3(VO_4)_2$ and $TaVO_5$ as minor secondary phases were detected on the Ta_2O_5 doped $ZnO–V_2O_5$ varistor ceramics.

The addition of Ta_2O_5 didnot improve the nonlinear coefficient, but decreased the breakdown field of $ZnO–V_2O_5$ varistor ceramics. Within experimental doping level and sintering temperature, the highest value of nonlinear coefficient 9.2 was obtained on the simple without Ta_2O_5 doping and sintered at 850°C. The more detail investigation is carrying on.

Refrences

[1] Clarke D 1999 *J. Am. Ceram. Soc.* **82** 485
[2] Gupta T 1990 *J. Am. Ceram. Soc.* **73** 1817
[3] Bueno P, Varela J and Longo E J 2008 *Eur. Ceram. Soc.* **28** 505
[4] Fan J and Freer R. 1993 *J. Mater. Sci.* **28** 1391
[5] Fan J and Freer R 1994 *J. Am. Ceram.Soc.* **77** 2663
[6] Fan J and Freer R 1995 *J. Appl. Phys.* **77** 4795
[7] Fan J, Poosimma P and Freer R 2015 *Adv. Appl. Ceram.* **214** 14
[8] Nahm C-W 2001 *Mater Letters* **47** 182
[9] Nahm C-W 2002 *J. Mater. Sci. Letters* **21** 201
[10] Hng H-H and Knowles K 1999 *J. Eur. Ceram. Soc.* **19** 721
[11] Hng H-H and Halim L 2003 *Mater Letters* **57** 1411
[12] Nahm C-W 2007 *J. Mat. Sci.* **42** 8370
[13] Nahm C-W 2010 Journal of Alloys and Compounds **505** 657
[14] Nahm C-W 2009 *Ceram. Int.* **35** 2679.

IOP Conf. Series: Materials Science and Engineering **359** (2018) 012020 doi:10.1088/1757-899X/359/1/012020

Chemical Composition Analysis of Extracts from Ficus Hirta Using Supercritical Fluid

S B Deng[1,2], J P Chen[3], Y Z Chen[1,2,*], C Q Yu[1,2,*], Y Yang[1,2], S H Wu[4] and C Z Chen[5]

[1] Guangdong Provincial Key Laboratory of Advanced Drug Delivery Systems, Guangzhou, China
[2] School of Pharmacy, Guangdong Pharmaceutical University, Guangzhou, China
[3] The First Affiliated Hospital, Sun Yat-sen University, Guangzhou, China
[4] Guangzhou Lvheyuan Biotechnology Co., Ltd. , Guangzhou, China
[5] Heyuan Jin Yuan green life Co. Ltd. , Heyuan, China

[*] Corresponding authors: (Y Z Chen), E-mail: pn111@163.com
(C Q Yu), E-mail: pn333@163.com

Abstract: Ficus hirta was extracted by supercritical carbon dioxide. The volatile chemical components of extracts were analyzed using gas chromatography-mass spectrometry (GC-MS). The percentage of products extracted by Supercritical Fluid Extraction(SFE) was 2.5%. Nineteen volatile compounds were identified. The main volatile components were Elemicin, Psoralen, Palmitic acid, Bergapten, α-Linolenic acid, Medicarpin, Retinoic Acid, Maackiain, and Squalene. The method is simple and quick, and can be used for the preliminary analysis of chemical constituents of supercritical extracts of Ficus hirta.

1. Introduction

Hairy fig (Ficus hirta Vahl.) is shrub or small tree, which belongs to the family of Moraceae. It mainly distributes in the temperate and subtropical zones of Asia, especially in southern China[1]. The root of hairy fig, known as Cantonese ginseng, has a pleasantly fragrance of coconut milk and has been used for the treatment of hepatitis and improving fatigue resistance by Hakka people in China[2]. Some research revealed that the roots of hairy fig exhibited bioactivities of hepatoprotective[3], antitumor[4], and antibacterial[5]. It has many functional constituents, including flavonoids, coumarins and saponins, such as psoralen, bergapten, luteolin, apigenin, vitexin and 3,5,4'-trihydroxy-3,7-dimethoxyflavone[6-9]. Wei et al. analyzed the Ficus hirta ethanol extract with

IOP Conf. Series: Materials Science and Engineering **359** (2018) 012020 doi:10.1088/1757-899X/359/1/012020

Gas chromatography-mass spectrometry(GC-MS) method[10]. Li et al. analyzed the Ficus hirta steam distillation liquid with GC-MS method[11]. Liu et al. analyzed the Ficus hirta ethanol extract and steam distillation liquid with GC-MS method[12]. Lin et al. analyzed the Ficus hirta ether extract and steam distillation liquid with GC-MS method[13]. No supercritical extraction of Ficus hirta and chemical composition analysis of its supercritical extracts have been reported. In this paper, GC-MS technique was used to analyze the chemical composition of supercritical extracts of Ficus hirta, aiming to provide the theoretical basis for its development and utilization.

2. Instruments and Reagents

2.1 Instruments

SFE-5(32) Supercritical extraction device (Guangzhou Lvheyuan Biotechnology Co., Ltd.); TRACE DSQ Gas chromatography-mass spectrometry (ThermoFinnigan);
YP 3001N Electronic balance (Shanghai Precision Scientific Instrument Co., Ltd.);
DFT-200 Portable high speed universal grinder(Wenling Lingda Machinery Co., Ltd.);
Sartorius One Over Ten-thousand Analytical Balance (Sartorius Scientifc Instruments (Beijing) Co., Ltd.).

2.2 Reagents

Ficus hirta (Provided by Heyuan Jin Yuan green life Co. Ltd., identified by Professor Liu Jizhu (Guangdong Pharmaceutical University) as the dried root of Moraceae Ficus hirta Vahl.); Ethyl acetate (Tianjin Zhiyuan Chemical Reagent Co., Ltd.); Carbon dioxide (Food grade).

3. Methods

3.1 Supercritical Fluid Extraction

Firstly, Ficus hirta were broke into pieces (20 mesh). Weigh accurately 2000.0g of the coarse powder, put into the extraction kettle. The extraction pressure was 28 MPa, the extraction temperature was 45℃, the CO_2 flow rate was 30 L/h, the separation pressure was 6 MPa, the temperature of separation kettle I was 55℃, and the extraction time was 2 h. The extract was yellow semi-solid oil with an extraction rate of 2.5%.

3.2 Test solution

Taking 400 mg of the supercritical extracts, accurately weighed, in a 5 mL volumetric flask, add ethyl acetate to volume and mix well, filter and discard the initial filtrate, preserved the successive filtrate in liquid vial, as a test solution for GC-MS analysis.

3.3 GC-MS Analysis

3.3.1 Gas Chromatography Conditions

The GC system used a fused capillary column DM-5MS, length 30m, internal diameter 0.25mm, and film thickness 0.25μm. Ultra-high purity helium (99.99%) was used as the carrier gas at a constant flow rate of 1mL/min. The column oven temperature was programmed at 50℃ under a holding time of

2 min, while the injection temperature was set at 250℃. The temperature program increased at a rate of 12℃/min to 240℃ under a holding time of 22 min. The injection mode was splitless. Injection volume was 1.0μL.

3.3.2 Mass Spectrometry Conditions

The acquisition mode was operated at 1000 scan speed with 1.0 s event time. Sample was run in the 15-380 amu range and the total ion chromatogram obtained was auto-integrated using ChemStation.

4. Results

4.1 Component Analysis of The Supercritical Extracts

The supercritical extracts of the Ficus hirta were analyzed by GC-MS to obtain the total ion chromatogram (Figure 1). The initial identification results of the target components were obtained by automatic retrieval of the NIST mass spectrometry, and the relative contents of the components were determined according to the peak area normalization method. According to the results of GC-MS analysis, there were nineteen compounds identified in the Ficus hirta supercritical extracts and ten compounds with relative content of more than 1.0%. The results were shown in Table 1.

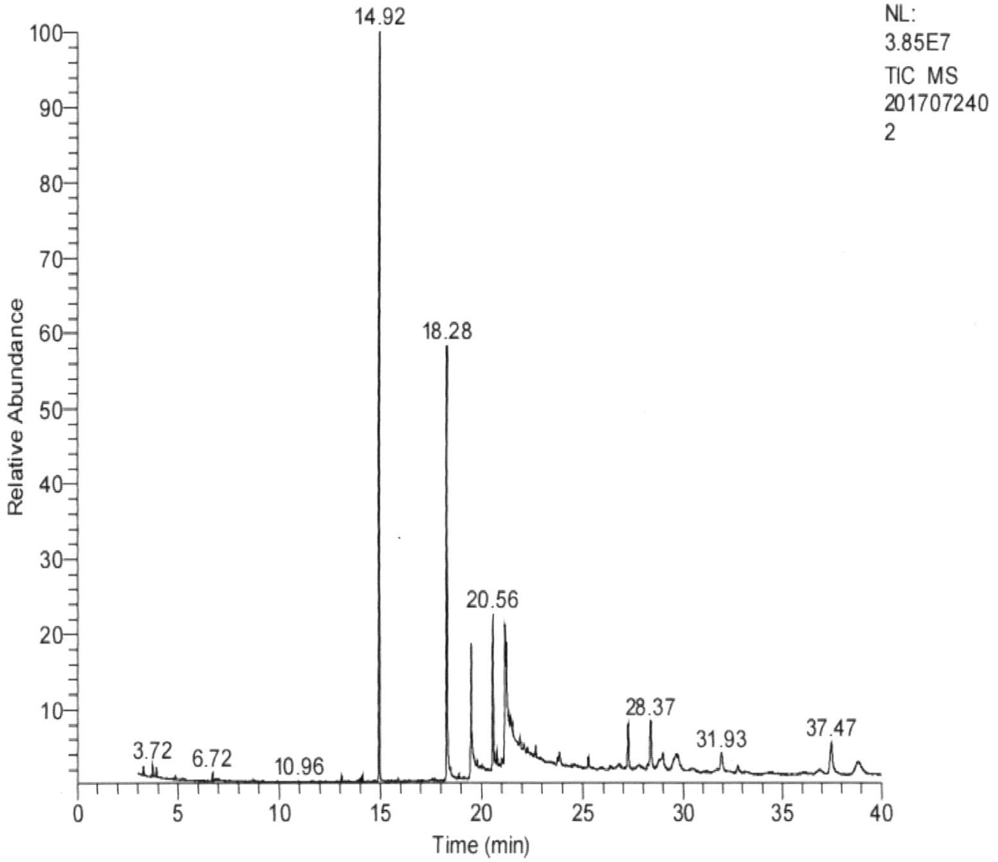

Figure 1. The total ion chromatogram of the Ficus hirta supercritical extracts

Table 1. Component analysis of the Ficus hirta supercritical extracts

Number	Retention time (min)	Compound	Molecular formula	Relative molecular mass	Relative content (%)
1	3.27	Toluene	C_7H_8	92	0.33
2	3.72	Hexanal	$C_6H_{12}O$	100	0.49
3	14.12	Tricyclo[6.3.0.0(2,4)]undec-8-ene, 3,3,7,11-tetramethyl-	$C_{15}H_{24}$	204	0.28
4	14.92	1,2,3-trimethoxy-5-(2-propenyl)-Benzene	$C_{12}H_{16}O_3$	208	23.11
5	18.28	7H-Furo[3,2-g][1]benzopyran-7-one	$C_{11}H_6O_3$	186	18.90
6	19.49	n-Hexadecanoic acid	$C_{16}H_{32}O_2$	256	8.34
7	19.80	Hexadecanoic acid, ethyl ester	$C_{18}H_{36}O_2$	284	0.34
8	20.55	7H-Furo[3,2-g][1]benzopyran-7-one, 4-methoxy-	$C_{12}H_8O_4$	216	6.85
9	20.76	2H,8H-Benzo[1,2-b:5,4-b']dipyran-2-one, 8,8-dimethyl-	$C_{14}H_{12}O_3$	228	0.93
10	21.15	9,12-Octadecadienoc acid (Z,Z)-	$C_{18}H_{32}O_2$	280	17.45
11	21.54	Cyclopropanebutanoic acid	$C_{25}H_{42}O_2$	374	0.57
12	21.91	Linoleic acid ethyl ester	$C_{20}H_{36}O_2$	308	0.61
13	22.68	[1,1'-Biphenyl]-4,4'-diamine, 3,3',5,5'-tetramethyl-	$C_{16}H_{20}N_2$	240	0.58
14	23.87	2,2',4,4',6-Pentamethyldiphenylsulfone	$C_{17}H_{20}O_2S$	288	1.06
15	25.28	Dibenz[a,c]cyclohexane,	$C_{18}H_{20}O_3$	284	0.84

		2,4,7-trimethoxy-			
16	28.37	6H-Benzofuro[3,2-c][1]benzopyran-3-ol, 6a,11a-dihydro-9-methoxy-, (6aR-cis)-	$C_{16}H_{14}O_4$	270	3.85
17	28.98	Retinoic acid	$C_{20}H_{28}O_2$	300	1.75
18	31.94	6a,12a-Dihydro-6H-(1,3)dioxolo(5,6)benzofuro(3,2-c)chromen-3-ol	$C_{16}H_{12}O_5$	284	2.41
19	37.47	Squalene	$C_{30}H_{50}$	410	4.14

4.2 Bioactivity Analysis of the Main Constituents of the Supercritical Extracts

Five compounds with relative content of more than 5% were identified in the Ficus hirta supercritical extracts. The content of compounds was arranged from high to low as: Elemicin (23.11%), Psoralen (18.90%), α-Linolenic acid (17.45%), Palmitic acid (8.34%), Bergapten (6.85%), which consisted the main components of the Ficus hirta supercritical extracts. Zhao et al. studied the volatile oil components and antioxidant activity of Dalbergia pinnata fragrance ingredient. It was found that Elemicin is the main component of the volatile oil from Dalbergia pinnata fragrance ingredient and the total volatile oil appears better antioxidant activity[14]. Tang et al. studied on Acorus tatarinowii Schott for cerebral diseases. The experiment showed the Elemicin, β-asarone and α-asarone of Acorus tatarinowii Schott which can pass through the rat's blood brain barrier (BBB) have direct effect on brain. Acorus tatarinowii Schott is a good herb for cerebral diseases on which the Elemicin, β-asarone and α-asarone have effect all together[15]. Psoralen showed strong antibacterial activity against gram positive bacteria and Staphylococcus aureus[16]. Psoralen also showed dose-dependent antitumor activity, which can significantly induce the apoptosis of tumor cells[17]. It can inhibit the bone metastasis of breast cancer cells in vivo, inhibit the growth of breast cancer cells, and regulate the function of osteoblasts and osteoclasts of tumor bearing mice in the bone microenvironment[18]. α-Linolenic acid is an essential fatty acid for the body to play its normal physiological function, which can modulate blood lipids, anti-thrombosis, regulating blood sugar and blood pressure, anti-tumor, anti-inflammatory, inhibiting allergic reactions, enhance intelligence and protect eyesight and so on[19]. Xiao found that Bergapten could promote the osteogenic differentiation of bone marrow mesenchymal stem cells into osteoblasts in the mice in vivo. It is beneficial to bone formation, so as to prevent and treat osteoporosis[20]. Huang et al. studied that Bergapten could suppresses lung cancer cells secreting soluble interleukin-2 receptor by inhibiting the binding and the transcriptional activity of STAT3 on the promoter of IL2RA[21]. In addition, there were also relatively high content of components such as Medicarpin (3.85%), Maackiain (2.41%) and Squalene (4.14%) in the Ficus hirta supercritical extracts. Li et al. found that Medicarpin could inhibit or kill human's liver cancer cells under the cultured condition[22]. Maackiain is a novel antiallergic compound that suppresses transcriptional upregulation of the histamine H1 receptor and interleukin - 4 genes[23]. Squalene is an important bioactive substance, which has anti-cancer, anti-tumor, anti-fatigue, anti-infection, immune and antioxidant effects, and has been widely used in medicine, food, cosmetics industry and other fields[24].

5. Discussion

Nineteen compounds in the samples of supercritical extracts were identified using GC-MS, accounting for 92.83% of the total chemical constituents. The main components were long chain organic acids (25.79%), coumarins (25.75%) and phenylpropanoids (23.11%). In addition, there were some unsaturated hydrocarbons, sulfones, isoflavones, dihydroflavones, terpenoids and so on. In the premise of the Ficus hirta from the same origin, the results of the Ficus hirta supercritical extracts were compared with those of the subcritical extracts, the ethanol extracts and the volatile oil extracted by the steam distillation. It was found that the relative content of Elemicin extracted by supercritical fluid extraction was the highest.

In this paper, GC-MS method was used to analyze the supercritical extracts of the Ficus hirta. In order to analyze the chemical constituents more detailed, other analytical methods, eg, High Performance Liquid Chromatography-Mass Spectrometry (HPLC-MS), should be used.

Acknowledgments

This study was sponsored by the joint project of Guangdong Provincial Academy of Chinese Medical Sciences and Guangdong Provincial Academy of Traditional Chinese Medicine (2016A020226038).

References

[1] Tao Y, Chen Q, He X, et al. Chemical quantification and antioxidant assay of four active components inFicus hirtaroot using UPLC-PAD-MS fingerprinting combined with cluster analysis[J]. *Chemistry Central Journal*, 2013, **7**(1):1-9.

[2] Au D T, Wu J, Jiang Z, et al. Ethnobotanical study of medicinal plants used by Hakka in Guangdong, China[J]. *Journal of Ethnopharmacology*, 2008, **117**(1):41-50.

[3] Cai Q Y, Chen H B, Cai S Q, et al. [Effect of roots of Ficus hirta on cocaine-induced hepatotoxicity and active components].[J]. *China Journal of Chinese Materia Medica*, 2007, **32**(12):1190.

[4] Qu G X, Wang J, Yao X S. Screening of anti-pyricularia oryzae activities of traditional Chinese herbal medicines[J]. *Journal of Shenyang Pharmaceutical University*, 2003, **20**(03):219-221.

[5] Chen Q, Ye S X, Yu J. Antibacterial Activity of RADIX FICI HIRTAE by Chromotest Microassay[J]. *Medicinal Plant*, 2012, **3**(09):13-16.

[6] Ji Y, Zhang X Q, Ying W, et al. Two new phenolic compounds from the roots of Ficus hirta.[J]. *Journal of Asian Natural Products Research*, 2010, **24**(7):621-625.

[7] Tao Y, Chen Q, He X, et al. Chemical quantification and antioxidant assay of four active components inFicus hirtaroot using UPLC-PAD-MS fingerprinting combined with cluster analysis[J]. *Chemistry Central Journal*, 2013, **7**(1):1-9.

[8] Li C, Bu P B, Yue D K, et al. [Chemical constituents from roots of Ficus hirta].[J]. *China Journal of Chinese Materia Medica*, 2006, **31**(2):131-133.

[9] Zhao L P, Di B, Feng F, et al. Chemical constituents from the roots of Ficus hirta[J]. *Pharmaceutical & Clinical Research*, 2008, **16**(01):5-7.

[10] Wei G, Liu C L, He J X, et al. Root of Ficus hirta characteristic components on the GC-MS[J]. *Chinese Archives Of Traditional Chinese Medicine*, 2005, **23**(07):1249-1251.

[11] Li J X, Hui J, Yang Y Y, et al. Component Analysis of Volatile Oil from Ficus hitra Vahl by GC-MS[J]. *Journal of Anhui Agri. Sci*, 2010, **38**(14):7281-7282.

[12] Liu C L, Wei G, He J X, et al. Analysis of Volatile Oils and Ethanol-Extracts from Different Parts of Radix Fici Hirtae[J]. *Journal of Guangzhou University of Traditional Chinese Medicine*, 2004, **21**(03):204-205.

[13] Lin L, Zhong X Q, Wei G, et al. GC-MS Analysis of Volatile Components in Root of Ficus Hitra[J]. *Journal of Chinese Medicinal Materials*, 2000, **23**(04):206-207.

[14] Zhao W B, Zhang D Y, Xu Z C, Mo W. Study on Composition and Antioxidant Activity of Volatile Oil from Dalbergia pinnata Fragrance Ingredient[J]. *Traditional Chinese Drug Research & Clinical Pharmacology*, 2017, **28**(05):659-662.

[15] Tang H M. A Study On Material Base of Acorus Tatarinowii Schott for Cerebral Diseases[J]. *Lishizhen Medicine And Materia Medica Research*, 2002, **13**(01):1-2.

[16] Khatune N A, Islam M E, Haque M E, et al. Antibacterial compounds from the seeds of Psoralea corylifolia[J]. *Fitoterapia*, 2004, **75**(2):228-230.

[17] Wang Y, Hong C, Zhou C, et al. Screening Antitumor Compounds Psoralen and Isopsoralen from Psoralea corylifolia L. Seeds[J]. *Evidence-Based Complementray and Alternative Medicine*, 2011, 1-7.

[18] Wu C, Sun Z, Ye Y, et al. Psoralen inhibits bone metastasis of breast cancer in mice.[J]. *Fitoterapia*, 2013, **91**(8):205-210.

[19] Li X H, Zhang J, Li C, et al. Bioactivity and Efficacy of Plant α-linolenic Acid[J]. *Health Vocational Education*, 2016, **34**(18):144-146.

[20] Xiao J J. An experimental study of the Bergapten effect on differentiation ability of the bone marrow mesenchymal stem cells into osteoblasts[D]. Southern Medical University, 2016.

[21] Huang X T, Chen T H, Qi M N, et al. Bergapten suppresses lung cancer cells secreting soluble interleukin-2 receptor[J]. *Chin J Clinicians(Electronic Edition)*, 2016, **10**(11):1567-1573.

[22] Li W D, Kan Y M, Hong M, et al. The inhibitory effects of homopterocarpin and medicarpin on human' s liver cancer cells in vitro[J]. *Journal of Shenyang Pharmaceutical University*, 2001, **18**(3):211-212.

[23] Mizuguchi H, Nariai Y, Kato S, et al. Maackiain is a novel antiallergic compound that suppresses transcriptional upregulation of the histamine H1 receptor and interleukin - 4 genes[J]. *Pharmacology Research & Perspectives*, 2015, **3**(5):f1-f2.

[24] Guan B, Zheng W C. Extraction and purification of squalene and its application[J]. *Science and Technology of Cereals,oils and Foods*, 2010, **18**(4):27-30.

IOP Conf. Series: Materials Science and Engineering **359** (2018) 012021 doi:10.1088/1757-899X/359/1/012021

Properties of Semi-dry Flue Gas Desulfurization Ash and Used for Phosphorus Removal

Y P Yu, Y Fang, S Y Chai and Z Z Zhuang

Materials Science and Engineering, Nanjing University of Technology,Nanjing 210009, China

E-mail: powder8817@njtech.edu.cn

Abstract: The composition of Semi-dry Flue Gas Desulfurization ash was as follows: $CaSO_3 \cdot 0.5H_2O$ and $Ca(OH)_2$. Most of the particle size was about 8 μm; particles were irregular and small, while the other part was rough and globular. $CaSO_3$ strongly oxidized at about 466℃ and $Ca(OH)_2$ decomposed at ~ 662℃. The maximum amount of phosphorus removal in Semi-dry Flue Gas Desulfurization ash was 79.898 mg/g.

1.Introduction

Buring coal to generate energy releases SO_2 caused acid rain and accelerated soil acidification. In order to reduce the impact of SO_2, many efforts have been made over the past years.Flue gas desulfurization technologies were usually classified as either wet or dry methods[1-2]. The semi-dry FGD process used lime as sorbent was the most used FGD technology in powder and thermal industry.[3]There were 43 million tons FGD gypsum produced in 2010(China Association Environment Protection industry)[4]. T Zaremba[5] reported the semi-dry adsorption process offered many advantages such as no waste water, straightforward operation,no or little reheating, low space requirements and lower investment, however the product had a complicated composition with higher content of $CaSO_3$, which affected the comprehensive utilization of flue gas desulfurization ash. The dry and semi-dry flue gas desulfurization were rich in a large number of active substances and activated substances,consists of $CaSO_3$、$CaSO_4$、$CaCO_3$、CaF_2 and fly ash,which had great potential application value in the field of building materials. Liming Chen al.reported that it was safe to grow plants in soil treated with $CaSO_3$ if the application was made at least three days to several weeks before planting[6]. The $CaSO_3$ could rapidly oxidize to gypsum in oxygenated environments[7]. Gypsum could improve the physical and chemical properties of soils and mitigating subsoil acidity and Al toxicity[8].

The FGD contained a large amount of hydroxyl ions, which could be used as heavy metal ions removal agent, such as Cu(Ⅱ)、Fe(III)、pb(Ⅱ)、Cd(Ⅱ).Yubo Yan al.[8] reported that the FGD gypsum generated from coal-fired power plants to adsorptively removed pb(Ⅱ) and Cd(Ⅱ) from wastewater.The kinetic analysis displayed that the adsorption processes both followed the pseudo-second order model well,and the FGD gypsum provided a higher sorption rate for Pb(Ⅱ). The maximum adsorption capacities of Pb(Ⅱ) and Cd(Ⅱ) were respectively calculated as 161.3 and 32.57 mg/g[9], which could compete with some other synthesized adsorbents. Phosphorus was a key factor of eutrophication and it could affect the people's production and life[10-11].Research on an efficient method for removal phosphorus was meaningful to improve the economic and social benefit.MTG Viana [12]modified the solar coral powder as adsorbent and the mount of phosphorus removal was 9.597 mg/g.C Han[13] took the slag as adsorbent and the amount of phosphorus removal was reached 89.9 mg/g. The Semi-dry contained a large number of $Ca(OH)_2$ and $CaCO_3$, all of which had good precipitation ability for phosphorus.It was a new attempt to use Semi-dry FGD for the phosphorus removal.The utility model not only reduced the waste and protected the environment,but also facilitated the reutilization of waste materials.

2.Properties of Semi-dry Flue Gas Desulfurization Ash

2.1Properties of Semi-dry FGD ash. A variety of analytical methods were used to obtain semi-dry FGD composition and properties.The chemical composition and content of the substance were determined by ARL-9800 X-ray fluorescence spectrometer(XRF).X-ray diffraction(XRD) patterns were recorded by the Rigaku SmartLab (made in japan) with the powdered samples less than 30 um,with Cu Kα radiation in the diffracton angle(2θ) range of 10-90°at a sweep rate of $10°/min^{-1}$(λ=0.154 nm,Voltage=40 KV,I=20 mA).The morphology of Semi-dry FGD before and after reaction was observed by scanning electronic microscope(JSM-6390A 15KV) and the size parameter was measured by Laser particle sizer(Mastersizer 2000) though wet measurement.The differential thermal and thermogravimetric analysis was carried out by means of a DTG-60H type derivatograph within the range of temperatures 20-1000℃ under air atmosphere.

Fig.1 was the X-ray diffraction diagram of semi-dry FGD and Table 1 was the chemical compositions of it.It consisted of $Ca(OH)_2$、$CaCO_3$、$CaSO_3 \cdot 0.5H_2O$、$CaSO_4$.The main chemical composition of semi-dry FGD were CaO and SO_3,The sum of both was over seventy percent.Fig3.showed that the size distribution ranged from 1 to 30 um.Most of the particles size were about 8 um and it contained small amounts of larger particles. Table 2 was specific parameters of each particle diameter.

Table 1 Chemical compositions of semi-dry FGD ash w%

$SiO_2Al_2O_3$ Fe_2O_3CaO		MgO	SO_3	LOI		
0.465	0.207	0.985	48.77	1.05	21.65	20.96

Table 2 Size parameter of Semi-dry FGD

Parameter	Semi-dry FGD ash	Obscuration（%）	Uniformity	Span	Dispersant
D_{50}/ um	3.038	15.05	2.41	9.465	Water
D[4,3] /um	8.796				
D[3,2] /um	1.864				
D_{16}* /um	0.955				
D_{26}* /um	1.445				
D_{74}* / um	7.586				
D_{84}* / um	17.378				
D_{90}* / um	30.200				

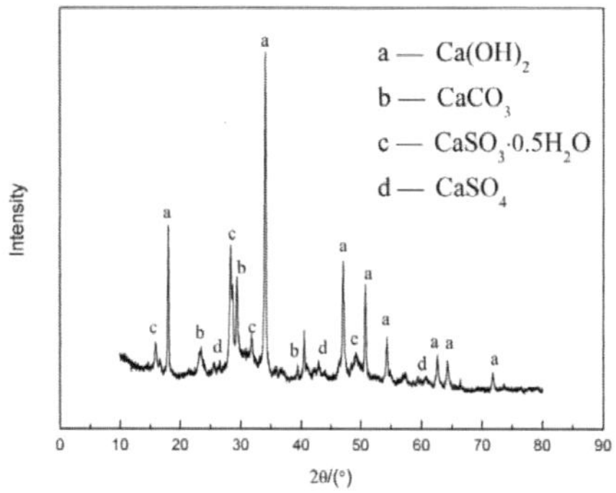

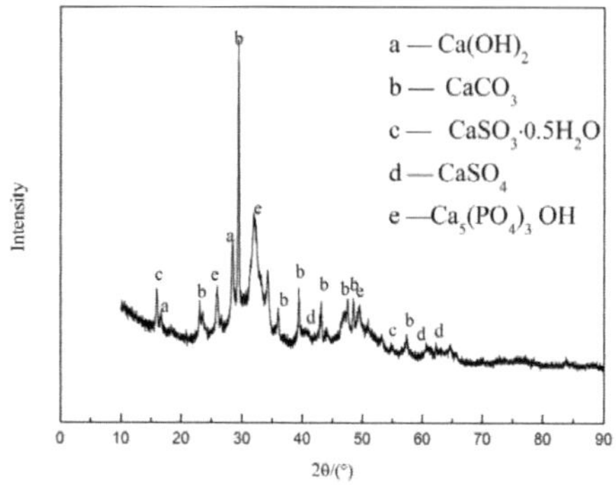

Figure.1 XRD of Semi-dry FGD ashFigure.2XRD of the reaction product

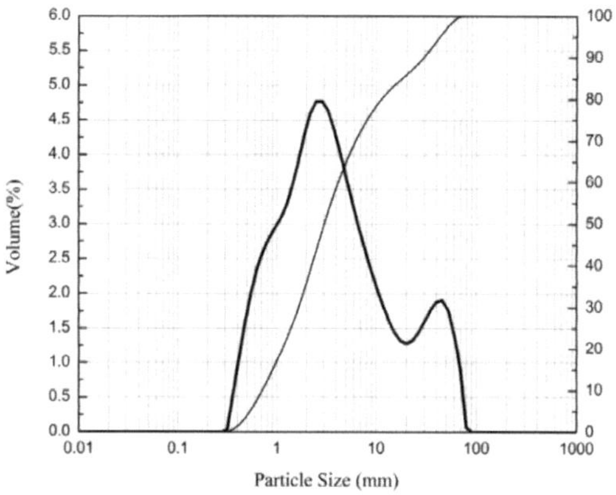

Figure.3 Size distribution of Semi-dry FGD ash

Fig.3was the TG-DTA of the semi-dry FGD.There were three endothermic peaks(351℃、414℃、662℃) and one exothermic peak(466℃) in the DTA curve.The endothermic peaks at 351 and 414 degrees centigrade were accompanied by obvious mass loss due to the loss of adsorbed water and the $CaSO_3 \cdot 0.5H_2O$ lost its crystalline water.The DTA curve showed a sharp exothermic peak at 466 ℃ and it was accompanied by an apparent in mass increase due to the $CaSO_3$ was oxidized. The DTA curve appeared an endothermic peak at 662 ℃,and there was obvious mass loss due to the decomposition of Ca（OH）$_2$ in the sample.

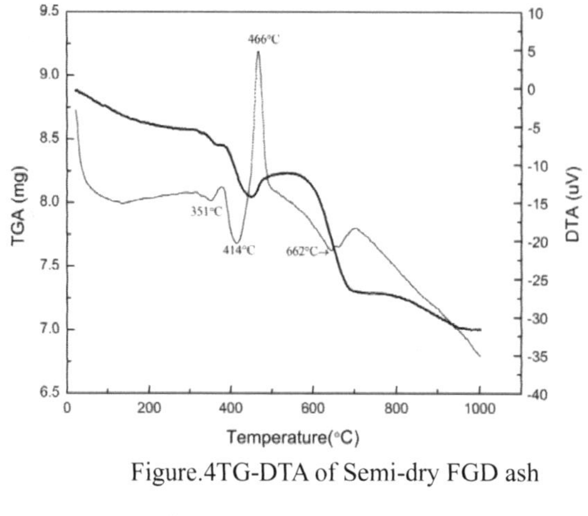

Figure.4TG-DTA of Semi-dry FGD ash

2.2Removal of phosphorus by Semi-dry FGD. The precipitation of calcium and phosphorus was carried out in two steps.The first was the genration of Amorphous Calcium Phosphate(ACP),then thehydroxyapatite (HAP) crystallization.

ACP: $3Ca^{2+}+2PO_4^{3-}+xH_2O \rightarrow Ca_3(PO_4)_2 \cdot x\ H_2O$

$Ca_3(PO_4)_2 \cdot xH_2O \rightarrow 3Ca^{2+}+2PO_4^{3-}+x\ H_2O$

HAP: $Ca_3(PO_4)_2 \cdot xH_2O+2\ Ca^{2+}+\ PO_4^{3-}+OH^- \rightarrow Ca_5(PO_4)_3OH+x\ H_2O$

Phosphorus concentration in wastewater was determined with molybdenum blue spectrophotometer method; Potassium dihydrogen phosphate (KDP) was used to simulate the phosphorus-containing wastewater,the initial phosphorus concentration 50 mg/L and the volume was 80 mL.All experiments were performed in 100 mL glass beakers and set the single variable (initial pH;add quantity;temperature;time).The linear regression equation was Y=1.95488X+0.00859,the R-Square=0.9999.Absorbance was measured by using 752 UV-Vis spectrophotometer. Fig.5was the curve of phosphorus concentration-absorbance.Phosphorus concentration was obtained by the curve of Phosphorus concentration-absorbance.

$$\text{P Elimination}(\%) = \frac{C0-C1}{C0} \quad (1)$$

Where C_1 represented the phosphorus concentration of wastewater solution and C_0 represented the initial concentration.

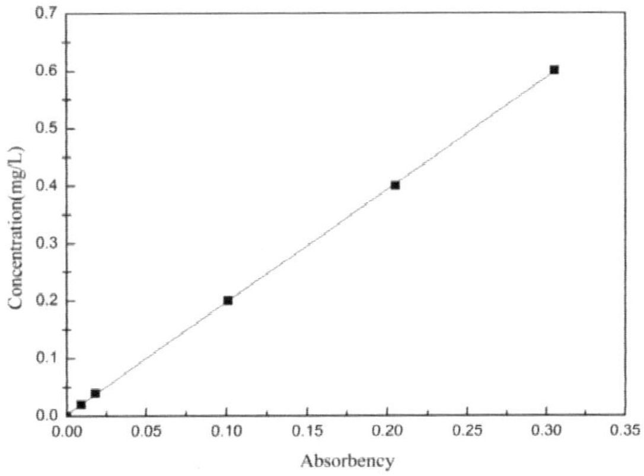

Figure.5 The curve of phosphorus concentration-absorbance

3.Results and discussions

Fig.2 was the XRD of the reaction product.The reaction product was $Ca_5(PO_4)_3OH$,Although the side effected consume Ca^{2+},the result $CaCO_3$ could be used as a weight agent to aid precipitation. L Montastruc[14] reported that the reaction pKs was 58.33 at 25℃.

Main reaction: $5Ca^{2+}+7OH^-+3HPO_4^{2-} \rightarrow Ca_5(PO_4)_3OH\downarrow+3H_2O$

Side reaction: $5Ca^{2+}+CO_3^{2-} \rightarrow CaCO_3$

According to Fig.8the amount of Semi-dry FGD was 0.05 g and the pH was 6.when the reaction time was 12h,the concentration of phosphorus was 0.099 mg/L.The optimum phosphorus removal time was 12 h. The P Elimination was 79.898 mg/g.According to Fig.7 Temperature had little effect on phosphorus removal efficiency.The P Elimination ranged from79.479-79.898 mg/g when the pH was 6 and temperature was 45 ℃。Fig.8 showed that the pH and Semi-dry FGD addition on phosphorus had great influence on the efficiency of phosphorus removal.When the pH was 2,the amount of phosphorus removal ranged from

70.458-72.898 mg/g.The range of phosphorus removal was 70.458-79.898 m/g when initial pH ranged from 2-10. The optimum phosphorus removal pH was 6.Excessive acid and alkali were not conductive to the efficient removal of phosphorus.

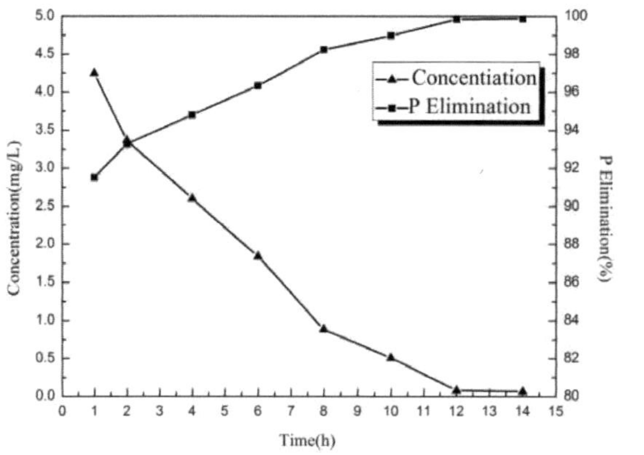

Figure.6 The time of phosphorus removal

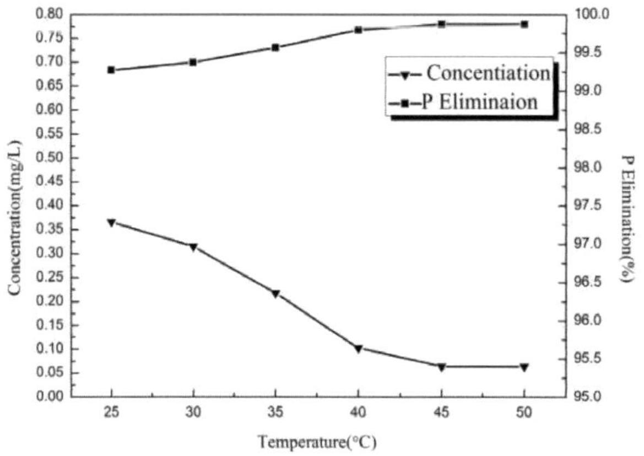

Figure.7 The temperature of phosphorus removal

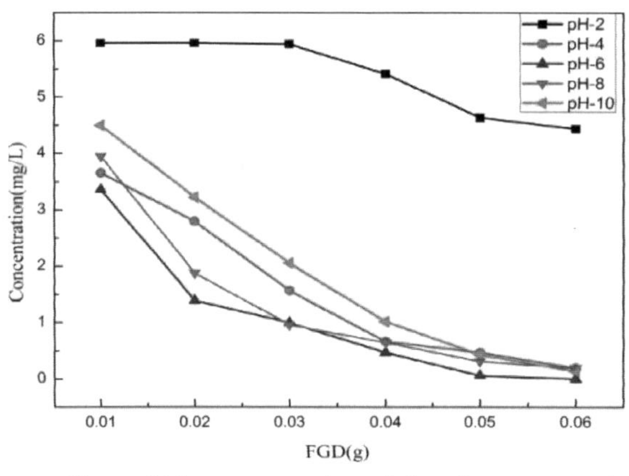

Figure.8 The concentration of phosphorus removal

Fig.9-Fig.10 was the scanning electron microscopy(SEM) photos of Semi-dry FGD.Some

IOP Conf. Series: Materials Science and Engineering **359** (2018) 012021 doi:10.1088/1757-899X/359/1/012021

particles were irregular and small,while the other part was rough globular.There were many holes in it.This was due to the formation of the Semi-dry FGD difficult to produced liquid phase when the gas-solid reaction happened at the high temperature.Although it could produce obvious diffusion of solid phase,it would not form strong densification,which would make its surface structure loose and porous. Fig.11-Fig.13 was the scanning electron microscopy(SEM) photos of phosphorus removal product.The particles were loosely spherical and larger pores.There were many small protrusions are formed on the surface of the particles.

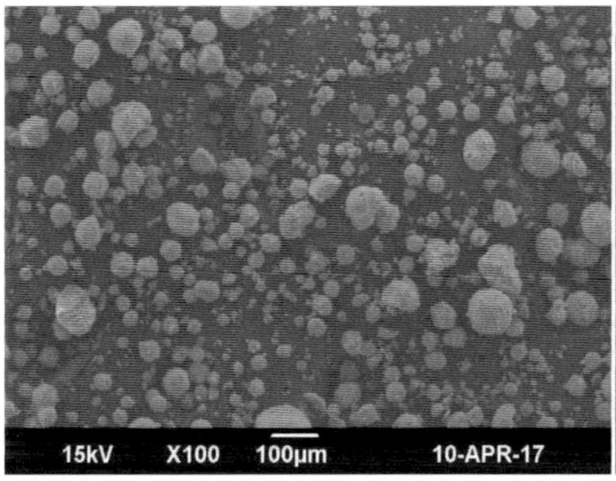

Figure.9 SEM of Semi-dry FGD ashFigure.10 SEM of Semi-dry FGD ash

Figure.11 SEM of Semi-dry FGD ashFigure.12 SEM of P removal product

Figure.13 SEM of P removal product

4.Summary

Semi-dry FGD consists of $Ca(OH)_2$、$CaCO_3$、$CaSO_3 \cdot 0.5H_2O$、$CaSO_4$ and the main chemical content was CaO and SO_3. The particles size were about 8 um and particles were irregularly spherical,the surface was and the structure was loose.$CaSO_3$ was strongly oxidized about 466 ℃and $Ca(OH)_2$ was decomposed at about 662 ℃.The optimum phosphorus removal time was 12 h and the pH was 2.Temperature had little influence on phosphorus removal efficiency and the main influenced factors of phosphorus removal efficiency were initial pH,Semi-dry FGD addition.The maximum amount of phosphorus removal was 79.898 mg/g.

Reference

[1] Takeshita M, Soud H N. FGD performance and experience on coal-fired plants[M]. London: IEA Coal Research, 1993.
[2] Soud H N. FGD installations on coal-fired plants[M]. IEA Coal Research, 1994.
[3] Xu G, Guo Q, Kaneko T, et al. A new semi-dry desulfurization process using a powder-particle spouted bed[J]. Advances in Environmental Research, 2000, 4(1): 9-18.
[4] China Association of Environmental Protection Industry (CAEPI), China Environ. Prot. Ind. 7 (2011) pp. 4–12 (In Chinese).
[5] Zaremba T, Mokrosz W, Hehlmann J, et al. Properties of the wastes produced in the semi-dry FGD installation[J]. Journal of Thermal Analysis and Calorimetry, 2008, 93(2): 439-443.
[6] Chen L, Ramsier C, Bigham J, et al. Oxidation of FGD-CaSO 3 and effect on soil chemical properties when applied to the soil surface[J]. Fuel, 2009, 88(7): 1167-1172.
[7] Pasiuk-Bronikowska W, Bronikowski T, Ulejczyk M. Mechanism and kinetics of autoxidation of calcium sulfite slurries[J]. Environmental science & technology, 1992, 26(10): 1976-1981.
[8] Shainberg I, Sumner M E, Miller W P, et al. Use of gypsum on soils: A review[M]//Advances in soil science. Springer US, 1989: 1-111.
[9] Yan Y, Li Q, Sun X, et al. Recycling flue gas desulphurization (FGD) gypsum for removal of Pb (II) and Cd (II) from wastewater[J]. Journal of colloid and interface science, 2015, 457: 86-95.
[10]Zhao G, Zhou Q, Xie L. Adsorption of phosphorus from aqueous solution on steel converter slags[J]. JOURNAL-TONGJI UNIVERSITY, 2007, 35(11): 1510.
[11]Dale B. Marine dinoflagellate cysts as indicators of eutrophication and industrial pollution: a discussion[J]. Science of the total environment, 2001, 264(3): 235-240.
[12]Vianna M T G, Marques M, Bertolino L C. Sun coral powder as adsorbent: Evaluation of phosphorus removal in synthetic and real wastewater[J]. Ecological Engineering, 2016, 97: 13-22.
[13]Han C, Wang Z, Yang H, et al. Removal kinetics of phosphorus from synthetic wastewater using basic oxygen furnace slag[J]. Journal of Environmental Sciences, 2015, 30: 21-29.
[14]Montastruc L, Azzaro-Pantel C, Biscans B, et al. A thermochemical approach for calcium phosphate precipitation modeling in a pellet reactor[J]. Chemical Engineering Journal, 2003, 94(1): 41-50.

IOP Conf. Series: Materials Science and Engineering **359** (2018) 012022 doi:10.1088/1757-899X/359/1/012022

AuRu/*meso*-Mn_2O_3: A Highly Active and Stable Catalyst for Methane Combustion

Z Han, J Y Fang, S HXie, J G Deng, Y X Liu and H X Dai*

Beijing Key Laboratory for Green Catalysis and Separation, Key Laboratory of Beijing on Regional Air Pollution Control, and Laboratory of Catalysis Chemistry and Nanoscience, Department of Chemistry and Chemical Engineering, College of Environmental and Energy Engineering, Beijing University of Technology, Beijing 100124, China

Email: hxdai@bjut.edu.cn

Abstract. Three-dimensionally ordered mesoporous Mn_2O_3 (*meso*-Mn_2O_3) and its supported Au, Ru, and AuRu alloy (0.49 wt% Au/*meso*-Mn_2O_3, 0.48 wt% Ru/*meso*-Mn_2O_3, and 0.97 wt% AuRu/*meso*-Mn_2O_3 (Au/Ru molar ratio = 0.98)) nanocatalysts were prepared using the KIT-6-templating and polyvinyl alcohol-protected reduction methods, respectively. Physicochemical properties of the samples were characterized by means of numerous techniques, and their catalytic activities were evaluated for the combustion of methane. It is found that among all of the samples, 0.48 wt% Ru/*meso*-Mn_2O_3 and 0.97 wt% AuRu/*meso*-Mn_2O_3 performed the best (the reaction temperature ($T_{90\%}$) at 90% methane conversion was 530–540°C), but the latter showed a better thermal stability than the former. The partial deactivation of 0.97 wt% AuRu/*meso*-Mn_2O_3 due to H_2O or CO_2 introduction was reversible. It is concluded that the good catalytic activity and thermal stability of 0.97 wt% AuRu/*meso*-Mn_2O_3 wasassociated with the high dispersion of AuRu alloy NPs (2–5 nm) on the surface of *meso*-Mn_2O_3 and good low-temperature reducibility.

1. Introduction

As the main component of natural gas, methane shows much higher greenhouse effect than CO_2[1], and more serious pollutants, such as CO, NO_x, and unburned hydrocarbons, are usually detected in the outlets of methane flame combustion [2]. Catalytic combustion is considered to be one of the most promising approaches for the suppression of the pollutants, in which the key issue is the availability of high-performance and stable catalysts. Hence, the development of novel high-efficiency and stable catalysts is of practical significance.

Supported noble metals (e.g., Au, Pd, Pt, and Ru) perform well at low temperatures, but they are poor in high-temperature stability. Although the Pd-based catalysts show good catalytic performance for methane combustion at low temperatures, these materials have some major drawbacks: (i) Pd-based catalysts often have a poor stability in the steady-state methane combustion processes [3,4]; and (ii) exposure of Pd-based catalysts to water or sulfur-containing compounds may cause a significant loss in activity via formation of inactive $Pd(OH)_2$ and stable Pdsulfate phases [5–7]. To overcome these problems, a second metal is expected to be added to palladium to generate a bimetallic alloy catalyst. For instance, the doping of Pt to Pd could increase low-temperature methane combustion activity,

hinder the growth of Pd or PdO particles [8], improve the catalyst thermal stability [9,10], and enhance the resistant ability of sulfur or water vapor poisoning [7].
Transition-metal oxides (e.g., MnO_x [11,12] and CoO_x [13]) have also been used as catalyst for methane combustion, among which MnO_x has widely been studied and considered as alterative catalysts due to their multiple oxidation states and oxygen vacancies [14]. For example, Han et al. [12] observed that the complete conversion of methane could be achieved over α-Mn_2O_3 at ca. 600 °C, and this catalyst exhibited ultrahigh stability. Three-dimensionally ordered mesoporous Mn_2O_3 (*meso*-Mn_2O_3) possesses a higher surface area and good redox ability, thus suitable for the use of catalyst and support. A high surface area is beneficial for the dispersion of active sites on the support surface, and a uniform pore size is favourable for the inhibition of agglomeration of noble metal nanoparticles (NPs), hence improving the high-temperature stability of the supported noble metal catalyst.
Herein, we report the preparation of *meso*-Mn_2O_3-supported AuRu alloy nanocatalysts and their catalytic performance for methane combustion. It is found that the AuRu/*meso*-Mn_2O_3 catalyst was highly active and stable under the adopted reaction conditions.

2. Experimental

2.1. Catalyst preparation
Mesoporous silica (KIT-6) was synthesized by adopting the procedures described in the literature [15]. Ordered mesoporous Mn_2O_3 was synthesized according to the method reported previously [16]. In a typical synthesis, 1.0 g of KIT-6 was suspended in 50 mL of *n*-hexane. After stirring at 65 °C for 0.5 h, 20.00 mmol of $Mn(NO_3)_2$ aqueous solution (50 wt%) was added under vigorous stirring. The mixture was further stirred at 65 °C for 3 h. Then, the powders were obtained after filtration and drying at room temperature (RT). The precursor@silica composite was put in a crucible, and then calcined in a muffle furnace at a ramp of 1 °C/min from RT to 600 °C and kept at this temperature for 6 h. The silica template was removed by etching twice with a hot NaOH aqueous solution (2.00 mol/L). The template-free mesoporous Mn_2O_3 (*meso*-Mn_2O_3) was obtained after centrifugation, washing with deionized water and ethanol, and drying at 80 °C.
The *meso*-Mn_2O_3-supported gold–ruthenium alloy nanocatalyst was prepared via a polyvinyl alcohol (PVA, $MW_{aver.}$ = 10,000 g/mol)-protected reduction route with $NaBH_4$ as reducing agent [17]. The typical preparation procedures were as follows: A desired amount of PVA was added to the $HAuCl_4$ and $RuCl_2$ mixed aqueous solution (100 mg/L, theoretical Au/Ru molar ratio = 1.0 : 1.0, noble metal/PVA mass ratio = 2.0 : 3.0) in an ice bath under vigorous stirring for 30 min. A certain amount of $NaBH_4$ aqueous solution (0.1 mol/L, noble metal/$NaBH_4$ molar ratio = 1.0 : 5.0) was quickly added to the above mixed solution, generating a dark brown noble metal sol. Then, a desired amount of *meso*-Mn_2O_3 powders was added to the noble metal sol (theoretical Au–Ru loading = 1.0 wt%) under stirring for 4 h. After that, the mixture was filtered, washed with deionized water, and dried at 80 °C for 12 h. The dried powders were then calcined in a muffle furnace at a ramp of 1 °C/min from RT to 600 °C and kept at this temperature for 6 h, thus obtaining the *x*AuRu/*meso*-Mn_2O_3 samples.
For comparison purposes, the *x*Au/*meso*-Mn_2O_3 (theoretical Au loading = 0.5 wt%) and *x*Ru/*meso*-Mn_2O_3 (theoretical Ru loading = 0.5 wt%) samples were also prepared using the above-mentioned method. The results of inductively coupled plasma atomic emission spectroscopic (ICP–AES) investigations reveal that the real loading (x) of noble metals was 0.97, 0.49, and 0.48 wt% in the *x*AuRu/*meso*-Mn_2O_3, *x*Au/*meso*-Mn_2O_3, and *x*Ru/*meso*-Mn_2O_3 samples, respectively; and the real Au/Ru molar ratio in the *x*AuRu/*meso*-Mn_2O_3 sample was 0.98.

2.2. Catalyst characterization
X-ray diffraction (XRD) patterns of the samples were recorded on a Bruker D8 Advance diffractometer with Cu $K\alpha$ radiation and nickel filter (λ = 0.15406 nm). Elemental analyses of the noble metal loadings were performed using the ICP–AES technique on a Thermo Electron IRIS Intrepid ER/S spectrometer. Transmission electron microscopic (TEM) images of the samples were

obtained using the JEOL-2010 equipment (operating at 200 kV). BET (Brunauer-Emmett-Teller) surface areas of the samples were measured via N_2 adsorption at −196 °C on a Micromeritics ASAP 2020 analyzer with the samples outgassed at 250 °C for 2.5 h under vacuum before measurement, and their pore size distributions were calculated based on the desorption branches of the isotherms.

Hydrogen temperature-programmed reduction (H_2-TPR) experiments were carried out on a chemical adsorption analyzer (Autochem II 2920, Micromeritics). Before TPR measurement, ca. 0.050 g of catalyst (40–60 mesh) was loaded to a quartz fixed-bed U-shaped microreactor (i.d. = 4 mm) and pretreated in an O_2 flow of 20 mL/min at 250 °C for 1 h. After being cooled at the same atmosphere to RT, then with helium flow of 30 mL/min purge for other 15 min. Finally, the pretreated sample was exposed to a flow (50 mL/min) of 5% H_2–95% Ar (*v*/*v*) mixture and heated from RT to 600 °C at a ramp of 10 °C/min. The alteration in H_2 concentration of the effluent was monitored online by the chemical adsorption analyzer. The reduction peak was calibrated against that of the complete reduction of a known standard powdered CuO (Aldrich, 99.995%).

2.3. Catalytic evaluation

Catalytic activities of the samples were evaluated in a continuous flow fixed-bed quartz tubular microreactor (i.d. = 6.0 mm). To minimize the effect of hot spots, 50 mg of the sample (40–60 mesh) was diluted with 0.25 g of quartz sands (40–60 mesh). Before the test, each sample was treated in an oxygen flow of 20 mL/min at 250 °C for 1 h. The reactant mixture was composed of 2.5 vol% CH_4 + 20 vol% O_2 + 77.5 vol% N_2 (balance), and the total flow was 16.6 mL/min, thus giving a CH_4/O_2 molar ratio of 1/8 and a space velocity (SV) of ca. 20,000 mL/(g h)). In the case of CO_2 introduction, 5.0 vol% CO_2 from a CO_2 cylinder (balanced with N_2) was introduced to the reaction system through a mass flow controller. Furthermore, in the case of water vapor introduction, 3.0 vol% H_2O was introduced by passing the feed stream through a water saturator at a certain temperature. Reactants and products were analyzed online by gas chromatography (GC-14C, Shimadzu) equipped with a flame ionization detector (FID) and a thermal conductivity detector (TCD), using a stabilwax® column (30 m in length) for methane separation and a Carboxen 1000 column (3 m in length) for permanent gas detection. The balance of carbon throughout the catalytic system was estimated to be 99.5%. Catalytic activities of the samples were evaluated using the temperatures ($T_{50\%}$ and $T_{90\%}$) required for achieving methane conversions of 50 and 90%, respectively. CH_4 conversion was defined as $(c_{inlet}-c_{outlet})/c_{inlet}\times$ 100%, where the c_{inlet} and c_{outlet} are CH_4 concentrations of the inlet and outlet feed stream, respectively.

3. Results and discussion

3.1. Crystal phase composition, morphology, pore structure, and surface area

Figure 1 shows the wide- and small-angle XRD patterns of the samples. By comparing to the XRD pattern (JCPDS PDF# 41-1442) of the standard Mn_2O_3 sample, one can realize that all of the Bragg diffraction peaks (figure 1A) of each sample could be well indexed to a cubic *meso*-Mn_2O_3 crystal structure. Furthermore, only cubic Mn_2O_3 phase was detected in the *meso*-Mn_2O_3, Au/*meso*-Mn_2O_3, Ru/*meso*-Mn_2O_3, and AuRu/*meso*-Mn_2O_3 samples. This result might be due to low noble metal loadings and good dispersion of Au, Ru or AuRu NPs on the surface of *meso*-Mn_2O_3. As seen from figure 1B, there were two peaks at $2\theta = 1.10^{\circ}$ and 1.80° in the small-angle XRD pattern of the *meso*-Mn_2O_3 and AuRu/*meso*-Mn_2O_3 samples, indicating formation of ordered mesopores in the two samples. According to the Scherrer equation, the calculated crystallite sizes of *meso*-Mn_2O_3 in the samples were in the range of 19.6–20.2 nm (table 1).

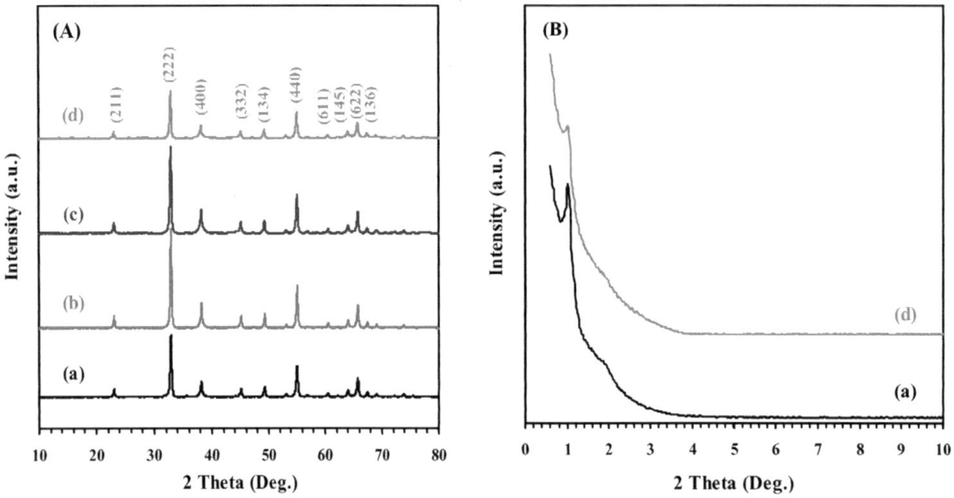

Figure 1. (A) Wide-angle and (B) small-angle XRD patterns of (a) *meso*-Mn_2O_3, (b) 0.49Au/*meso*-Mn_2O_3, (c) 0.48Ru/*meso*-Mn_2O_3, and (d) 0.97AuRu/*meso*-Mn_2O_3.

Table 1. BET surface areas, pore volumes, average pore sizes, Mn_2O_3 crystallite sizes (D_{Mn2O3}), noble metal particle sizes, real Au or Ru contents, and real Au/Ru molar ratios of the samples.

Sample	Surface area (m^2/g)	Pore volume (cm^3/g)	Average poresize (nm)	$D_{Mn2O3}{}^a$(nm)	Real Au/Ru content[b] (wt%)	Real Au/Ru molar ratio[b] (mol/mol)
meso-Mn_2O_3	104.8	0.36	11.0	19.6	–/–	–
0.49Au/*meso*-Mn_2O_3	100.6	0.34	10.1	20.2	0.49/–	–
0.48Ru/*meso*-Mn_2O_3	99.7	0.34	10.0	20.1	–/0.48	–
0.97AuRu/*meso*-Mn_2O_3	98.8	0.35	9.9	19.9	0.63/0.33	0.98

[a] Data were calculated according to the Scherrer equation using the FWHM of the (222) line of Mn_2O_3;

[b] Data were determined by the ICP–AES technique.

Figure 2 shows the TEM images of *meso*-Mn_2O_3, 0.49Au/*meso*-Mn_2O_3, 0.48Ru/*meso*-Mn_2O_3, and 0.97AuRu/*meso*-Mn_2O_3. It can be clearly seen that all of the samples displayed an ordered mesoporous structure with an average pore size of 10–11 nm. Since the method for the synthesis of AuRu NPs was the same as that for the synthesis of AuPd NPs with an alloy structure [17], the obtained AuRu NPs would also possess an alloy structure. From the TEM images of the supported noble metal samples, one can observe that the noble metal particles were 2–5 nm in diameter. Figure 3 shows the N_2 adsorption–desorption isotherms and pore-size distributions of the samples. The *meso*-Mn_2O_3 and its supported noble metal samples displayed a type IV isotherm with a hysteresis loop in the relative pressure range of 0.5–1.0 (figure 3A), indicating generation of a mesoporous structure in these samples. From the pore-size distribution curves (figure 3B), one can observe two peaks in the ranges of 2–6 and 10–16 nm, respectively. As summarized in table 1, surface area of *meso*-Mn_2O_3 was 104.8 m^2/g. After loading of noble metal NPs, surface areas (98.8–100.6 m^2/g) of the 0.49Au/*meso*-Mn_2O_3,

0.48Ru/*meso*-Mn_2O_3, and 0.97AuRu/*meso*-Mn_2O_3 samples decreased only slightly. Pore volumes (0.34–0.36 cm^3/g) of all of the samples were also similar.

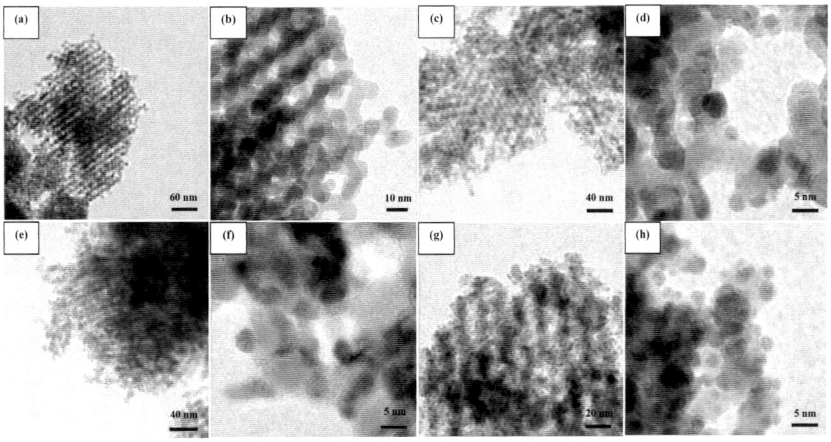

Figure 2. TEM images of (a, b) *meso*-Mn_2O_3, (c, d) Au/*meso*-Mn_2O_3, (e, f) Ru/*meso*-Mn_2O_3, and (g, h) AuRu/*meso*-Mn_2O_3.

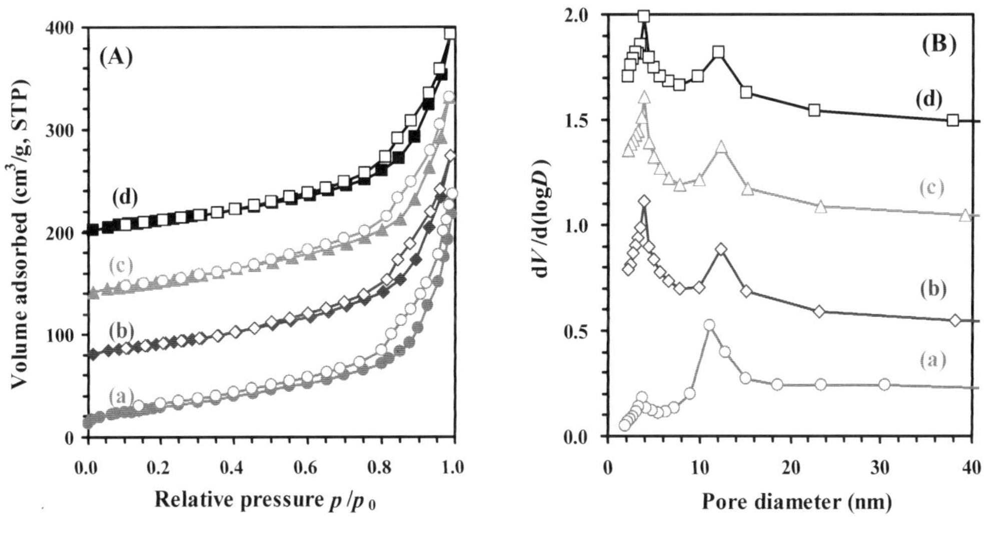

Figure 3. (A) Nitrogen adsorption–desorption isotherms and (B) pore-size distributions of (a) *meso*-Mn_2O_3, (b) 0.49Au/*meso*-Mn_2O_3, (c) 0.48Ru/*meso*-Mn_2O_3, and (d) 0.97AuRu/*meso*-Mn_2O_3.

3.2. Reducibility

It is well known that, the reduction of Mn_2O_3 can be divided by the successive processes [18,19]: $Mn_2O_3 \rightarrow Mn_3O_4 \rightarrow MnO$. The reduction signals in the H_2-TPR profile may be detected due to the reduction of species at different local environments [19]. Figure 4 illustrates the H_2-TPR profiles of the samples and their H_2 consumptions are summarized in table 2. There were two main reduction peaks centered at 372 and 524 °C for the *meso*-Mn_2O_3 sample, attributable to the reduction of Mn_2O_3 to Mn_3O_4 and Mn_3O_4 to MnO [18–20], respectively. After loading of the noble metals, the reduction

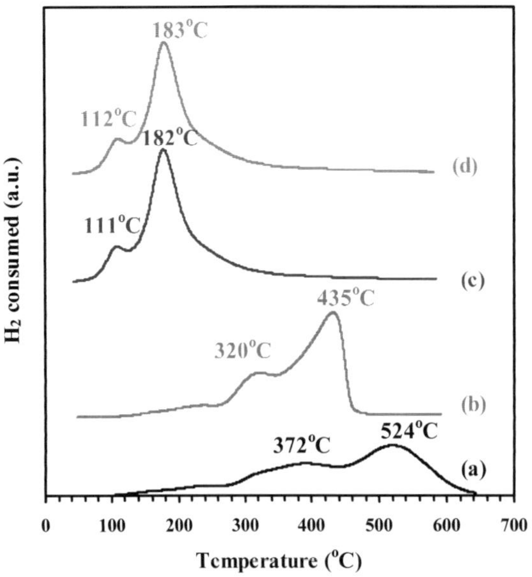

Figure 4. H_2-TPR profiles of (a) *meso*-Mn_2O_3, (b) 0.49Au/*meso*-Mn_2O_3, (c) 0.48Ru/*meso*-Mn_2O_3, and (d) 0.97AuRu/*meso*-Mn_2O_3.

peaks of the samples shifted to low temperatures, especially for the 0.48Ru/*meso*-Mn_2O_3 and 0.97AuRu/*meso*-Mn_2O_3 samples (111–112°C and 182–183°C, respectively), which was a result due to the strong interaction between the noble metal NPs and the *meso*-Mn_2O_3 support [21]. The H_2-TPR study demonstrates the improvement in low-temperature reducibility, which was due to the reduction of Mn_2O_3 promoted by the activated H atoms on the noble metals. The 0.48Ru/*meso*-Mn_2O_3 and 0.97AuRu/*meso*-Mn_2O_3 samples exhibited the best low-temperature reducibility. The total H_2 consumption (14.80 mmol/g_{cat}) of the *meso*-Mn_2O_3 sample was slight higher than those (13.71–14.25 mmol/g_{cat}) of the *meso*-Mn_2O_3-supported noble metal samples.

Table 2. H_2 consumptions and catalytic activities of thesamples at SV = 20,000 mL/(g h).

Sample	H_2 consumption (mmol/g_{cat})	Methane combustion activity	
		$T_{50\%}$ (°C)	$T_{90\%}$ (°C)
meso-Mn_2O_3	14.80	486	> 600
0.49Au/*meso*-Mn_2O_3	13.71	480	580
0.48Ru/*meso*-Mn_2O_3	13.84	465	530
0.97AuRu/*meso*-Mn_2O_3	14.25	470	540

3.3. Catalytic performance

In the blank experiment (only quartz sands were loaded in the microreactor), no conversion of methane was detected at SV = 20,000 mL/(g h) and below 600 °C. In other words, there were no gas-phase reactions in the catalytic system under the adopted conditions. Fig. 5A and B shows the catalytic activities of the samples and the effect of SV on methane conversion over the 0.97AuRu/*meso*-Mn_2O_3 sample for methane combustion, respectively. It is observed that methane conversion monotonously increased with the rise in temperature over *meso*-Mn_2O_3; however, methane conversion first increased, then decreased, and finally increased over the supported noble metal samples, which might be due to the active oxidized noble metal (AuO_x and/or RuO_x) on the surface of the sample decomposed to less active metallic noble metal (Au^0 and/or Ru^0) at 380–390°C and the metallic noble metal was re-oxidized to the oxidized noble metal above 400°C. Obviously, the loading of noble metal improved the catalytic performance of *meso*-Mn_2O_3. As shown in figure 5A and table 2, the *meso*-Mn_2O_3-supported Ru or AuRu sample performed the best: $T_{50\%}$ = 465°C and $T_{90\%}$ = 530°C over 0.48Ru/*meso*-Mn_2O_3; $T_{50\%}$ = 470°C and $T_{90\%}$ = 540°C over 0.97AuRu/*meso*-Mn_2O_3 at SV = 20,000 mL/(g h). From figure 5B, one can see that methane conversion decreased with the rise in SV, a result due to the shorting in contact time of the reactants with the catalyst.

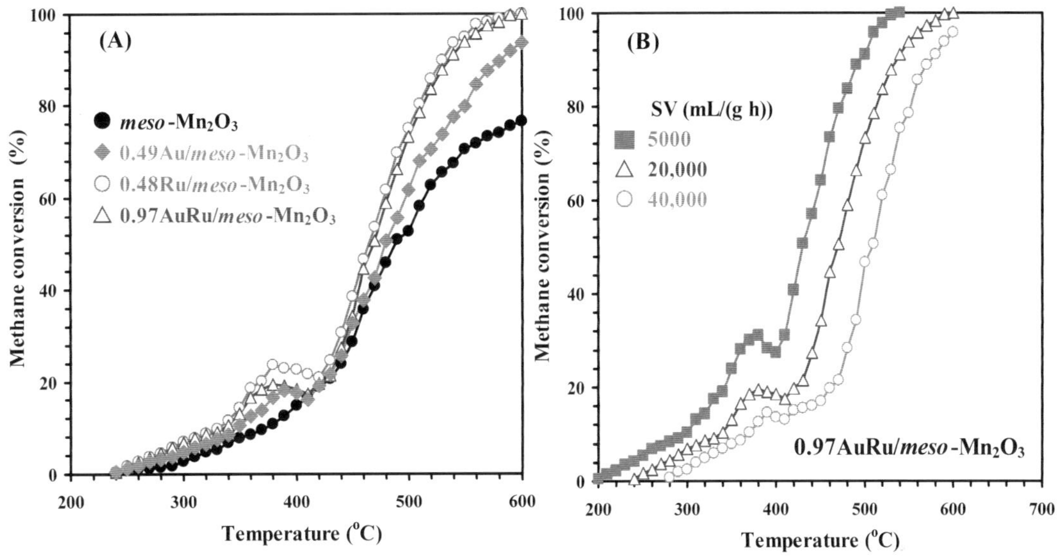

Figure 5. (A) Methane conversion as a function of temperature over the samples at SV = 20,000 mL/(g h) and (B) effect of SV on methane conversion over the 0.97AuRu/meso-Mn_2O_3 sample.

Table 3. Comparison on methane oxidation activity of the samples reported in the literature.

Catalyst	Reaction condition	$T_{50\%}$ (°C)	$T_{90\%}$ (°C)	Ref.
0.48Ru/*meso*-Mn_2O_3	2.5 vol% CH_4, SV = 20,000 mL/(g h)	465	530	Present work
0.97AuRu/*meso*-Mn_2O_3	2.5 vol% CH_4, SV = 20,000 mL/(g h)	470	540	Present work
$La_{0.7}Ag_{0.3}MnO_3$	1 vol% CH_4, SV = 10,000 mL/(g h)	690	750	[22]
$La_{0.7}Sr_{0.3}MnO_3$	1 vol% CH_4, SV = 10,000 mL/(g h)	720	780	[22]
$La_{0.7}Ce_{0.3}MnO_3$	1 vol% CH_4, SV = 10,000 mL/(g h)	740	810	[22]
1.0 wt% Pd/Al_2O_3	5 vol% CH_4, SV = 30,000mL/(g h)	640	710	[23]
1.0 wt% Pd/ZrO_2	5 vol% CH_4, SV = 30,000mL/(g h)	600	650	[23]
1.1 wt% Pt/3DOM $Ce_{0.6}Zr_{0.3}Y_{0.1}O_2$	2 vol% CH_4, SV = 30,000 mL/(g h)	479	598	[24]

In the past years, a number of works on methane combustion over the Mn-based and supported noble metal catalysts have been reported in the literature, and their results are summarized in table 3. Apparently, the $T_{50\%}$ and $T_{90\%}$ over 0.48Ru/*meso*-Mn_2O_3 or 0.97AuRu/*meso*-Mn_2O_3 were much lower than those over $La_{0.7}Ag_{0.3}MnO_3$ [22], $La_{0.7}Sr_{0.3}MnO_3$ [22], $La_{0.7}Ce_{0.3}MnO_3$ [22], 1.0 wt% Pd/Al_2O_3 [50], 1.0 wt% Pd/ZrO_2 [23], and 1.1 wt% Pt/3DOM $Ce_{0.6}Zr_{0.3}Y_{0.1}O_2$ [24].

Over the Pd-based catalysts, methane oxidation can proceed via the Mars–van Krevelen redox mechanism [25–27]. At lower temperatures, methane could react with the chemisorbed oxygen to produce CO_2 and H_2O; at higher temperatures, however, lattice oxygen of the oxidized noble metals (AuO_x and/or RuO_x) involves in the oxidation of methane. O_2 molecules could be activated in the oxygen vacancies of *meso*-Mn_2O_3 or near the interface between the AuO_x or RuO_x and *meso*-Mn_2O_3 to active oxygen species, replenishing the lattice oxygen in AuO_x or RuO_x consumed in methane oxidation and re-oxidizing the reduced catalyst.

3.4. Catalytic stability and effects of water and carbon dioxide on catalytic activity

Figure 6 shows the conversion of methane as a function of on-stream reaction time over the 0.48Ru/*meso*-Mn_2O_3 and 0.97AuRu/*meso*-Mn_2O_3 samples at SV = 20,000 mL/(g h). It is observed that methane conversion over the 0.48Ru/*meso*-Mn_2O_3 sample decreased from 90 to 67% after 30 h of on-stream reaction, however, there was no significant loss in catalytic activity over the 0.97AuRu/*meso*-Mn_2O_3 sample within 30 h of on-stream methane combustion. This result demonstrates that the 0.97AuRu/*meso*-Mn_2O_3 sample was catalytically more stable than the 0.48Ru/*meso*-Mn_2O_3 sample. Alloying of Ru with Au could significantly promote the adsorption and activation of oxygen, possibly enhancing the migration of activated oxygen to the noble metal alloy–Mn_2O_3 interface, hence stabilizing the catalytic activity. The excellent catalytic stability of 0.97AuRu/*meso*-Mn_2O_3 might be associated with its enhanced oxygen activation ability and better low-temperature reducibility.

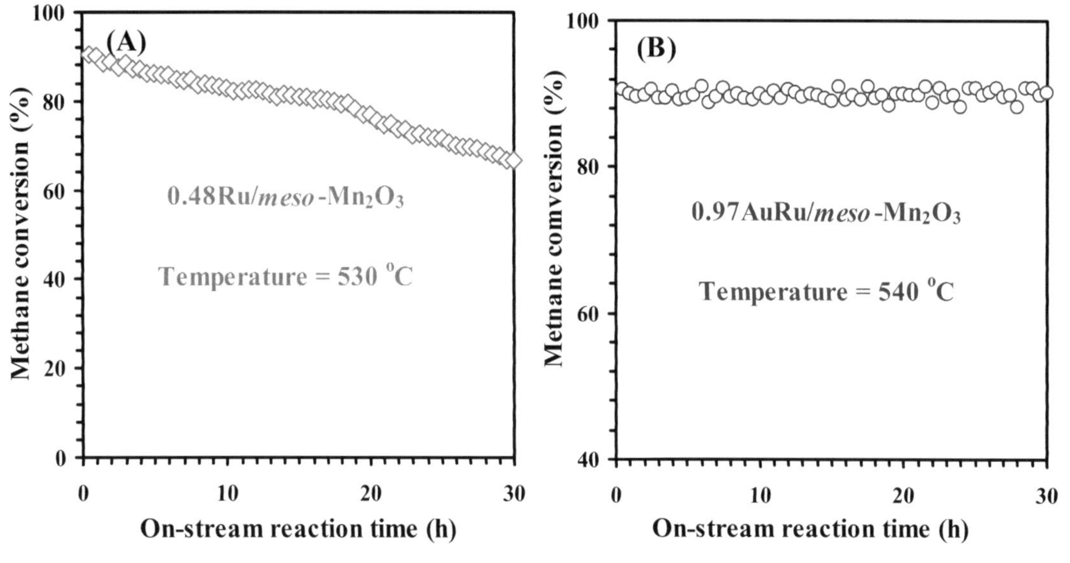

Figure 6. Methane conversion versus on-stream methane oxidation time over (A) 0.48Ru/*meso*-Mn_2O_3 and (B) 0.97AuRu/*meso*-Mn_2O_3 at SV = 20,000 mL/(g h).

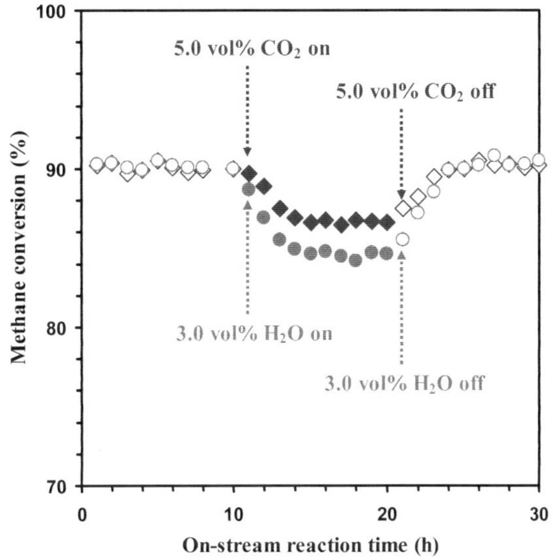

Figure 7. Effect of 3.0 vol% H_2O or 5.0 vol% CO_2 on methane conversion over the 0.97AuRu/*meso*-Mn_2O_3 sample at SV = 20,000 mL/(g h).

To examine the effects of water vapor and carbon dioxide on catalytic activity, we conducted the combustion of methane over 0.97AuRu/*meso*-Mn_2O_3 sample in the presence of 3.0 vol% H_2O or 5.0 vol% CO_2, and the results are shown in Fig. 7. It is clearly observed that with the introduction of 3.0 vol% H_2O or 5.0 vol% CO_2 to the reaction system, methane conversion over 0.97AuRu/*meso*-Mn_2O_3 decreased by 6 or 4%, respectively; after 3.0 vol% H_2O or 5.0 vol% CO_2 was cut off, however, methane conversion was restored to their original value in the absence of H_2O or CO_2. The partial deactivation of this sample was due to formation of hydroxyls or carbonates on the surface of the *meso*-Mn_2O_3 support that could hinder the oxygen exchange between the support and active sites [28]. The above results indicate that the partial deactivation of the 0.97AuRu/*meso*-Mn_2O_3 sample due to H_2O or CO_2 addition was reversible.

4. Conclusions

3D ordered *meso*-Mn_2O_3 and its supported Au, Ru, and AuRu alloy (0.49Au/*meso*-Mn_2O_3, 0.48Ru/*meso*-Mn_2O_3, and 0.97AuRu/*meso*-Mn_2O_3) nanocatalysts could be prepared via the KIT-6-templating and PVA-protected reduction routes, respectively. The *meso*-Mn_2O_3 support possessed a cubic crystal structure and an ordered mesoporous architecture. The noble metal nanoparticles were highly dispersed on the surface of *meso*-Mn_2O_3, and the AuRu in 0.97AuRu/*meso*-Mn_2O_3 was possibly present in an alloy. Among all of the samples, 0.48Ru/*meso*-Mn_2O_3 and 0.97AuRu/*meso*-Mn_2O_3 showed the highest catalytic activity ($T_{90\%}$ = 530–540°C), but the latter exhibited a better thermal stability than the former. Furthermore, the partial deactivation of 0.97 wt% AuRu/*meso*-Mn_2O_3 due to H_2O or CO_2 addition was reversible. It is concluded that high dispersion of AuRu alloy NPs (2–5 nm) on the surface of *meso*-Mn_2O_3 and good low-temperature reducibility were responsible for the good catalytic activity and thermal stability of 0.97 wt% AuRu/*meso*-Mn_2O_3.

Acknowledgements

This work was supported by the National Natural Science Foundation of China (21377008 and 21677004), the National High Technology Research and Development Program ("863" Program) of

China (2015AA034603), and the Foundation of the Creative Research Team Construction Promotion Project of Beijing Municipal Institutions.

References

[1] Yang Z, Grace JR, Lim CJ and Zhang L 2011 *Energy Fuel* **25** 975
[2] Graham LA, Rideout G, Rosenblatt D and Hendren J 2008 *Atmos. Environ.* **42** 4665
[3] Ersson A, Kusar H, Carroni R, Griffin T andJärås S 2003 *Catal. Today* **83** 265
[4] Persson K, Ersson A, Jansson K, Iverlund N andJärås S 2005*J. Catal.* **231** 139
[5] Ciuparu D, Lyubovsky MR, Altman E, Pfefferle LD and Datye A 2002*Catal. Rev.* **44** 593
[6] Choudhary TV, Banerjee S andChoudhary VR 2002 *Appl. Catal. A* **234** 1
[7] Gélin P, Urfels L, Primet M andTene E 2003 *Catal. Today* **83** 45
[8] Kinnunen NM, Hirvi JT, Suvanto M andPakkanen TA 2012 *J. Mol. Catal. A* **356** 20
[9] Yoshida H, Nakajima T, Yazawa Y and Hattori T 2007 *Appl. Catal. B* **71** 70.
[10] Barrera A, Fuentes S, Diaz G, Gomez-Cortes A, Tzompantzi F, Molina JC 2012 *Fuel* **93** 136
[11] Han YF, Chen LW, Ramesh K, Widjaja E, Chilukoti S, Surjami I-K and Chen JS 2008*J. Catal.* **253** 261
[12] Ramesh K, Chen LW, Chen FX, Zhong ZY, Chin JH, Mook HW and Han YF 2007 *Catal. Commun.* **8** 1421
[13] Li JH, Liang X, Xu SC, Hao JM 2009 *Appl. Catal. B* **90** 307
[14] Morales MR, Barbero BP andCadús LE 2006 *Appl. Catal. B* **67** 229
[15] Kleitz F, Choi SH andRyoo R 2003 *Chem. Commun.* 2136
[16] Sun XH, Shi YF, Zhang P, Zheng CM, Zheng XY, Zhang F, Zhang YC, Guan NJ, Zhao DY andStucky GD 2011*J. Am. Chem. Soc.* **133** 14542
[17] Xie SH, Deng JG, Zang SM, Yang HG, Guo GS, Arandiyan H andDai HX 2015*J. Catal.* **322** 38
[18] Santos VP, Pereira MFR,Órfão JJM and Figueiredo JL 2010 *Appl. Catal. B* **99** 353
[19] Gac W 2007 *Appl. Catal. B* **75** 107
[20] Xu R, Wang X, Wang DS, Zhou KB and Li YD 2006 *J. Catal.* **237** 426
[21] Xie SH, Liu Y X, Deng JG, Zhao X T, Yang J, Zhang K F, Han Z and Dai HX 2017*Appl. Catal. B* **206** 221
[22] Teng F, Qu W, Wen GD, Wang ZM, Tian ZJ, Yang XM, Xu P, Zhu YF and Xiong GX 2008*Microporous Mesoporous Mater.* **111** 620
[23] Schwartz WR, Ciuparu D and Pfefferle LD 2012*J. Phys. Chem. C* **116** 8587
[24] Ordóñez S, Bello L, Sastre H, Rosal R and Dìez FV 2002*Appl. Catal. B* **38** 139
[25] Ordóñez S, Paredes JR and Díez FV 2008 *Appl. Catal. A* **341** 174
[26] Ren L, Wang P, Han Y, Hu C and Wei B 2009*Chem. Phys. Lett.* **476** 78
[27] Sasaki K, Tanaike O and Konno H,1998 *Can. Mineral.* **36** 1225
[28] Gremminger AT, de Carvalho HWP, Popescu R, Grunwald J-D and Deutschmann O 2015 *Catal. Today* **258** 470

Effects of Coal Gangue on Cement Grouting Material Properties

J Y Liu and H X Chen

School Of Material Science and Engineering, Chang'an University, Xi'an, Shaan Xi 710054, P.R.China

Corresponding author's email: 306190545@qq.com

Abstract. The coal gangue is one of the most abundant industrial solid wastes and pollute source of air and water. The use of coal gangue in the production of cement grouting material comforms to the basic state policy of environment protection and the circular using of natural resources. Through coal gangue processing experiment, coal gangue cement grouting materials making test, properties detection and theoretical analysis, the paper studied the effects of coal gangue on the properties of cement grouting materials. It is found that at the range of 600 to 700 ℃, the fluidity and the compressive and flexural strengths of the cement grouting materials increase with the rising up of the calcination temperatures of coal gangue. The optimum calcination temperature is around 700 ℃. The part substitution of cement by the calcined coal gangue in the cement grouting material will improve the mechanical properties of the cement grouting material, even thought it will decrease its fluidity. The best substitution amount of cement by coal gangue is about 30%. The fluidity and the long term strength of the ordinary silicate cement grouting material is obviously higher than that of the sulphoaluminate cement one as well as that of the silicate-sulphoaluminate complex cement one.

1. Introduction

The coal gangue is one of the most abundant industrial solid wastes [1]. The emission and pile up of coal gangue not only occupy the cultivated soil, but also pollute the air and river [2], The burning explosion and the collapse of gangue mountain is also one of the safety threats which seriously affects peoples' production and life [3]. It has become the common concerned problem for many researchers to reduce the impact of coal gangue to the environment, and and to convert the industrial waste into available resources [4-6].

Cement grouting material is a kind of slurry cementing building materials, based on cement as the main binding material, characterised by fast hardening, high strengths, high liquidity and micro-expansion [7]. It is widely used in equipment installation, civil engineering, transportation and other engineering reinforcement and repair fields. As the ecological and environmental problems growing, the development of green ecological building materials has become a social consensus. The use of coal gangue in the production of cement grouting material conforms to the basic state policy of environment protection and the circular use of natural resources.

IOP Conf. Series: Materials Science and Engineering **359** (2018) 012023 doi:10.1088/1757-899X/359/1/012023

Through the coal gangue processing experiment, coal gangue cement grouting materials making test, properties detection and theoretical analysis, the paper studied the effects of coal gangue on the cement grouting materials properties, hoping to provide some technical support for the application of coal gangue in cement grouting material.

2. The raw materials, instruments and methods of test

2.1 Main raw materials

2.1.1 Coal gangue

The coal gangue used is from Shaanxi Tongchuan which is in black lump shape (fig. 1). The raw ore sample's X-ray diffraction is shown in figure 2.

Fig.1 photo of coal gangue

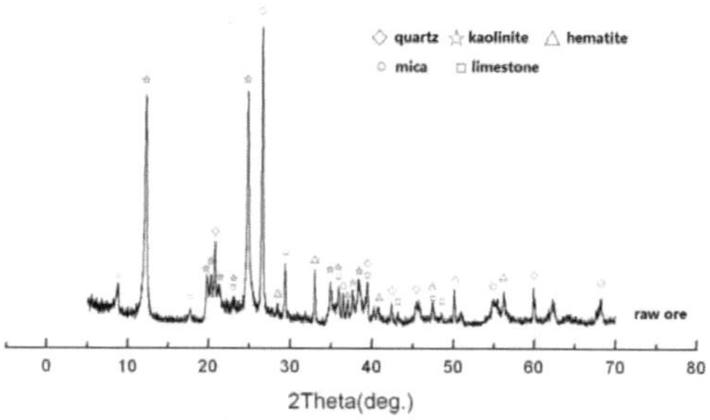

Fig.2 XRD of coal gangue

Its chemical composition is listed in table1,and the mineral constituents are shown in table2.

Table 1 Chemical composition of coal gangue (%)

composition	SiO_2	Fe_2O_3	Al_2O_3	CaO	MgO	SO_3
Coal gangue	42.31	25.69	23.28	2.69	0.7	0.69

Table 2 Minaral constituents of coal gangue (%)

minaral	quartz	kaolinite	hematite	limestone	mica
Coal gangue	53.39	26.41	5.59	9.79	trace

2.1.2 Cement

The silicate cement and sulphoaluminate cement were used in the test. The silicate cement is the Qinling brand 42.5 ordinary silicate cement from Shaanxi province. Its technical indexes are in accordance with the relevant provisions of the current national standards (GB175-2007). The sulphoaluminate cement is a low alkalinity rapid hardening cement from Zhengzhou Cement Co., LTD. The quality of the product meets the requirements of GB20472-2006.

2.1 .3 Aggregate

The river sand was used, which was sieved into coarse sand in the size of 4 ~ 5mm.

2.1.4 Other raw materials

Other raw materials include naphthalene type water reducing agent, calcium sulphoaluminate expanding agent, thickening agent carboxymethyl cellulose and additives silica fume. All the materials are the commercial products from the market.

2.2 Experimental methods and instruments

2.2.1 Coal gangue processing

Crushed the coal gangue in the jaw crusher to 3 ~ 5mm, then put into the roll type crushing-screening machine to get the granularity of 1mm. Followed was to calcined the materials in the box type furnace at 600 ~ 700 °C for 4 hours. After cooling, ground them up to 200 meshes in the mill.

2.2.2 Grouting test specimen preparing

Coal gangue powder (600 ~ 700 ℃ calcined and fine grinded), cement, aggregate, expansive agent and additives were blended together after weighing to get the dry mixtures. The water reducer, thickener and water are mixed to get the water reducing liquid. Mix the dry mixtures and the water reducing liquid to get the cement grouting material. Measure the fluidity of the grouting material with the flowmeter of cement mortar. Meanwhile, pour the grouting material to the molds to get the test specimen. Maintained the specimen until the provisions of age. Detected the mechanical properties on Electric Bending Machine and the Compression Test Machine.

2.3 The basic proportions of the ingredients of the test

See Table 3.

Table 3 Basic mix proportion of test

cement	aggregate	water	Water reducer	thickener	Swelling agent	additives
1	3	0.45	0.01	0.01	0.1	0.03

2.4 Research factors and levels

See Table 4.

Table 4 Test factors and levels

factors	Level 1	Level 2	Level 3
Calcining tempereture of Coal gangue ,℃	700	650	600
Proportion of coal gangue replacing cement	20%	30%	40%
Cement types	Ordinary silicate 1	Sulphoaluminate 1	Ordinary 0.8+ Sulphoaluminate 0.2

3. Experimental results and analysis

3.1 The effection of coal gangue calcining temperature

The replacement of cement by calcined coal gangue was 20%. Aggregate was the coarse sand. Other components used the basic mix proportion.

3.1.1 Effects of calcining temperature of the coal gangue on the fluidity of cement grouting

Effects of coal gangue calcining temperature on the cement grouting material are shown in figure 3.

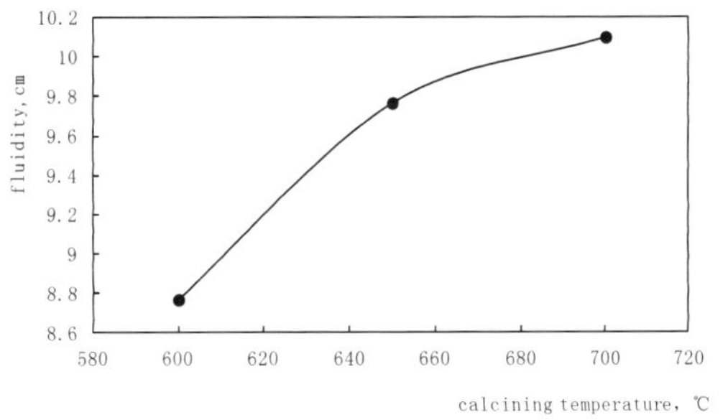

Fig.3 Effects of calcining temperature of coal gangue on the fluidity of cement grouting material

Figure 3 shows, that as the temperature of the coal gangue calcination goes up, the liquidity of the cement grouting rises up. This is because the temperature increases, the residual carbon in the coal gangue burns more sufficiently, and the quantity of amorphous material in the mineral structure increases, which makes the material water absorption lower and the liquidity higher.

3.1.2The effect of coal gangue calcining temperature on the strength of cement grouting

The impact of coal gangue calcining temperature on the compressive strength and flexural strength of cement grouting is shown in figure 4 and figure 5.

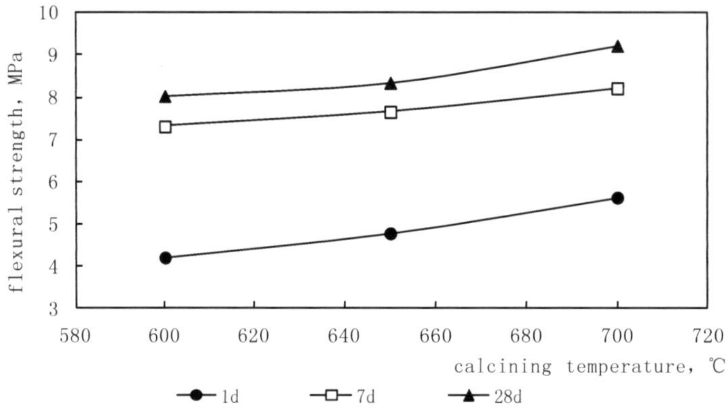

Fig.4 Effects of calcining temperature of coal gangue on the flexurel strengths of cement grouting material

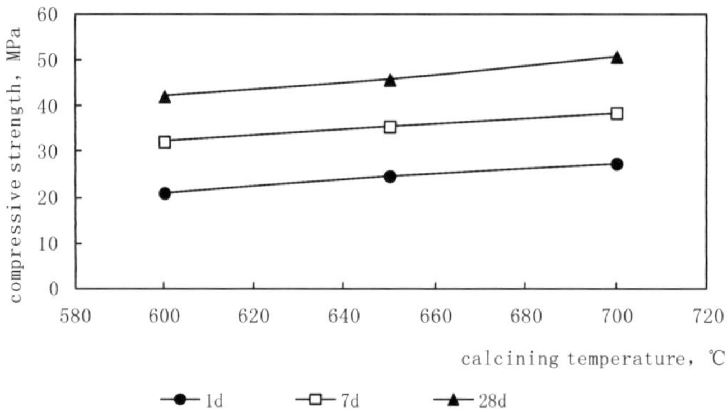

Fig.5 Effects of calcining temperature of coal gangue on the compressive strengths of cement grouting material

As shown in figure 4 and figure 5, the temperature of the coal gangue has a large effect on the strength of the cement grouting. Both the flexural and the compressive strength are all increased with the increase of calcining temperature. Calcination at 700 ℃ is better than that at 650 ℃ and 600 ℃.

Analyze the reason, under the temperature 700 ℃, more residual carbon is burned out, more structure water in the mineral is discarged, and more amorphous SiO_2 and Al_2O_3 are produced from the mineral structure, which increases the volcanic activity of the calcined coal gangue, and improves the strength of the grouting material. While at 600 ℃ ~ 650 ℃, we can imagine, the calcined coal gangue has more carbon residues and less amorphous material and lower volcanic activity, so they has lower strengths.

According to grouting material fluidity and strength test results, coal gangue calcined at 700 ℃ was selected as raw material for the following test.

3.2 Effects of quantity of coal gangue

Taken the 700 ℃ calcined coal gangue and coarse sand in the subsequent test. For other components, the basic mix proportions were used.

3.2.1 The effect of the amount of coal gangue on the flow degree of grouting material

The influence of coal gangue dosage on the flow degrees of grouting is shown in figure 6.

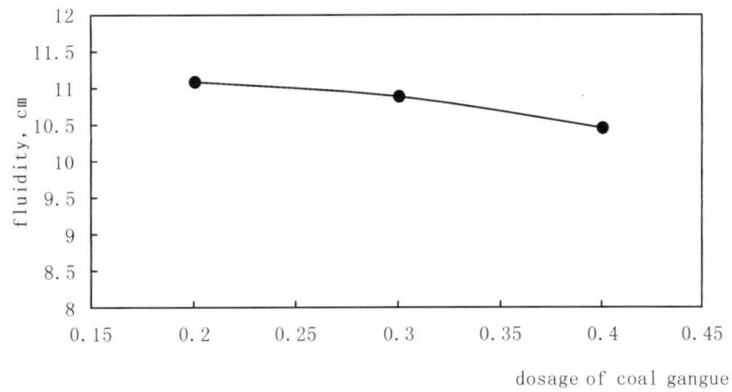

Fig.6 Relationship between coal gangue dosage and the fluidity of cement grouting material

Figure 6 shows, when the dosage of coal gangue is small, the fluidity of the grouting material is high, while the amount of coal gangue is large, the liquidity of the grouting material is low. The flow degree of the grouting material becomes lower with the coal gangue dosage increases.

Thinking about the reason, the calcined coal gangue contains a certain amount of porosity, The water absorption rate of calcined coal gangue is greater than that of cement, as a result, the more the coal gangue replaces the cement, the lower the liquidity of the grouting material.

3.2.2 The effect of coal gangue dosage on grouting strength

The effect of coal gangue dosage on the compressive and flexural strengths is shown in figure 7 and figure 8, respectively.

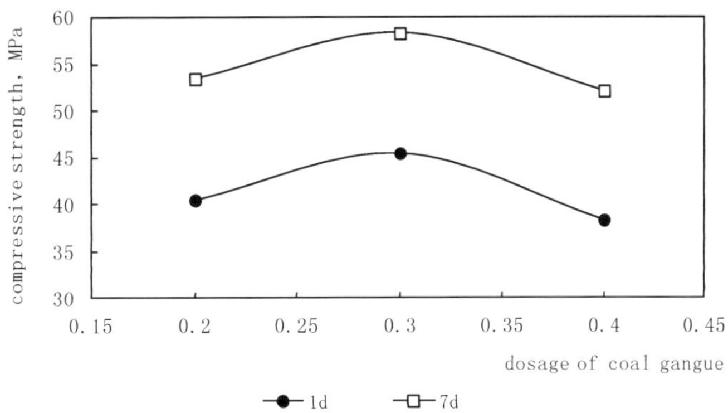

Fig.7 Relationship between coal gangue dosage and the compresive strength of grouting material

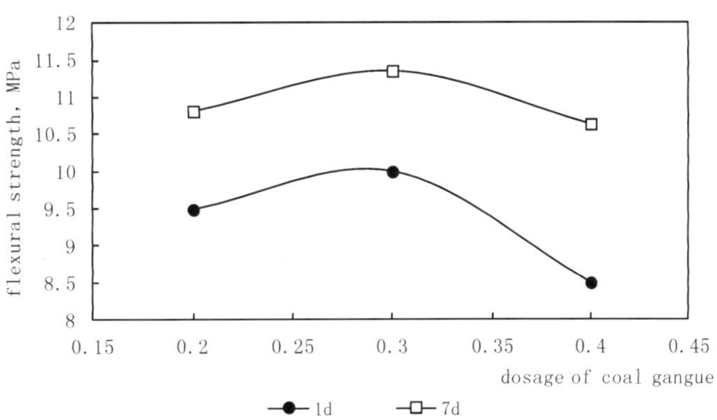

Fig.8 Relationship between coal gangue dosage and the flexural strength of cement grouting material

We can find in figure 7 and figure 8, the influence of coal gangue on the compressive strength and that of the flexural strength are similar to each other. With the increase of coal gangue,both the compressive and the flexural strength of grouting are improved at first. However, when the amount of coal gangue continues to add, and the compressive and the bendig strength decrease. The compressive and bending strength of the grouting materials reaches highest at 30% cement replacement by coal gangue.

Thinking about the reason, the calcined coal gangue, having certain volcanic activities and containing amorphous SiO_2 and Al_2O_3, can react with the calcium hydroxide produced in cement hydration. The silicate gel will be produced to increase the strength of the grouting. The possible chemical reaction is as follows:

$$SiO_2+xCa(OH)_2+ mH_2O=xCaO\cdot SiO_2\cdot mH_2O \quad (1)$$

$$Al_2O_3+3Ca(OH)_2+3H_2O=3CaO\cdot Al_2O_3\cdot 6H_2O \quad (2)$$

In addition, the coal gangue powder has a fine aggregate effect. It can fill the pores in the grouting material system and increase the grouting material compactness, which also increase the strength of the grouting material. Therefore, when the dosage of coal gangue is increased, the strengths of grouting is increased. Nevertheless, since the coal gangue itself contains certain pores, its water absorption rate is higher than the cement, when the replace quantity of coal gangue continues to increase to more than 30%, the liquidity of the grouting will reduce (Fig.6), and the density of the grouting material will also cut down, causing the strengths decrease of the material.From the above test result, 30% cemen replacement by 700 ℃ calcined coal gangue was adopted for further test.

3.3 The effect of the concrete variety

Taken 30% cement replacement by 700 ℃ calcined coal gangue and coarse sand in the subsequent test. For other components, the basic mix proportions were used.

3.3.1 The effect of cement variety on the flow degree of grouting

Figure 9 is the grouting material fluidity test result with 42.5 ordinary silicate cement, sulphoaluminate cement, and complex cement of ordinary silicate cement 0.8 + sulphoaluminate cement 0.2, respectively.

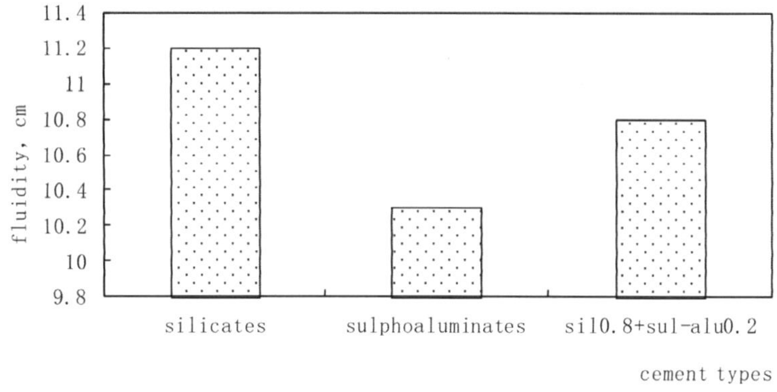

Fig.9 Relationship between cement types and the fluidity of cement grouting material

Figure 9 shows that, the fluidity of silicate cement grouting is significantly higher than that of sulphate aluminium cement and the complex cement. This may be related to the rapid hardening property of the sulphate aluminium cement. Sine after mixing, the latter coagulates quickly and reduces the liquidity. This tells us that the silicate cement is more suitable for the condensation time of the grouting material.

3.3.2 Effect of Cement Varieties on Strength of Grouting

Figure 10 and Figure 11 shows the influence of cement varieties on compressive strength and flexural strength of groutings.

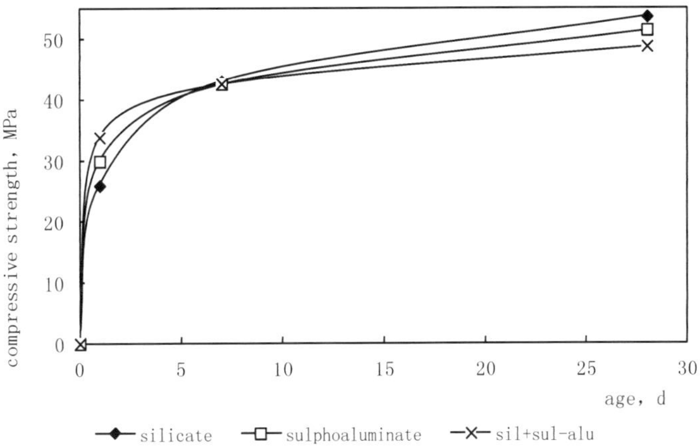

Fig.10 Development of compressive strength of cement grouting material with the age of different cement types

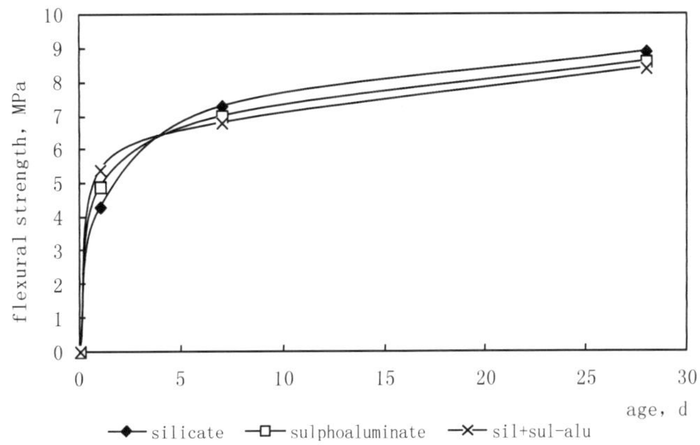

Fig.11 Development of flexural strength of cement grouting material with the age of different cement types

It can be seen from Figure 10 and Figure 11, that the strength of different types of cement grouting material has a similar development rule with age, whether compressive or flexural strength. In the early strength, complex cement grouting material> sulphoaluminate cement grouting material> ordinary silicate cement grouting material. But in the long-term strength, the strengths of the grouting material are in the order of the opposite. That is, complex cement grouting material <sulphoaluminate cement grouting material < ordinary silicate cement grouting material. In the medium-term strength, the three varieties of cement are similar. The compressive strengths are basically the same at 7d, and the flexural strengths are basically the same at 4d.

In principle, the early strength of sulphoaluminate cement is high because of its fast hydration speed and the formation of ettringite ($C_3A \cdot 3CaSO_4 \cdot 32H_2O$). While in the complex cement system, not only sulphoaluminate cement forms ettringite in early hydration, but also calcium gypsum and aluminate existed both in sulphoaluminate cement and ordinary silicate cement reacts to produce ettringite, promoting the growing of early strength of grouting material. The reaction is as follows:

$$3CaO\cdot Al_2O_3\cdot 6H_2O + 3CaSO_4 + 26H_2O = 3CaO\cdot Al_2O_3\cdot 3CaSO_4\cdot 32H_2O \quad (3)$$

Therefore, the early strength of the complex cement grouting material has the highest strength.

However, in the long-term strength, due to the anhydrous sulphoaluminate in sulphoaluminate cement has been consumed by the reaction at early age, its hydration products become less in the latter period, and the strength growth becomes low, so its long-term strength is less than ordinary silicate cement grouting material.

From the early strength of silicate cement, the compressive strength of 1d has reached 25MPa, which can meet the requirement of the grouting. Considering the long-term strength and cost of grouting(The price of sulphoaluminate cement is higher than silicate cement), the ordinary silicate cement was choosed first.

4. Conclusions

(1) In the temperature range 600～700 ℃,with the increase of coal gangue calcination temperature, the fluidity of cement grouting material increase.

(2) In the temperature range 600～700 ℃,with the increase of coal gangue calcination temperature, the compressive and flexural strengths of cement grouting material increase. The best calcination temperature is about 700℃。

(3) With the increase of the dosage of coal gangue, the fluidity of cement grouting material shows a trend of decline.

(4) With the increase of the dosage of coal gangue, the compressive strength and flexural strength of grouting material goes up first, then goes down. The top strengths appear at 30% amount of cement replaced by coal gangue.

Acknowledgement
This project funded by Shaanxi major scientific innovation project（No：2010ZGKG-58）

References

[1] Yong Yang, zhizheng Zhang, dehai Zhang. Introduction to resource comprehensive application of coal gangue [J].Shandong coal technology, 2010, (5) :19-22.
[2] Yunxi Chang, Xueshun Zhu, Changbin Song, et al. The harm of coal gangue and prevention [J]. Journal of geologic disasters prevention and control of China, 2001, 12(2): 39-43.
[3] Yuliang Qiu, Xiaoxu Zhang, Kaiping Liu, et al, Research on mechanical behaviour of cement mortar with high-volume coal gangue[J]. Advanced Materials Research, 2011, 263(3):685-689.
[4] G.Kakali, T.Perraki, S.Tsinilis, et al. Thermal treatment of kaolin the effect of mineralogy on the activity[J]. Applied Clay Science,2001,(20) :73-80.
[5] Haihong Zhang, Kaiping Liu, Lixin He, et al． Research on Application of Activated Coal Gangue in Cement Mortar[J]. Non-Metallic Mines, 2009, 32(1):45-44.
[6] Armesto L, Merino J L. Characterization of some coal combustion solid residues[J], Fuel, 1999, 78(5): 613-618 .
[7] Yanhui Hu, Qingrui Yue, Yaokai Zhang. Application of high strength grouting material in the structure reinforcement and repair technology [J]. industrial architecture,1999,(6) :17-19.

IOP Conf. Series: Materials Science and Engineering **359** (2018) 012024 doi:10.1088/1757-899X/359/1/012024

Cetyltrimethyl Ammonium Bromide as Corrosion Inhibitor for Zinc Used in Hydrochloric Acid

C X Sun[1,2], J J Du[1], Z W Ma [1],C S Huang[2] and J Y Wu[2]
1.College of Mechanical and Electrical Engineering, Qing Dao Binhai University, West Jialingjiang Road, Development Zone of Qingdao, Shandong. 266555 China
2.Institute of Energy Research, Henan Academy of Science, No.29 Garden Road, Zhengzhou R.P. China.450008

Corresponding email: caixiasun@alum.imr.ac.cn

Abstract. A compound inhibitor composed of cetyltrimethyl ammonium bromide (CTAB) and bromohexadecyl pyridine was tested as corrosion inhibitor for zinc in hydrochloric acid. The results of static coupon test show that the compound inhibitor can effectively protect zinc from corrosion and the best concentration ratio is CTAB 50 mg/L and bromohexadecyl pyridine 200 mg/L. The polarization results show that the compound inhibitor will cause a negative shift of E_0 of zinc in hydrochloric acid. The EIS (electrchemical impedance spectra) results show that the inhibitor leads to a bigger radius and has one time constant. SEM results show that the CTAB and bromohexadecyl pyridine form a uniform and compact membrane on the surface of zinc that can protect zinc from corroding effectively.

1.Introduction

Galvanized sheet can be found used in different industries fields. While the galvanized metals used in water-circulating systems usually produce white rust, at the same time scale and slime formed on the surface of cooling tower packing. The scale and slime will reduce the heat transfer effect. so the method to prevent or remove white rust, scale and slime is important.[1] The effectively and quickly method to remove the rust, scale and slime is chemical cleaning. The zinc on sheet will be reaction off, so the use of inhibitors is one of the most piratical method for protection zinc against corrosion especially in acidic media.[2-3]traditional inhibitor for metal such as nitrite and chromate have adverse effects on the environment[4].Organic inhibitors with electron-donating groups such as nitrogen, sulphur and oxygen play an important part in prevent zinc from corroding.[5-7]

This work aims to study the properties of compound inhibitor composed of CTAB and bromohexadecyl pyridine used for zinc in 0.5m HCl.

2.Experimental methods

2.1Materials

Cetyltrimethyl Ammonium Bromide(CTAB); BromohexadecylPyridine; Hydrochloric Acid.

2.2. Inhibitors Prepared

1:Dissolved 4.0g CTAB in water at 70°C, then set volume to 100ml for use. 2:dissolved 1.5g bromohexadecyl pyridine at 50°C, then set volume to 250ml for use, in order to get compound inhibitors just by adding A and B to the solution to get the excepted inhibitor solutions.

Compound: CTAB 50~ 400 mg/L,bromohexadecyl pyridine 200 mg/L;

2.3. Instrument
Potentiostat/galvanostat CS 2305, JSM-6510 scanning electron microscopy (SEM), BT224s Electronic Balance.

2.4 Experimental Procedure
*2.4.1 Weight loss test of zinc.*The corrosion of zinc in inhibitors was evaluated by using the static hanging piece test, which referenced HG/T2387-2007. The corrosion inhibition test was operated at 35°C for 96h. The basic testing solution is 0.5m hydrochloric acid. The experimental solution composed 0.5m HCl and inhibitors. The corrosion rate was calculated refereed as the followed formulas (Eq. 1)

$$K = \frac{W_1 - W_2}{S \cdot T} \quad (1)$$

K—corrosion rate, $g/cm^2 \cdot h$;
W_1— weight of zinc plate before test, g;
W_2— weight of zinc plate after test, g;
S— area of plate, cm^2;
T— time of experimental time, h.
*2.4.2Pre-treatment of plate.*Before add hydrochloric acid into solution play the plate in inhibitor solution for a time. the weight loss of Zinc were tested after pre-treament refered above experimental procedure. The compound inhibitor solution composed of CTAB 300mg/L and Bromohexadecyl Pyridine 100mg/L.
2.4.3 Electrochemical test of zinc. Zinc electrodes were sealed in epoxy resin except the exposed cross-section as working surface. The zinc electrodes were embedded in epoxy resin with an exposed cross-sectional area of 1cm2. The counter electrode in the electrochemical test was platinum while saturated calomel electrode (SCE) was used as the reference electrode. The electrochemical studies were operated on the Potentiostat/galvanostat CS2305. Linear polarization measurement were carried out with a potential perturbation of ±100mV around the open circuit potential (OCP). Electrolyte was prepared by adding different inhibitors into 0.5m hydrochloric acid solution with different concentration.
2.4.4. SEM test. The zinc plates after testing were observed on SEM JSM6510. The operation parameters were set at 20 kV acceleration voltage.

3.Results and discussion

3.1 Corrosion test results
Figure 1 gives the weight loss results of Zn in inhibitor solution. The corrosion rate of Zn (figure a) in different inhibitor solutions show that the influence of concentration of CTAB on corrosion rate is not obvious within the experimental range. The influence of pre-treatment time on the corrosion rate results (figure 1b) show before expose Zn to 0.5m HCl immerse it in inhibitor solution for 1d can effectively prevent Zn from corrosion. The results of pre-treatment derived from the adsorption of inhibitor solution on the An plate surface. Adsorption film can protect the corrosive process between Zn and 0.5m HCl.

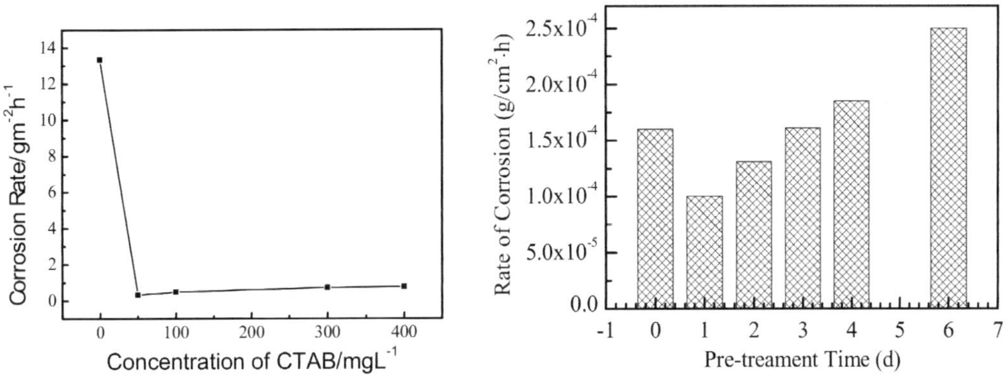

(a) (b)

Figure 1.(a) corrosion rate of Zinc in inhibitor solutions,(b) influence of pre-treatment time on the corrosion rate of Zinc

3.2 Electrochemical test results

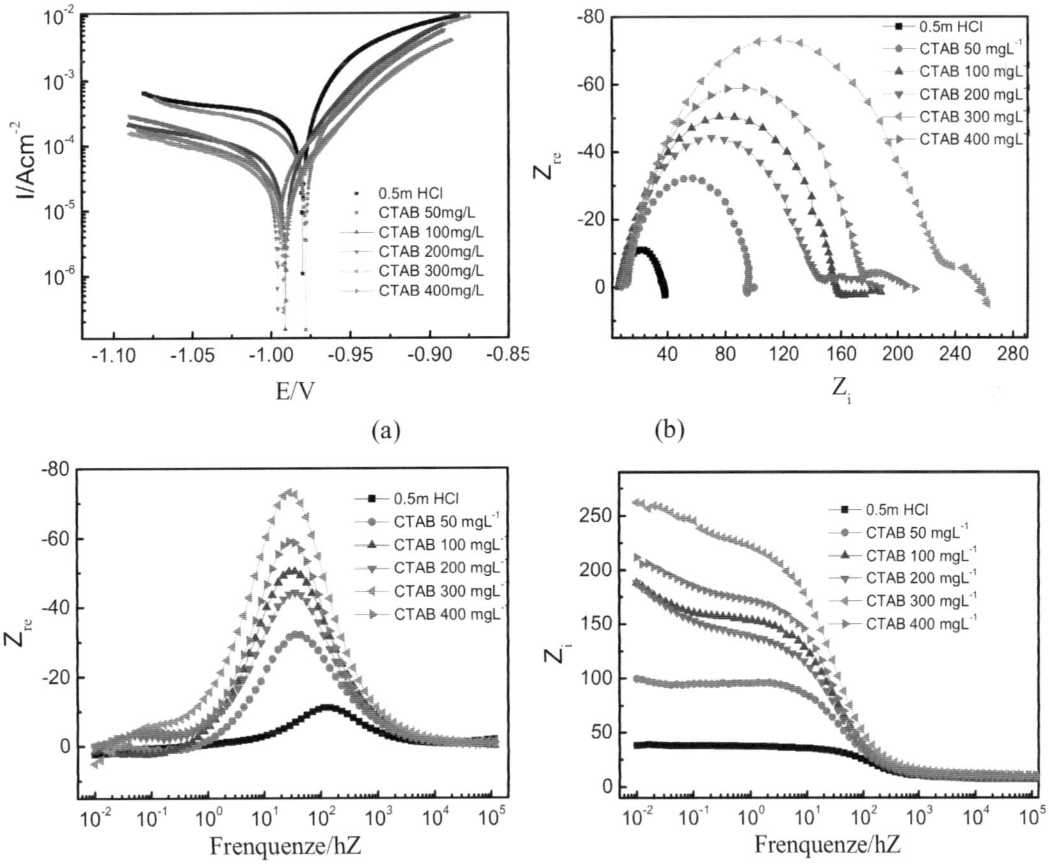

(a) (b)

(c) (d)

Figure 2.(a)polarization plot of Zinc in inhibitor solution,(b) EIS plot of Zinc in inhibitor solution,(c)and(d)nyqiust plots

Figure 2a gives the polarization results of Zinc in different inhibitor solution. The results show that the compound inhibitor make the E_0 of Zn in 0.5m HCl negative shift. So the compound inhibitor in the experimental range is cathodic corrosion inhibitor. Figure 2b shows the EIS results of Zn in 0.5mHCl with different concentration inhibitor addition amount. The compound inhibitor makes the radius of EIS bigger which indicate the addition of inhibitor enhance the anti-corrosion property of Zn in 0.5m HCl. Table 1 shows the parameters of Zn in different inhibitor solutions. The value of R_p change bigger with the increase of inhibitor concentration when the concentration below 300mg/L. R_p indicate the anti-corrosion property.

In summary the corrosion test and electrochemical test results show that the compound inhibitor has better anti-corrosion performance on Zinc in 0.5m HCl.

Table 1. paraments of Zinc obtained from EIS results

Concentration of CTAB(mg/L)	R_s	CPE-T	CPE-P	R_p
0	8.496	$8.86x10^{-5}$	0.8946	15.2
50	7.302	$1.76x10^{-4}$	0.7492	90.5
100	6.634	$1.18x10^{-4}$	0.7736	156.2
200	8.286	$1.48x10^{-4}$	0.7620	144.9
300	10.89	$9.86x10^{-5}$	0.7630	228.6
400	10.79	$9.77x10^{-5}$	0.8070	173.2

3.3 Results of SEM test

Figure.3 shows the SEM photos of zinc in different compound Inhibitor solutions for 96h at 35°C. Figure.3a shows that there are many corrosion pit on the Zn surface when the addition concentration

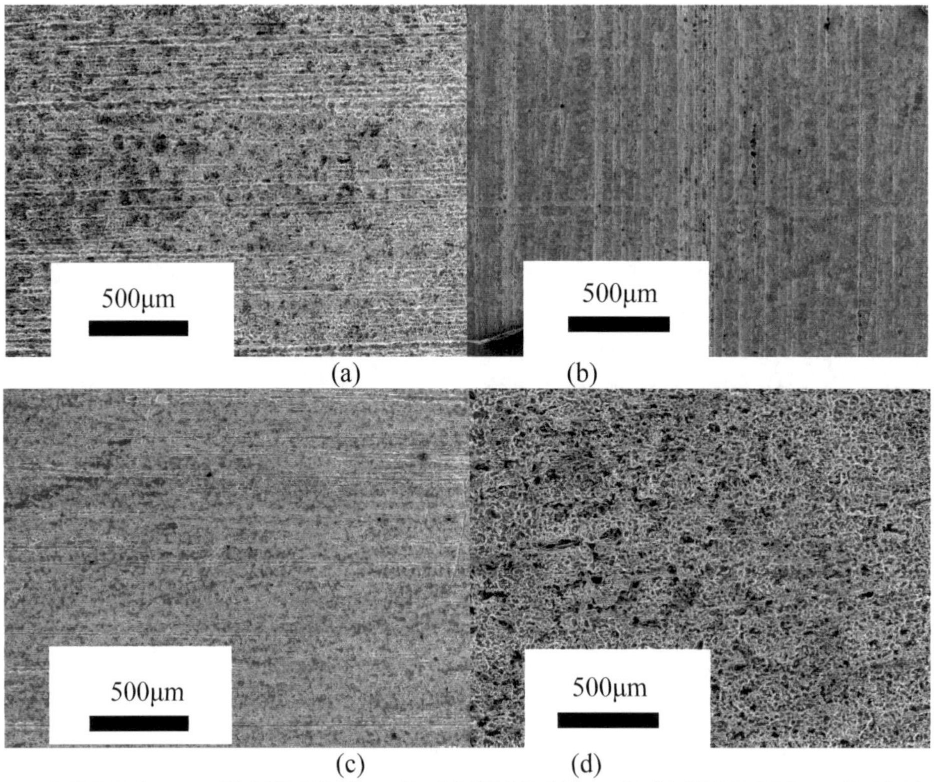

Figure 3 SEM photos of (a)CTAB 50mg/L, (b)CTAB 100mg/L,(b)CTAB 300 mg/L,(d)CTAB 400mg/L

of CTAB is 50mg/L. While the compound inhibitor with the concentration of CTAB is 100mg/L(figure 3b) and 300mg/L(figure 3c) the film on the surface of Zn is more uniform. But when the concentration of CTAB is 400mg/L(figure 3d) the film formed on the surface has many defects.

4.Conclusion

1.Weight loss results show that CTAB had no evident effect on the weight loss results which indicate the macro corrosion of Zn in the 0.5m HCl. The pre-treatment of Zn plate in compound inhibitor solution had big effect on the weight loss results. So when the compound inhibitor solution used in practical can circulate it in the tower system before add 0.5m HCl.
2.The polarization plots and EIS results indicate that when CTAB concentration addition is 300mg/L the inhibitor had most performance anti-corrosion property.
3.SEM results show that the inhibitor solution can form a film on the surface of Zn which can protect it from corroding.

Acknowledgments
This authors would like to acknowledge financial support from the Natural Science Foundation of Shandong (Grant Nos. ZR2017MEM017), Research project of Qingdao Binhai University (Grant Nos.2017KY01), Shandong Higher School Superior Discipline Talent Team Development Program of China. A Project of Shandong Province Higher Educational Science and Technology Program [Grant Nos J17KB 005]

Reference

[1] Fontana M W 5 Yale Ave., Reverside, ClaymontDE (US)2001 US *patent 6183649 B1*
[2] Abiolaa O K and JamesA O, 2010*Corro. Sci.*52661-664
[3] Trabanelli G1991*Corros.*47 410
[4] JiangJ Z, Meng F G, Tang Y G 2003 *Nonferrous Metals*5521-24
[5] Abdallah M 2003*Corros. Sci.*452705–2716
[6] Gadallah A G and Moustafa H M 1992 J. Appl. Electrochem.22 644–648
[7] Morad M S 1999 *J. Appl. Electrochem.*29619–626

IOP Conf. Series: Materials Science and Engineering **359** (2018) 012025 doi:10.1088/1757-899X/359/1/012025

Development, Problems and Countermeasures of Chinese Racing Car Industry

J J Yang[1 2]

[1] School of Automotive Engineering, Wuhan University of Technology, Wuhan, 430070, China

[2] School of Mechatronics Engineering, Wuhan Business University, Wuhan, 430056, China.

E-mail: yangjianjun2004@163.com

Abstract. In recent years, motor car racing has developed rapidly in China. However, under the background of maximum vehicle production and car ownership in China, the racing car industry has a long way compared with that of the developed countries. The paper analyzes the current situation and summarizes the problems of Chinese racing car industry with supporting documentation and review of the literature. The future trend of the development of car industry in China is discussed. On the basis of the analysis and prediction, the strategies to respond to the future racing car industry in China are presented.

1.Introduction

The racing car industry is the related industrial chain from motor vehicle manufacturing, research and development, event organization, catering, entertainment, transportation, to the car derivatives (motor training school, racing clothing, supplies, etc.). Complete industrial chain is helpful to promote the regional economic development, promote the development of the automotive industry and set up the city brand image, etc.

Since the date of birth, motorsport and automotive industries is mutually dependent. The racing car industry is developed with the development of automobile industry, which is one of the symbols of the advanced auto industry. The racing car industry is wing of the auto industry, which has positive influence.

2. The developing history of Chinese motor sports and the present situation of the racing car industry.

A Chinese motor sport was firstly held in 1985, the session of which was from Hong Kong to Beijing. Till now, projects include rally, cross-country, track racing and karting race, etc. and the different sizes of autodromes were built in Beijing, Shanghai, chengdu, zhaoqing, ordos, zhuhai and macau. Motor sport is flourishing. the racing car industry also flourishes. The development of China's car industry probably has experienced three stages:

2.1 starting stage. In 1985 in the first session of rally from Hong Kong to Beijing before China, there were almost no car sports in China, let alone racing car industry. From 1985 to about 2004 years, Chinese racing car industry came to its first stage. the main sign of the stage are: Beijing-Hong Kong rally, Beijing cross-country rally, such as Beijing and Paris, construction of zhuhai circuit, the establishment of Chinese motor sports federation, a group of folk motor sports association and the establishment of the club and the car sport of world championships, etc. in this stage, the development of Chinese auto industry was rapid. Motor racing is a rich sport. motor racing organization system gradually perfected. Racing talent was in scarcity. Racing drivers mainly came from motorcycle and other sports related industries.

2.2 Rapid Development Stage. From 2004 to 2013, the racing car industry came to rapid development phase. The main sign of the phase are: formed seven championships, club management specification, construction of different size of circuit in Beijing, Shanghai, chengdu, zhaoqing, ordos and other places, the release of " reform and development compendium of Chinese motor sports federation from 2013 to 2022". In this stage, Chinese automobile industry developed in a very fast speed. China became the biggest car production and sales country. The population of participating motor sport increased obviously. Keen and in the number of, management capacity and profitability of clubs were strengthened. The competition and training school have trained a large number of racing talents. Participations of high level race in domestic and foreign increased significantly.

2.3 rapid growth stage. With the development of automobile industry and the increase of car ownership, the pursuit of cultural products is more and more colorful. Motorsport flourish and industrialization are an inevitable trend. Thus, Chinese car industry came into the fast growth stage. This phase (2013-2013) is characterized by a steady growth of the auto industry, the similar rate of number of people involved in motor sports compared with the developed countries, motor manufacturer valuing investment on motor sport.. Complete racing industrial chain of manufacturing, research and development, event organization, catering, entertainment, transportation and racing car derivatives, will be formed.

3. Problems of Chinese Racing Car Industry.

Because Chinese auto industry started relatively late and per capita car ownership is low, the car industry's development in China also has a lot to do compared to the auto industry in developed countries. There are a number of problems with Chinese car industry development:

3.1 short of racing car talents. Racing industry chain is very long. all kind of talents are needed. so the shortage of car industry talents is all-around in China, including race car driver and referee, car maintenance personnel and related service personnel. Lack of talent is performed on the insufficient number of people and low quality of the personnel. Taking race car driver as example, "car population" is very important to motor sport. racer usually starts kart racing in their 4-5 years old in developed countries, and gradually promoted to entry-level formula. With relatively mature and perfect training system, racer can obtain broad participation, training and competition, which laid a solid foundation for racing talents reserves. From the realistic conditions of our country, it is impossible to train racer in from the age of 4 to 5.

3.2 bad mass base of the racing car industry. Bad mass base lies in following aspects. on the one hand, the car race is known as expensive sport. Participation ratio is very low on the condition of low life level and no possession of cars; On the other hand, car racing is associated with the auto industry. our country's industrial civilization is not developed. So attention to car race is little. What's more, motor sports' unbalanced regional development leads to few motor racing in less developed areas.

3.3 brand effect of racing car industry isn't obvious and competitive. Auto manufacturers have insufficient understanding of the car industry. They haven't realized the problems of the racing car industrialization from the aspects of industry, culture and market as a whole to know the car industry. They don't care about the car industry, or just for the sake of the car race. They do not to combine motor racing with the auto industry development. So with motorsport, there's no motor sports brand. the car's marketing function is not well used.

4. Countermeasure of Chinese racing car development.

4.1 bring the government macro management function into play. From successful racing experience in Shanghai, zhuhai, Beijing, we can see carrying out car race is the government behavior. The government's core roles lie in not only investment of money and land but also policy support, management, coordination, etc. as for motor racing in the future, the government need to fully recognize the correlation between car industry and racing car industry, the promoting function of racing car industry. Government should encourage the development of car industry in order to realize the auto industry's fast development.

4.2 Enhancing training of the racing industry talent. Talent training of the car industry is a comprehensive system engineering, which need to use the advantages of university discipline group, such as automobile, tourism, management. All disciplines should involve in motor racing to train talents of car design and manufacture, operation management for sporting events, race car driver, the referee, car maintenance, the service of the event, etc. Auto club need to play role of training talents of racing and cultivating the students' practical ability. The mature experience of developed countries needs to be drawn. For instance, British racing car industry ranks no.1 in the world. Its car industry has stood on the high ground. There are a lot of experience can be used for, such as through cooperation with British colleges and universities, joint training racing talents, etc.

4.3 Improving and standardizing the motor sports club to raise the level of event organization and operation. On the basis of the existing management framework, we will continue to improve the construction about cars, including the driver, the referees, coaches, sites, the vehicle registration and management system, etc. To strengthen the construction of the standardization of the racing club, draw lessons from foreign car club success experience, in order to establish perfect rules and regulations, ensure the lawful rights and interests of the driver, the club and the sponsors. All measure above are important for the healthy development of motor sport in China.

4.4 Cultivating the car culture and promoting the development of car industry. With the rapid development of automobile industry in China in recent years, car culture has been formed. Based on this, the car culture needs to be promoted greatly. Promotion ways include: firstly, making full use of the motor sports column of the TV, radio, magazines, websites, video introduces the history and culture of motor sports; secondly, carry out the interesting motor sports, such as around the pile, racing, assembled sexual activity and all kinds of safe driving training activities, in order to let more automotive enthusiasts to participate in the movement, reduce vehicle movement into the threshold.

4.5 Encouraging domestic car to carry out motor sport especially self-owned brand car. Improve the domestic car brands' participation, especially the self-own brand car, to form a good interaction of domestic automobile industry and motor sports. Guide domestic auto especially self-own car brands, to become famous brands. These are necessary way to becoming famous brand. In recent years, there are a growing number of local brand cars in 7 big race, which is a good phenomenon. Motor sport management institutions and relevant government departments need to provide more platforms to attract local brands to join, to realize the benign interaction of the domestic automobile industry and vehicle movement.

5.Summary

Looking forward to the future, as continuing growth of Chinese auto industry, motor sports culture is deeply rooted in the hearts of people. car manufacturer pay more attention to motor sport unprecedentedly. Chinese auto sports will become more popular, and the car industry will become the wings of boosting the development of auto industry.

6.Acknowledgments
This work was supported by the The humanities and social sciences research project of Hubei Provincial Department of Education: The research on the impact of the development of the racing industry on improving the competitiveness of Wuhan City.

References

[1] State general administration of sports car motorcycle sports management center. Chinese motor sports association 2013 annual work report, 2013.12. http://www.autosports.org.cn/

[2] Feng Wang. Study of Development Status and Countermeasures of Chinese Motor Sport [J]. Sports Culture Guide, 2014, (4): 24-26.

[3] Lei Shi. China Circuit Championship State of the Art [M]. Beijing, Capital Institute of Physical Education, 2008.

[4] Yuan Gao. China auto field championship business strategy research [M]. Wuhan, Wuhan Institute of Physical Education, 2008.

[5] Yongxiang Tang. Analysis on National Automobile Market Prospect at the Angle of F1 Match in Shanghai [J]. Journal of Nanjing Institute of Physical Education, 2005, 19(4): 43-45.

[6] James Meredith. Study of Combustion Performance of Biodiesel for Potential Application in Motorsport[A]. .International Journal of Intelligent Information and Management Science(Volume 5,Issue 2,April 2016)[C],2016:8.

[7] Zhang Keke. The Design and Optimization of FSC Racing Suspension Rocker System[A]. Chinese Construction Machinery Society （CCMS）.Proceedings of the 2015 International Conference on Advances in Construction Machinery and Vehicle Engineering(ICACMVE'2015)[C].Chinese Construction Machinery Society （CCMS）,2015:6.

[8] Bin Zhu. Design and Optimization of FSAE Race Car suspension system[A].Proceedings of 2015 3rd International Conference on Mechanical Engineering and Intelligent Systems(ICMEIS 2015)[C].2015:4.

[9] Jun Ni. The Suspension Optimization of FSAE Racing Car Based on Virtual Prototyping Technology[A]. Society of Automotive Engineers of China(SAE-China)、International Federation of Automotive Engineering Societies(FISITA).FISITA 2012 World Automotive Congress Proceedings --Volume 7:Vehicle Design and Testing[C].Society of Automotive Engineers of China(SAE-China)、International Federation of Automotive Engineering Societies(FISITA):,2012:10.

IOP Conf. Series: Materials Science and Engineering **359** (2018) 012026 doi:10.1088/1757-899X/359/1/012026

Relationship between Maceral of Coal and Coal-bed Methane adsorption ability in Sihe Coalmine of Qinshui Basin, China

M S Wang[1], G G Zou[2,*] and R B Zhu[2]

1,School of Mechanics and Architectural Engineering, China University of Mining and Technology (Beijing), Beijing, 100083;

2,College of Geoscience and Surveying Engineering, China University of Mining and Technology (Beijing), Beijing, 100083

*cumtzgg@foxmail.com

Abstract. Maceral components and its content of coal were divided based on the microscopic characteristics of coal. The Langmuir volume and the Langmuir pressure were tested, and the Langmuir volume represents the adsorption capacity of coal. The formation of coal bed methane is affected by the partition of the maceral components in coal. Therefore, the relationship between maceral composition and coal bed methane adsorption capacity of coal was analyzed. The results show that the maceral components of coal are dominated by vitrinite and inertinite in the study area, and the content of inertinite is below 32%. The vitrinite group has a negative linear correlation with the Langmuir volume, and the inertia composition has a positive linear correlation with it. The cellular structures in the inertinite are the main site of coalbed methane enrichment. The microstructure of coal affects the coal bed methane content and the stage of hydrocarbon generation in coal. This indicates that the microstructure of coal is one of the important factors influencing the adsorption capacity of coal seam..

1. Introduction

Coalbed methane is produced from coal and stored in coal seams, it is a kind of unconventional natural gas mainly in an adsorbed state. The main component is methane (CH4), which was seemed as a new, clean and safe energy.Coalbed methane development and utilization can help to reduce the effective emissions of methane in coal mines, while improve coal mine production safety and efficiency.

The development of coalbed methane in China began in the late 20th century, and the Qinshui Basin is one of the coalbed mehane enrichment regions in the China, due to its unique geological features and resource reservation. The coalbed methane adsorption was studiedfrom different perspectives, such as the adsorption capacity, permeability, crack, stress and the another properties of coal, but the study on the adsorption capacity of the macerals of coal is still ambiguous. The coalbed methane was deposited in the coal seam with adsorption state, dissociate state and dissolved state. For the anthracite, it is mainly composed of the adsorption state gas. Meng Zhaoping[1,2]and Bustin[3]think that there is a certain relationship between maceral components and adsorption ability.Su Xianbo through the analysis of the maceral-components of coal samples in different areas, it was concluded thatthe adsorption capacity of the vitrinite was the strongest, the inertia was the second, the exinite was weaker[4]. He Xiao-guang analyzes the content and type of maceral-components in coal, and thinks that the maceral-components have a great influence on the adsorption of coalbed methane[5]. Some

IOP Conf. Series: Materials Science and Engineering **359** (2018) 012026 doi:10.1088/1757-899X/359/1/012026

scholars believed that the maceral of coal is not directly related to the content of coalbed methane.Zhang Feiyan collectedgeological information on different exploration areas in China, and found that the Langmuir volume and vitrinite content did not change continuously, and the reflectivity of vitrinite was closely related to the Langmuir volume. When the vitrinite reflectivity is large, the inertia group has a higher effect on the adsorption of coalbed methane, and the content of the vitrinite has no effect on the adsorption of coalbed methane[6]. We studied the linear relationship between the maceral components of coal and the isothermal adsorption parametersto study its effect on the methane adsorption capacity of coal.

2. Experimental samples and methods

2.1 Experimental principles and methods for maceral composition

The experimental sample is from coal #3in Sihe Coalmine of the Shanxi Qinshui Basin, which is a kind of anthracite with high metamorphic and simple structure. According to"Method of preparing coal samples for the coal petrographic analysis(GB/T16772-2008)", the pulverized coal-fired powder were producedas shown in Fig. 1.

Fig. 1 15 samples of coal seam #3 from Sihe Coalmine, China

The pulverized coal sample was placed under the polarized light microscope. In the case of polarized or incomplete orthogonal polarizations, we divide the maceral components and inorganic minerals, and count thevolume fraction of maceral componentsusing the statistical method.

(1)The microscope and the computer were turn on, the sample was put on the stage and fixed.The reflected light was turn on, and lens were focused and corrected to the center.The objective lens were adjusted to 20 times, the eyepiece 110 times.the light source and the visual threshold aperture were adjusting to make a more clear image.

(2)It was ensured that noless than 500 effective measuring points are evenly covered with the whole film. Counting was started from one end of the line, when the line was finished and then changed to a new line. The cross point of the eyepiece should fall at the center of the object to be viewed as a valid point.

(3) When the cross fell on the boundaries of different ingredients, it was started from the upper right quadrant, the apparent threshold fully filled withthe maceral components was selected in the clockwise order.

2.2 Isothermal adsorption experiment

Through the experiment of isothermal adsorption curve, we can assess the adsorption capacity of coal[7].According to the geological conditions of this study area, the test temperature is 30 ℃ and the methane concentration is 99.9%. The isothermal adsorption test was carried out under constant temperature. The experimental pressure range was 0 ~ 7 MPa.Adsorption equilibrium time is not less than 12hours. Before the isothermal adsorption test, the coal samples were treated with equilibrium moisture.

The experimental steps are as follows:

(1) Experimental pressure point distribution: the equilibrium pressure was not less than 8MPa, the experimental pressure pointswere not less than six.

(2) Inflammation: Placed the tested sample in the cylinder, opened the regulating valve and the reference cylinder valve, filled the system with methane , adjust pressure in the reference cylinder to the target pressure.

(3) Data acquisition: when the target pressure and the temperature were stable, opened the acquisition program.Time, pressure, temperature and other related data were collected, and the data were recorded as files.

(4) Adsorption equilibrium time: when the methane adsorption value does not change, it is stable adsorption, the adsorption time were determined.

(5) Repeat (3) to (4) step, Experiment were carried out one by one pressure point until the last pressure point experiment ends.

3. Experimental results and analysis

3.1 Observation and analysis of macerals in coal

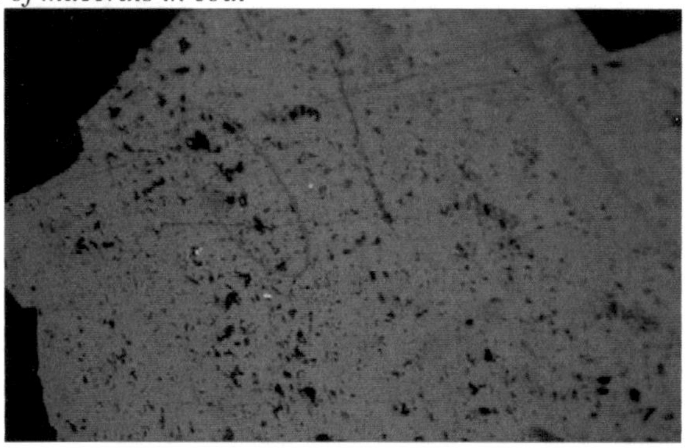

Fig.2 Inertinite and sulfide minerals in coal sample No. 8from Sihe Coalmine, China

Fig.2 shows inertinite with bright-white or yellow-white under the reflection light.It mainly derived fromplant stem, roots, branches of xylem, which were suffered from the carburization[8,9]. In Fig.2the minerals in yellow were the sulfide minerals. It were pale yellow-white with the protruding surface, sometimes it also shown the "sunflower" petal-shaped in the surface under the microscope[10]. The clay minerals, as shown in Figure 2, are gray or dark gray under the microscope, the protrusions are not obvious, the surface is particle, and sometimes uneven.

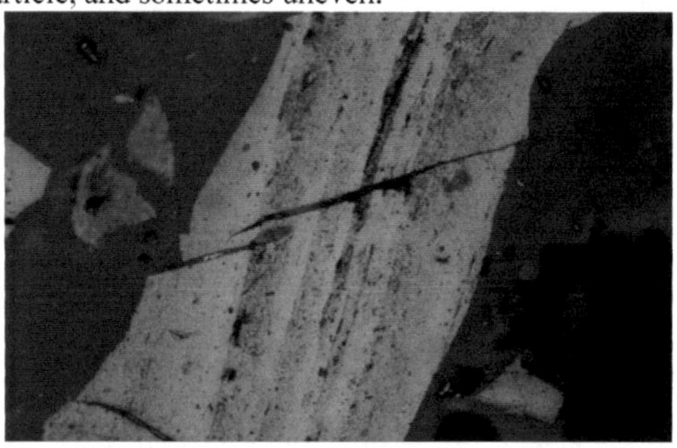

Fig.3 Vitrinite in coal sample No. 8from Sihe Coalmine, China

The vitrinite wasone of the most common and important maceral components of coal. It was formed by the condensation of the roots, stems and leaves of the plant under the condition of water reduction[8,9]. Vitrinite is gray to dark gray, and sometimes it is difficult to distinguish between vitrinite and semifusinite in the high rank coal. Fig. 3 shown the sample is in highly metamorphic, and homoeollinite is banded or lenticular appearance with clear outline, sometimes it is perpendicular to the surface of the cracks. One of the reasons why homoeollinite did not show the cellular structure was that it is difficult to distinguish the refractive index and color between the humic gel filling the cell cavity and the gel cell wall[11].

Table.1 Quantitative statistics of maceral components of different coal samples from Sihe Coalmine, China

Coalsamplenumber	Organic Component (%)		
	Vitrinite	Inertinite	Subtotal
1	73.1	26.9	100
2	82.0	18.0	100
3	87.2	12.8	100
5	81.2	18.8	100
7	88.6	11.4	100
8	86.6	13.4	100
10	91.5	8.50	100
11	78.3	21.7	100
12	85.5	14.5	100
13	87.6	12.4	100
14	71.1	28.9	100
15	68.9	31.1	100
18	75.5	24.5	100

3.2Experimental results and analysis of isothermal adsorption

As shown in Fig.4,The adsorption capacity of samples from coal seam #3 had an obvious difference. It can be seen from Table 1 that the vitrinite content of coal sample No. 13 and No. 14 is different.In Fig.4, coal samples No. 13 and No. 14 shown gradually increasing methane adsorption ability with the increase of pressure. This indicates that the adsorption amount of methane increases with the pressure increase, and this rule still exists in the case of different vitrinite.

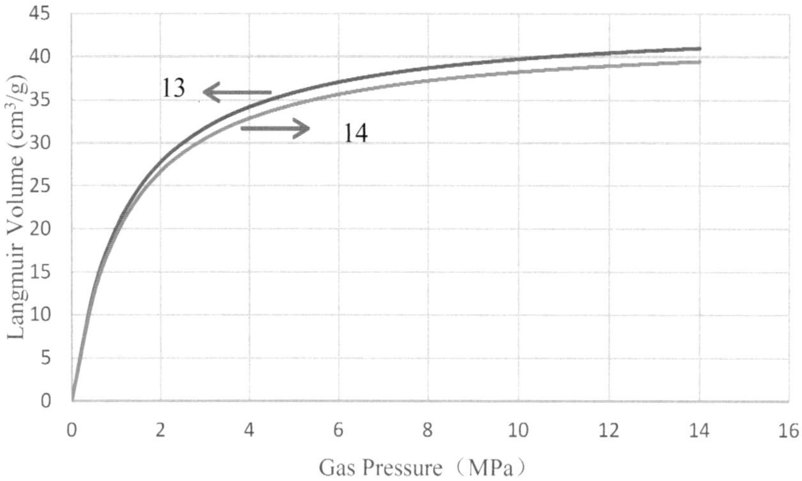

Fig. 4 Isothermal adsorption curves of coal samples 13 and 14

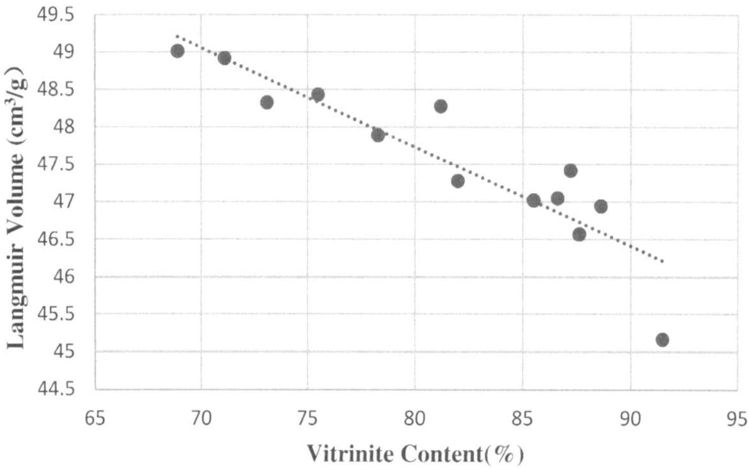

Fig.5 Linear relationship between Langmiur volume and vitrinite content

InFig5. It can be concluded that: with the increase of vitrinite content, the adsorption capacity of coal bed methane decreases gradually, which is negatively correlated.In the vitrinite content of 68.9% ~ 91.5%, the vitrinite content were almost linearly changing with the Langmuir volume, which range change is 4.21cm^3/g. For the same coal sample, the more the vitrinite content, the weaker the adsorption capacity.

Coalbed methane is mainly adsorbed on coal matrix surface. In the high rank coal in Sihe coal mine, it can be seen from Table 1 thatthe vitrinite content is higher than the inertia content. Because of the higher metamorphic degree, the non-filled cavities in the inertia were increased. The methaneadsorption capacity of coal is negatively correlated with the content of vitrinite, and basically conforms to the linear relationship.

4. Conclusion

Taking the high rankcoal #3 from the Shihe Coal Mine in Qinshui Basin as an example, the maceral components of coal were obtained under the microscope. The results shown that the vitrinite content in the coal seam #3 of the Sihe Coalmine is dominant, because it is more than 68%.The average of the inertia is about 16% and the maximum is not more than 32%. The inertia contained a large number of fusinite. While the exinite only accounted for a small proportion of coal, so it is

IOP Conf. Series: Materials Science and Engineering **359** (2018) 012026 doi:10.1088/1757-899X/359/1/012026

negligible.The vitrinite group has a negative linear correlation with the Langmuir volume, and the inertia compositionhas a positive linear correlation with that. The cellular structures in the inertiniteare the main site of coalbed methane enrichment. The maceral content of coal affects the coalbed methanecontent, and the stage of hydrocarbon generation in coal.

This indicates that the maceral content of coal is one of the important factors influencing the adsorption capacity of coal seam in Sihe coalmine.This study is of great significance to the exploration and developmentof coalbed methane. Especially, the metamorphic degree of coal is relatively high in China, which is different from that of low and middle rank coal.

References

[1] Meng Z, Tian Y, Li F. Geology theory and method of coalbed methane development[M]. Science Press, 2010.

[2] Meng Z, Liu S, Wang B. Adsorption Properties and Pore Structure Characteristics of Different Structure Coal[J]. Journal of Coal Science, 2015.

[3] BustinRM,ClarksonCR. Geological Controls on Coalbed Methane Reservoir Capacity and Gas Content[J] . International Journ al of Coal Geology,1998,38 (1/2) :3-26.

[4] Su X,Lin X. Coalbed methane geology [M]. Coal Industry Press, 2009.

[5] He X. Study on coal maceral-component in coalbed methane resource prediction [J]. Yunnan coal, 2012.

[6] Zhang F, Cheng W, Ji L, et al. Effect of maceral composition of coal on adsorption capacity of coal reservoir[J]. China coalbed methane, 2011, 8 (2): 29-31.

[7] Zhang X, Sang S, Qin Y, et al. Isotherm adsorption of coal samples with different grain size[J]. Journal-China University of Mining and Technology-Chinese edition, 2005, 34(4): 427.

[8] Han D. Coal Petrology of China[M]. China University of Mining and Technology Press,1996.

[9] Yang Y, Ren D. Atlas for coal petrography of China[M].China University of Mining and Technology press,1996.

[10] Zhou F. Study on petrophysical and physical simulation of coalbed methane reservoir in Qinshui basin[D]. PhD thesis, 2013.

[11] Fu X, Qin Y, Wei Ch. Coalbed methane geology[M].China University of Mining and Technology Press,2006.

IOP Conf. Series: Materials Science and Engineering **359** (2018) 012027 doi:10.1088/1757-899X/359/1/012027

Microstructural evolution and mechanical properties of a low alloy high strength Ni-Cr-Mo-V steel during heat treatment process

C Wu[1*] and S Han[2]

[1]National-Local Joint Engineering Laboratory of Intelligent Manufacturing Oriented Automobile Die & Mould, School of Mechanical Engineering, Tianjin University of Technology and Education, NO.1310 Dagu South Road, Tianjin 300222, China

[2] School of mechanical engineering, Yanshan University, Qinhuangdao, 066004, China

E-mail:zjkwuchuan@126.com

Abstract. In order to obtain an optimal heat treatment for a low alloy high strength Ni-Cr-Mo-V steel, the microstructural evolution and mechanical properties of the material were studied. For this purpose, a series of quenching and temper experiments were carried out. The results showed that the effects of tempering temperature, time, original microstructure on the microstructural evolution and final properties were significant. The martensite can be completely transformed into the tempered lath structure. The width and length of the lath became wider and shorter, respectively with increasing temperature and time. The amount and size of the precipitates increased with temperature and time. The yield strength (*YS*), ultimate tensile strength (*UTS*) and hardness decreased with temperature and time, but the reduction in area (Z), elongation (*E*) and impact toughness displayed an opposite trend, which was related to the morphological evolution of the lath tempered structure.

1.Introduction

Recently, with the rapid development of ship building, drilling platform of offshore oil & gas, and aerospace manufacture, the low alloy high strength Ni-Cr-Mo-V steels have been used widely due to their promising mechanical properties [1]. In these fields, many components have to be possessed with proper microstructures, excellently mechanical properties and good dimensional stability [2]. In order to satisfy these requirements, the materials usually have to be subjected to a series of heat treatments. It is well-known that the microstructure takes a complex evolution during the processes. Moreover, the microstructural evolution directly affects the mechanical properties. Therefore, it is necessary to investigate the effects of heat treatment parameters on the microstructural evolution and mechanical properties.

In the past years, many authors have paid attention on this subject. Tao et al [3] investigated the effect of cooling method on the microstructural characteristics and mechanical properties of a as-cast high-strength low-alloy steel. Peng et al[4] studied the microstructural evolution and strength of 9Cr-3W-3Co steel under different aging conditions by experiments. Shin et al [5] investigated the microstructural evolution and the tensile behavior of precipitation hardened martensitic steel after aging treatment. Maji et al [6] carried out some experiments to ascertain the effect of cooling rate on microstructure and mechanical properties of an eutectoid steel subjected to a heat treatment.

Although these studies investigated the microstructural evolution and mechanical properties of steels, some conclusions, obtained from these studies, may not be valid for the present material. Therefore, it is necessary to investigate the microstructural evolution and its effects on the mechanical properties of the Ni-Cr-Mo-V alloy during heat treatments.

2.Material and experimental procedure

A 10Ni5CrMoV alloy was used in this study. The chemical composition of the steel is listed in Table 1. Its initial microstructure was tempered martensite and the grain number was 8.5, as shown in Fig.1(a). The material was subjected to quenching and tempering treatments. This procedure has been showed in Fig.1(b).

The heat-treated material was subsequently subjected to a standard metallographic grinding, polishing and etching (by a solution of 97% alcohol and 3% nitric acid). The material was shot in the center zones by an optical microscope (Axiovert 200 MAT) with magnifications from 50 to 1000 times. Three to five micrographs at each condition were used to determine the grain size, area percentage and lath width (ASTME112-10).

The hardness, tensile and impact tests were performed. A Vicker's hardness testing machine (Tukon 2100B) using 3-kgf load was used. The standard tensile specimens were prepared and conducted on a tensile testing equipment (CSS-44300) with a load of 10 kN and loading strain rate of 10-3 s-1. The Charpy V-notch impact specimens with 10 mm×10 mm×55 mm size were prepared according to GB/T229-2007 [10]. The impact specimens were performed at temperatures of -120℃and -80℃on an impact equipment with a capacity of 500 J.

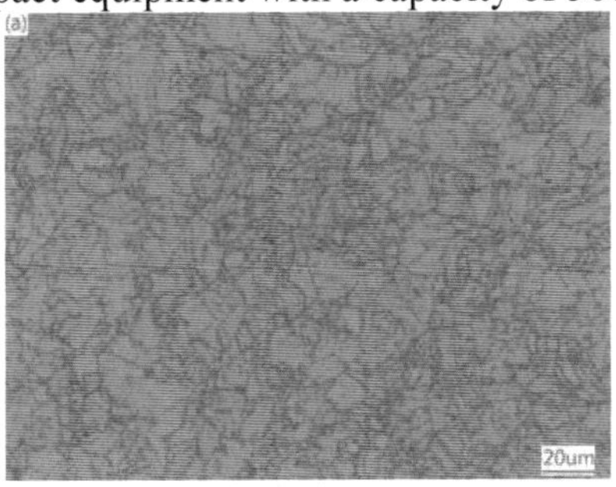

(a). The initial microstructure

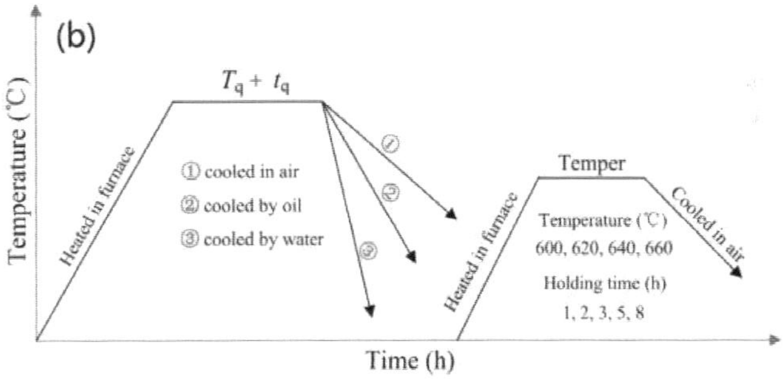

(b) heat treatment procedure

Figure 1 Material and experimental procedure

Table 1. The chemical composition of the material

Element	C	Si	Mn	Cr	Ni	Mo	V	Al	Co
wt. (%)	0.09	0.30	0.50	0.55	4.55	0.50	0.05	0.021	0.1
Element	S	P	H	O	N	Fe			
wt. (%)	0.002	0.005	0.147	0.0024	0.0073	balanced			

3.Results and discussions

3.1 Microstructural evolution during quenching

The material was subjected to a quenching (including water-quenched, oil-quenched and air cooled) and tempering process. During the process, the microstructure had a complex evolution, as shown in Fig.2. It can be seen that a complete lamellar martensitic structure can be obtained through the water-quenched treatment (as shown in Fig.2(a)). For the lamellar structure, the length and width were very sensitive to the cooling rate. The sizes of primary lamellar structure were larger than that of the secondary structure, because the former had more time to grow than the latter. When the material was subjected to the oil quenched, the original structure was replaced by a large amount of lath martensite and a little of upper bainite, as shown in Fig.2(b). The sizes of the structure were obviously larger than those of lamellar

IOP Conf. Series: Materials Science and Engineering **359** (2018) 012027 doi:10.1088/1757-899X/359/1/012027

structure Fig.2(c) showed the optical micrograph of the material subjected to air cooled. The final microstructure comprised a little martensite and much bainite.

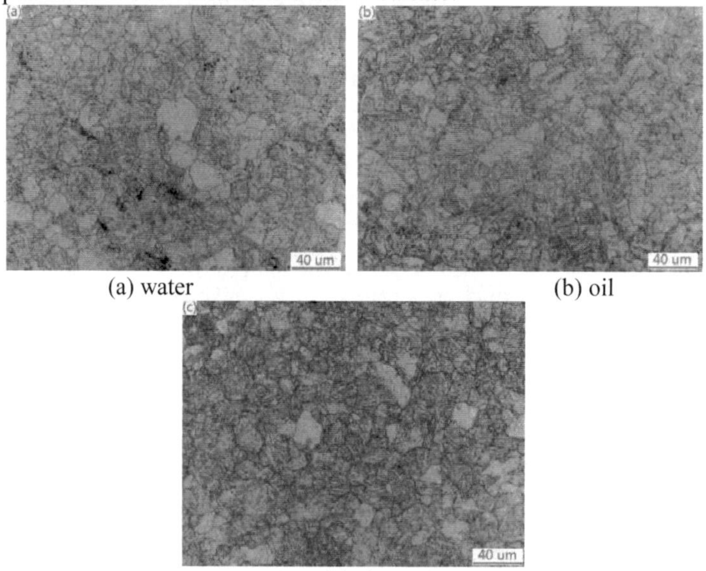

(a) water (b) oil

(c) air

Figure 2. The microstructural evolution of the material subjected to different quenching method.

3.2 Microstructural evolution during tempering

After the water-quenched treatment, the material was subjected to a tempering treatment. The tempering temperatures were 600℃, 620℃, 640℃ and 660℃ and the time was from 1 to 8 hours. As shown in Fig.3 (at 600℃), a tempered structure can be obtained. Clearly, the tempered structure gradually became coarsening with the increasing time. According to Dong's results [7], this phenomenon can be attributed to the dislocation annihilation and recombination of two subgrain boundaries. At 620℃, as shown in Fig.4, the size of the tempered structure was larger than that at 600℃. When the tempered temperature increased to 640℃, as shown in Fig.5, a large amount of carbide precipitates can be observed in the microstructure at 3 hours. The needle-like precipitates were much tiny (less than 0.1 um). The size and amount of the precipitates increased with the time. The tiny needle-like structure gradually developed into the short rod-like or equiaxed. The length of the tempered lath decreased while its width increased with the temperature and time, which resulted in the decreasing of length-width aspect ratio of the lath. This phenomenon has been explained by Guimarães et al [8]. They believed that the aspect ratio was proportional to the elastic strain energy of the lath. With the increasing of temperature and time, the elastic strain energy was gradually relaxed by the plastic deformation within and the transformation of self-accommodating groups of martensite units. When material was tempered at 660℃, as shown in Fig.6, a large number of precipitates were formed in the tempered lath since 2 hours.

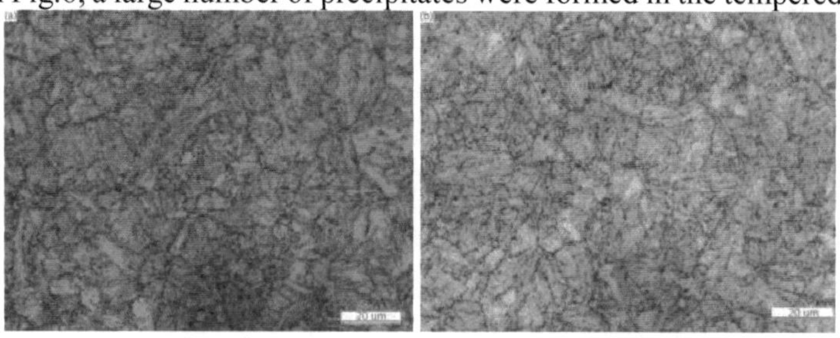

(a) 1 hour (b) 2 hours

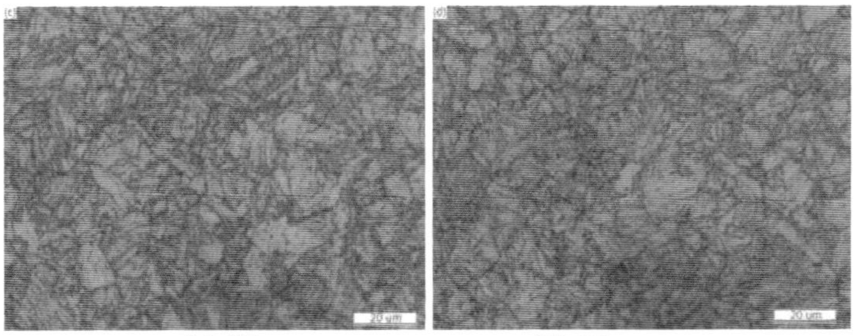

(c) 5 hours (d) 8 hours

Figure 3. The microstructural evolution of the material tempered at 600℃ for different time

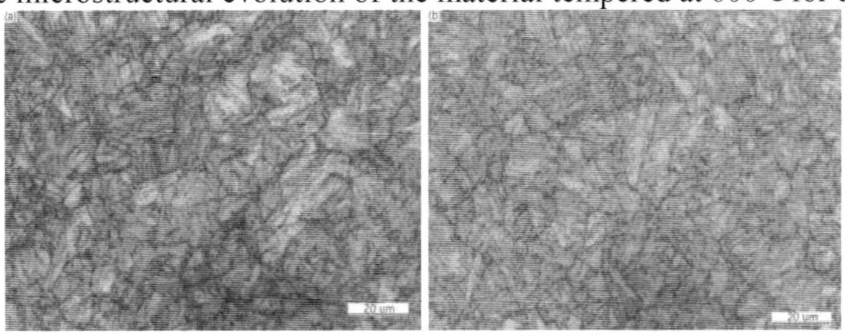

(a) 1 hour (b) 2 hours

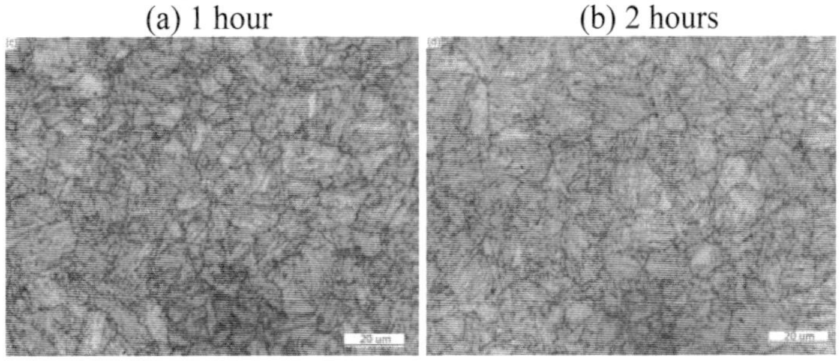

(c) 5 hours (d) 8 hours

Figure 4. The microstructural evolution of the material tempered at 620℃ for different time.

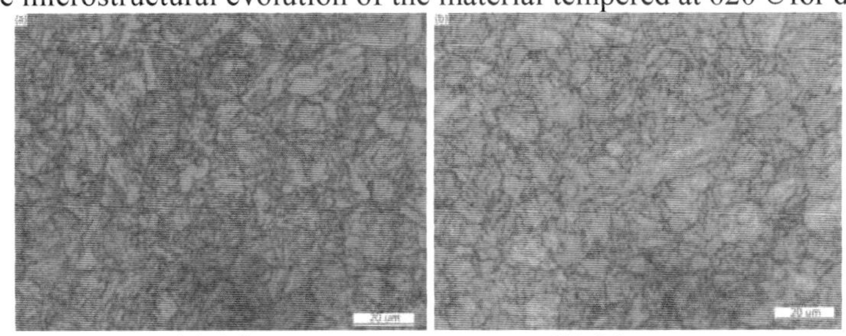

(a) 1 hour (b) 2 hours

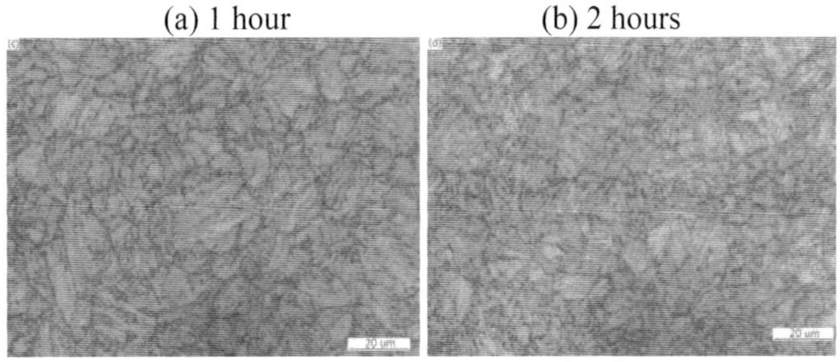

(c) 5 hours (d) 8 hours

Figure 5. The microstructural evolution of the material tempered at 640℃ for different time.

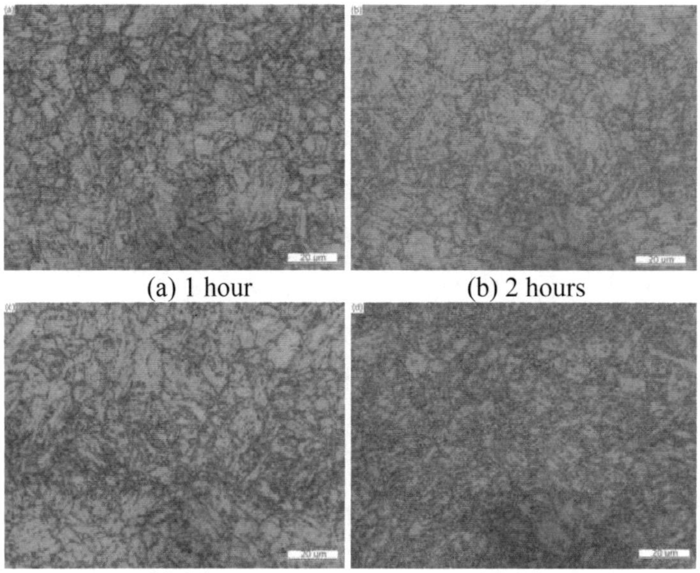

(a) 1 hour (b) 2 hours

(c) 5 hours (d) 8 hours

Figure 6. The microstructural evolution of the material tempered at 660℃for different time.

3.3 Mechanical properties

3.3.1 Hardness. The variation of hardness value of the material with the tempering temperature and time was shown in Fig.7. It can be seen that the value was sensitive to the time. A general trend was that the value decreased with the increasing tempering temperature and time. At 600℃, the value was always larger than the original value (310 HV), but the hardness value became smaller than the original value at 660℃. This indicated that it was not proper for the material to be tempered at these temperatures. At 620℃ and 640℃ for holding 2-3 hours, the hardness value (located in the red dotted rectangle) was close to the original value of the material. Thus, the proper tempering conditions for the material may be at 620-640℃ for holding 2-3 hours.

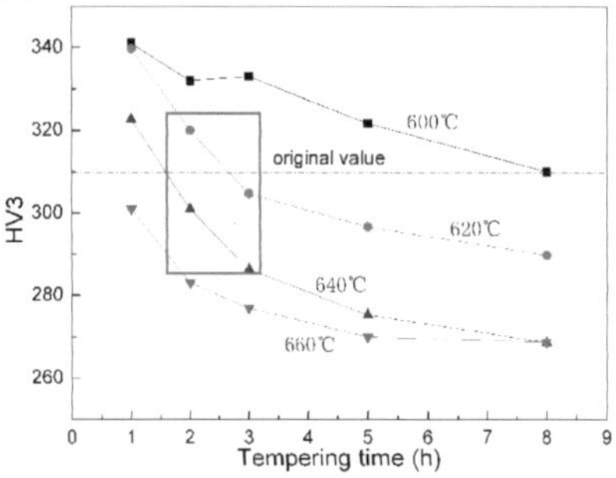

Figure 7. The variation of hardness of the material at 620℃with tempering time.

3.3.2 Tensile properties. The variation of the ultimate tensile strength (UTS) and yielded strength (YS) with the tempering time was displayed in Fig. 8(a). Clearly, both the YS and UTS decreased with the time. Zhang et al's [9] study had shown that the martensitic lath was the key factor controlling the strength. The martensitic lath width and dislocation density decreased with the time, which led to the decrease of strength. In addition, the solution strengthening effect became weaker with time. However, the precipitation strengthening effect was minor. Therefore, the strength of the material decreased with the time.

Fig.8(b) showed the variation of reduction of area (*Z*) and elongation (*E*) with time. It can be seen that both of the *Z* and *E* increased with the time. It has been reported that the decrease of ductility was attributed to the decrease of aspect ratio of martensite lath [10]. During the tempering process, the aspect ratio of the tempered lath decreased with the time. Therefore, the values of *Z* and *E* increased with the holding time.

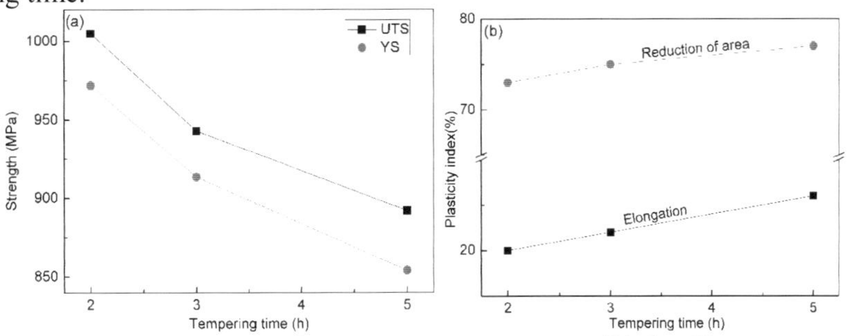

(a) strength (b) ductility

Figure 8. The variation of mechanical properties of the material tempered at 620℃with time.

3.3.3 Impact toughness. The variation of impact toughness of the material with time was showed in Fig. 10. It can be seen that the impact toughness increased with the time. Two reasons can explain this phenomenon. The first was the softening of the base metal due to the precipitation of interstitial atoms [11]. The second was the change of crystallographic orientation and crack direction. Therefore, in order to ensure a good combination of strength and plasticity, the temper condition should be selected as 620℃+ 2~ 5 hours.

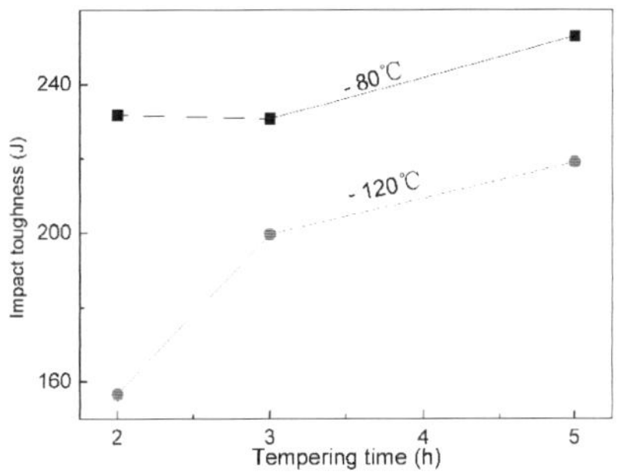

Figure.9 The variation of impact toughness at 620℃with tempering time.

4.Summary

The width and length of the tempering lath became wider and shorter with the temperature and time. The amount and size of carbide precipitates increased with the temperature and time. The beginning time of the precipitation was shorter for a martensite microstructure a mixed microstructure. The yielded strength, ultimate tensile strength and hardness decreased with the tempering temperature and time, while the reduction of area, elongation and impact toughness increased with the temperature and time. The variation of these properties was closely related to the variation of lath width and length. The optimal quenching and tempering treatment was obtained. The material was subjected to the water quenched and then was tempered at 620℃ for 2-3 hours, which can make sure a good combination of strength and toughness.

5.Acknowledgments

This work was supported by the Scientific Research Foundation of Tianjin University of Technology and Education (KYQD1801) and the National Natural Science Foundation of China (51301121) and the Natural Science Foundation of Tianjin City of China (No.13JCYBJC38900).

References

[1] Castan C, Montheillet F and Perlade A 2013 Dynamic recrystallization mechanisms of an Fe-8% Al low density steel under hot rolling conditions *Scripta. Mater.* **68** 360-364.
[2] Wang Q, Liu X S, Wang P, Xiong X and Fang H Y 2017 Numerical simulation of residual stress in 10Ni5CrMoV steel weld-ments *J. Mater. Process. Technol.* **240** 77-86.
[3] Tao P, Yu H, Fan Y F and Fu Y 2014 Effects of cooling method after intercritical heat treatment on microstructural characteristics and mechanical properties of as-cast high-strength low-alloy steel *Mater. Des* **54** 914-923.
[4] Peng Y, Liu Z D, Bao H S, Weng Y Q and Liu W 2013 Effect of microstructural evolution on high-temperature strength of 9Cr-3W-3Co martensitic heat resistant steel under different aging conditions *Mater. Sci. Eng. A.* **588** 22-28.
[5] Shin J H, Jeong J S and Lee J W 2015 Microstructural evolution and the variation of tensile behavior after aging heat treatment of precipitation hardened martensitic steel *Mater. Charact.* **99** 230-237.
[6] Maji S, Subhani A R, Show B K and Maity J 2017 Effect of cooling rate on microstructure and mechanical properties of eutectoid steel under cyclic heat treatment *J. Mater. Eng . Perform.* **26** 3058-3070.
[7] Dong J, Zhou X S, Liu Y C, Li C, Liu C X and Guo Q Y 2017 Carbide precipitation in Nb-V-Ti microalloyed ultra-high strength steel during tempering *Mater. Sci. Eng. A.* **683** 215-226.
[8] Guimarães J R C and Rios P R 2015 Microstructural path analysis of martensite dimensions in FeNiC and FeC alloys *Mater. Res.* **18** 595-601.
[9] Zhang C Y, Wang Q F, Ren J X, Li R X, Wang M Z, Zhang F C and Yan Z S 2012 Effect of microstructure on the strength of 25CrMo48V martensitic steel tempered at different temperature and time *Mater. Des.* **36** 220-226.
[10] Zhan K, Yong Q L, Sun X J, Li Z D, Zhao P L and Chen S D 2014 Effect of tempering temperature on microstructure and mechanical properties of high Ti-microalloyed directly quenched high strength steel, Acta. Metall. Sin. **50** 913-920 (in Chinese).
[11] Yang J J, Yu H C, Yin J, Gao M, Wang Z M and Zeng X Y 2016 Formation and control of martensite in Ti-6Al-4V alloy produced by selective laser melting *Mater. Des.* **108** 308-318.

IOP Conf. Series: Materials Science and Engineering **359** (2018) 012028 doi:10.1088/1757-899X/359/1/012028

Effective Identification on Adulteration of Polyethylene With Post-consumer Ones

S Zhao[1, a], W B Qin[1, b], J F Guo[1], J Liu[1], Y L Wang[1], W Zhang[1], X Y Zhao[2] and L Wang[*1, c]

[1] Hebei Research Institution for Product Quality Supervision and Inspection, Shijiazhuang, 050000, China

[2] Department of Materials Science and Engineering, Hebei University of Science and Technology, Shijiazhuang, 050000, China

E-mail: [a] 930575554@qq.com, [b]1920489166@qq.com,

*[1,c]282932157@qq.com (corresponding author)

Abstract. This paper mainly describes the effective identification of the adulteration of polyethylene with post-consumer ones. Degradation would be happened when multiple processings occurred. The melt flow index (MFI) analysis, thermal gravimetric analysis (TGA) ,differential scanning calorimeter (DSC) were used to characterize the processability and thermal stabilities of virgin polyethylene and recycled polyethylene which adulterated post-consumer PE. The results indicated that MFI of PE increased with the increasing doping content. Adulterating reclaimed PE had effects on the thermal stability of PE, which led in lower thermal decomposition temperature. Melting peak of recycled LLDPE varied from merely single to double, which differently compared with virgin LLDPE. Besides, with the doping content of post-consumer LDPE, the melting temperature had a decreasing tendency.

1. Introduction

Plastics are widely used in food and drug packaging industry due to their intrinsic good properties such as light weight, excellent chemical stability, convenient processing etc.[1~3] Food outer packaging, food inner packaging films, beverage bottles, drug bottles and infusion bags are commonly used plastic materials. Virgin plastic granules should be chosen to be used to manufacture food and drug packaging in order to ensure the food and drug safety. Recycled plastics are strictly forbidden to product the food and drug packaging. Because the ingredient of reclaimed PE was sophisticated, and the properties of those reclaimed ones had a reduction.[4~6]. For this reason, research on the effects on the structure and properties of food-grade reclaimed plastics has significance[7~8]. In this thesis, multiple reprocessings were performed by using twin-screw extruder. A series of recycled PE samples which adulterated a few post-consumer PE were prepared by high-speed commixture and melt extrusion. The aim of this work was to identify the adulteration of polyethylene with post-consumer ones.

2. Experimental

2.1 Materials. Virgin PE and recycled ones were prepared in this study. The recycled PE were carefully washed and dried. The materials are presented in Table 1.

Table 1. Descriptions of the materials in the study

Type	Commercial grade	Manufacturer
LLDPE	1001KW	Exxon Mobil
LDPE	Q210	SINOPEC
LLDPE	Recycled 1001KW/1002KW/1018FA...	Exxon Mobil
LLDPE	Recycled	Xiongxian, Hebei Province
LDPE	Recycled	Suning, Hebei Province

2.2 Material reprocessing. Before the mechanical recycling process, virgin and recycled polyethylene should be put in an oven at 60℃ for 4h in order to avoid hydrolysis reactions. Then high-speed commixture of adulterated post-consumer PE was needed in order to make virgin and post-consumer ones mixed homogeneously. Subsequent reprocessing was performed by using a CTE35 twin-screw extruder with 35mm diameter and a L/D ratio of 44(CKC, Nanjing). The temperature profile used was 160, 210, 211, 216, 219, 220, 220, 220, 220 and 220℃. The die temperature was controlled to be 210℃.

3. Results and Discussion

3.1 MFI analysis. Fig.1 depicted the melt flow indices of LLDPE and LDPE with different doping contents. The reclaimed LLDPE were chosen from the Xiongxian city. It was observed that both MFI of PE increased with the increasing adulteration, which indicated the rise of processability. As shown in Fig.1, a predominant reaction of chain scission had occured in the thermal-mechanic degradation, which resulted from poor free radicals' recombination in the blend of virgin PE and post-consumer ones.

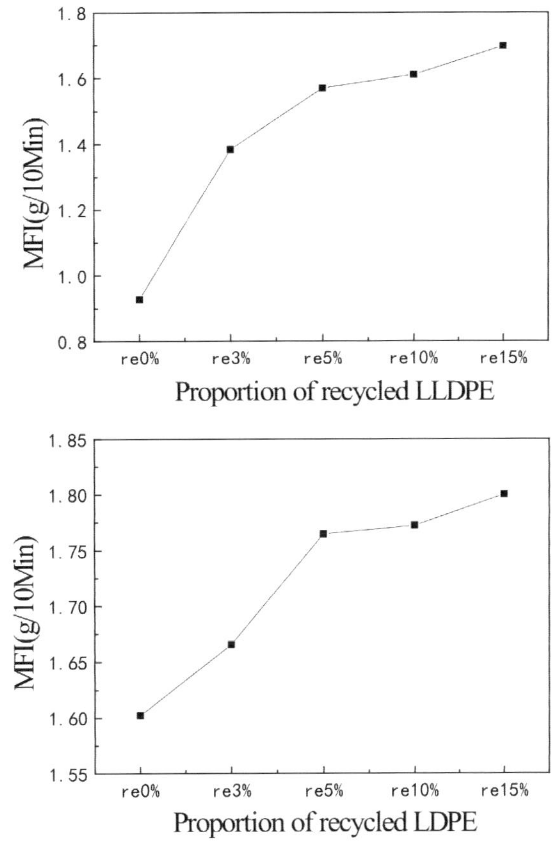

Fig.1 Melt flow indices of LLDPE and LDPE with different doping contents

3.2 Thermo-gravimetric analysis. The TG curves of LLDPE with different doping content were shown in Fig.2. The reclaimed LLDPE was chosen from the Xiongxian city. It had been suggested that the virgin LLDPE had excellent thermal properties due to the high decomposition temperature, while the thermal temperature had a decreasing tendency with the increasing doping content, which was much lower than virgin LLDPE.

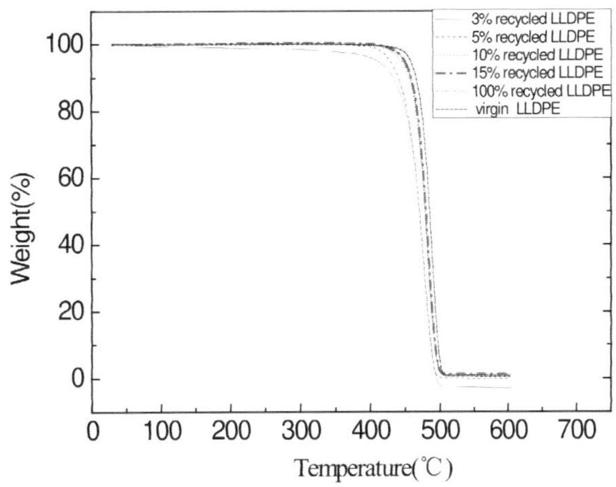

Fig.2 TG curve of LLDPE with different doping contents

The results indicated virgin LLDPE had better thermal stability due to its single structure. However, the ingredient of the post-consumer LLDPE was prone to be complicated when they suffered from reprocessing, reserve and daily-use. Therefore there existed an adverse effect on the thermal stability of the ones which adulterated with post-consumer polyethylene ones.

3.3 DSC Analysis. The melting temperature(T_m) of virgin LDPE and the recycled LDPE with doping was illustrated in Fig.3. The reclaimed LDPE was chosen from the Suning city. As shown in Fig.3, the melting temperature of virgin LDPE was 111℃. After adulterating post-consumer LDPE, the melting peak had no effective variation, but the melting temperature decreased. From the perspective of thermodynamics, when the temperature of polyethylene got to the melting temperature, the crystalline phase and amorphous phase reached to thermodynamic equilibrium. The melting temperature was calculated by using the relationship:

$$T_m = \frac{\Delta H}{\Delta S} \qquad (1)$$

where ΔH was related to the strength of molecular force; ΔS was related to the flexibility of molecular chain.

Due to suffering from high temperature and high shear force, the molecular chain of recycled LDPE with doping cracked, and the intermolecular forces receded, which made ΔH decreased. This was the one reason that resulted in the decrease of melting temperature. The other reason is that the backbone of LDPE which doped with post-consumer LDPE had breakage. The recombination of free radicals made the symmetry and the regularity of molecular chain destroyed. ΔS had an increasing tendency so that the melting temperature decreased.

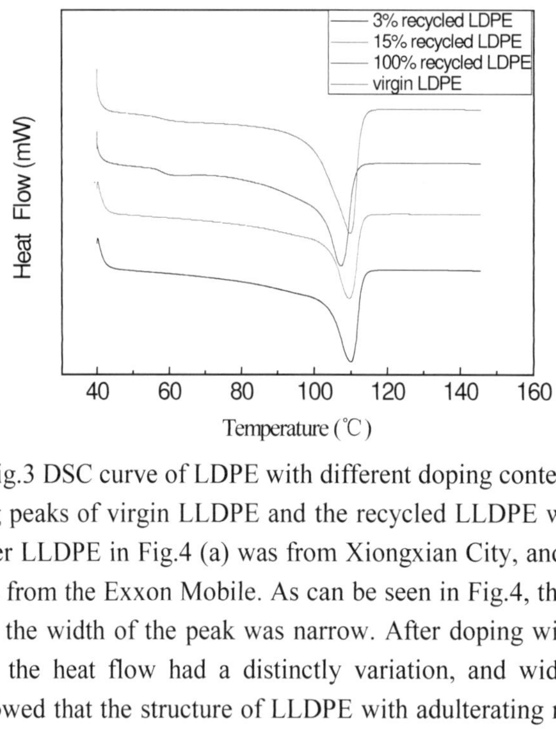

Fig.3 DSC curve of LDPE with different doping contents

The variation of melting peaks of virgin LLDPE and the recycled LLDPE with doping was shown in Fig.4. The post-consumer LLDPE in Fig.4 (a) was from Xiongxian City, and the other post-consumer LLDPE in Fig.4 (b) was from the Exxon Mobile. As can be seen in Fig.4, the virgin LLDPE had only single melting peak and the width of the peak was narrow. After doping with different proportion of post-consumer LLDPE, the heat flow had a distinctly variation, and widen or bimodal peak had appeared. The result showed that the structure of LLDPE with adulterating reclaimed LLDPE was no longer single. Because of complicated ingredient of post-consumer LLDPE and kinds of factors after reprocessing, the molecular chain cracked and the molecular structure varied. The polydispersity increased which resulted in the variation of melting peak of the LLDPE with doping post-consumer LLDPE.

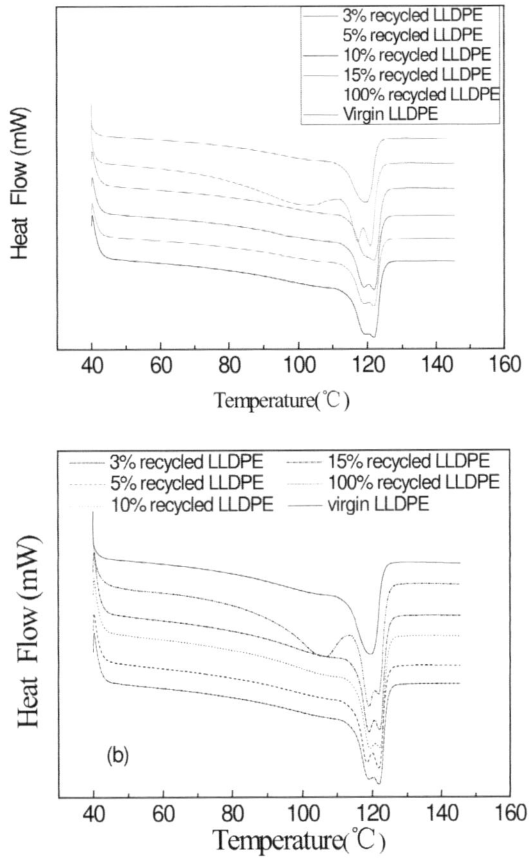

Fig.4 DSC curve of LDPE with different doping contents

4. Conclusion

In summary, due to the sophistication of the ingredient of post-consumer PE, the simplicity of the molecular structure of PE materials which adulterated different proportion of reclaimed PE was poor. The MFI increased along with the increasing doping content, which illustrated the improved processing fluidity and the decreased molecular weight. The thermal degradation temperature of adulterated post-consumer PE decreased slightly in comparison with virgin ones. Besides, after adulteration of reclaimed PE, the melting peak had a significant variation: recycled LLDPE appeared bimodal or widened melting peak. The melting temperature of recycled LDPE decreased with the increasing of the doping content of reclaimed LDPE.

Acknowledgments

The Project was supported by the National Special Found for Scientific Research in the Public Interest of China (Grant No. 201310151)

IOP Conf. Series: Materials Science and Engineering **359** (2018) 012028 doi:10.1088/1757-899X/359/1/012028

References

[1] Xie L G, Sun H M and Jin S H 2011 Anal. *Chim. Acta* Vol. 706, p. 312-320

[2] Jin H Y, Gonzalez-Gutierrez J, Oblak P and et al 2012 Polym. *Degrad. Stabil* Vol. 97, p. 2262-2272

[3] Zare Y 2013 *Waste Manage* Vol. 33, p. 598-604

[4] Camacho W, Karlsson S 2002 Polym. *Degrad. Stabil* Vol. 78, p. 385-391

[5] Tzankova Dintcheva N, La Mantia F P, Acierno D and etal 2001 Polym. *Degrad. Stabil* Vol. 72, p. 141-146

[6] Sun L, Zhao X Y, Sun Z Y 2014 Adv. *Mater. Res* Vol. 1003, p. 96-99

[7] Cruz S A and Zanin M 2003 Polym. Degrad. Stabil Vol. 80, p. 31-37

[8] Choudhury A, Mukherjee M, Adhikari B 2005 *Thermochim Acta* Vol. 430, p. 87-94

IOP Conf. Series: Materials Science and Engineering **359** (2018) 012029 doi:10.1088/1757-899X/359/1/012029

Study on Corrosion-induced Crack Initiation and Propagation of Sustaining Loaded RCbeams

X P Zhong, Y Li, C B Yuan, Z Yang and Y Chen

College of Civil Science and Engineering, Yangzhou University, Yangzhou, Jiangsu, China

Corresponding author. Email: zhongxiaoping@zju.edu.cn

Abstract. For 13 pieces of reinforced concrete beams with HRB500 steel bars under long-term sustained loads,at time of corrosion-induced initial crack of concrete,and corrosion-induced crack widths of 0.3mm and 1mm, corrosion of steel bars and time-varying behavior of corrosion-induced crack width were studied by the ECWD (Electro-osmosis - constant Current – Wet and Dry cycles) accelerated corrosion method. The results show that when cover thickness was between 30 and 50mm,corrosion rates of steel bars were between 0.8% and 1.7% at time of corrosion-induced crack,and decreased with increasing concrete cover thickness;when corrosion-induced crack width was 0.3mm, the corrosion rate decreased with increasing steel bar diameter,and increased with increasing cover thickness;its corrosion rate varied between 0.98% and 4.54%;when corrosion-induced crack width reached 1mm, corrosion rate of steel bars was between 4% and 4.5%;when corrosion rate of steel bars was within 5%, the maximum and average corrosion-induced crack and corrosion rate of steel bars had a good linear relationship.The calculation model predicting the maximum and average width of corrosion-induced crack is given in this paper.

1. Introduction

Many researchers have studied corrosion-induced crack process of concrete cover.Most of these focused on two aspects:1)corrosion amount of steel bars at time of corrosion-induced crack on the surface of concrete cover,or time of corrosion-induced crack on the surface of concrete cover;2)development law of crack width on the surface of concrete after cracking.In regards to corrosion amount of steel bars at time of corrosion-induced crack on the surface of concrete cover,Andrade *et al.*[1]conducted accelerated corrosion experiments on 4 group different specimen of steel bars position,steel bars diameter and cover thickness,found that reinforcement corrosion depth of the first visible surface crack(crack width was 0.05~ 0.10mm)was usually 10~50μm;Oh *et al.*[3]judged the crack by strain value,thought when the monitored strain value reached concrete crack strain,cover surface appeared initial crack,thus the relationship between corrosion rate of steel bars at time of cover surface crack and cover thickness was established.For study of surface crack width,Vidal *et al.*[4] studied 2 pieces reinforced concrete beams exposed under chlorine salt environment for 14 years and 17 years respectively,measuring calculated reinforcement corrosion at different positions by weightlessness method,and measured surface crack width of the corresponding location,set up the relationship between crack width and corrosion cross-section loss area of reinforcement;Zhang *et al.*[6] also built relationship between the average cross-section loss area of reinforcement and corrosion-induced crack width by experimental study.However,the effect of long-term sustained load on corrosion-induced crack and crack width of cover has not been considered in above experimental studies.In fact,the actual structure is working under load and environment interaction,so,studying corrosion-induced crack behavior of reinforced concrete members under load has very important practical value for durability design and service life prediction of structure.

2. Experimental investigation

2.1. Experimental material

Ordinary Portland cement 42.5MPa was used as concrete raw material, coarse aggregate particle size is 5~20mm of rubble; natural river sand was used as fine aggregate,fineness modulus is 2.5; II grade fly ash was used,its dosage is 15%;JM-B type naphthalene series water reducing agent was used as high efficiency water reducing agent.Concrete mix proportion is shown in table 1.

Table 1.mix proportion of concrete

w/c ratio	Water	Cement	Fineaggregate	Coarseaggregate	Fly ash	waterreducing agent
		kg/m^3	kg/m^3	kg/m^3	kg/m^3	kg/m^3
0.44	215	416	573	1217	73	8.8

2.2. Specimen design

Taking cover thickness,steel bar diameter as the main parameter,13 pieces reinforced concrete beams were designed and made.Test beam parameters are shown in table 2,reinforcement is as shown in Figure1. In order to make chloride ion electricity permeability to surface of longitudinal reinforcement, a piece of 30mm ×1300mm × 0.2mm stainless steel sheet was put in the middle layer of each test beam.

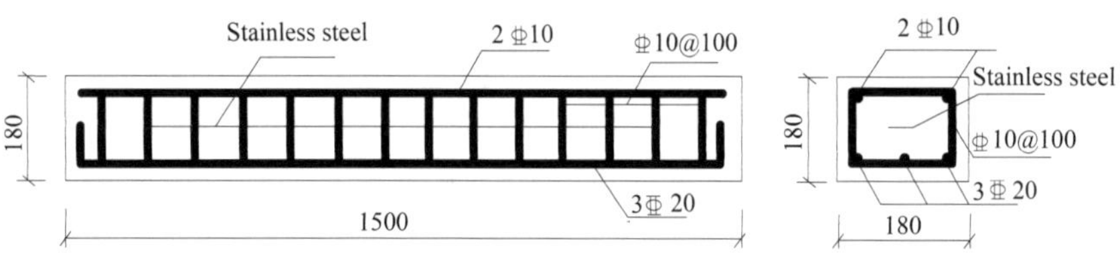

Figure 1. Details of the tested specimens

2.3. Load level

Considering the actual structure durability degradation is under the coupled action of load and environment,so the accelerated corrosion experiment under long-term sustained load was done for B1-B13 beams.Each two beams were anchored each other, force was pressed by tightening nuts.Load level is subject to member tensile surface crack width in 0.2～0.25mm as a control condition.

2.4. Accelerated corrosion method

The accelerated corrosion can be divided into two phases:

1) In the electro-migration phase, the embedded stainless steel sheet in the concrete was connected with the positive of direct current dc power source by wire,external stainless steel net was connected with the cathode of power by wire,opening the dc power supply,maintaining a constant voltage of 30v, the chloride ion was migrated to the surface of the corroded reinforcement. Theelectro-migration time calculation of each beam is shown in Table 2.

2)The wet-dry cycle phase,constant current power phase.After the electro-migration experiment,a current density of $0.2mA/cm^2$ was applied through the flexural tensile reinforcing steel in concrete beams(acting as the anode)and the stainless steel net(acting as the cathode)to accelerating corrosion.Taking surface area of corroded reinforcement as combined surface area of reinforcement within flexural tensile area 1m length,the required corrosion current sizes of each beam can be obtained by calculation,as shown in Table 2.A wet-dry cycle was used during accelerated corrosion,each dry-wet cycle is for 14 days,the ratio of dry and wet is 1:1.The preset corrosion degree of each beam was subject to corrosion-induced crack width on concrete surface as control condition,as shown in Table 2.The accelerated corrosion experiment ended after the preset corrosion-induced crack width was achieved.

Table 2. Test beam sizes,corrosion parameters and practically measured corrosion-induced cracking widths

Beam designation	$l \times b \times h$	d	c	t [e]	I [a]	$w_{c,de}$ [b]	ρ [d]	$w_{c,ex}$ [c]
	mm	mm	mm	d	A	mm	%	mm
B1	1500×168×180	16	30	5.32	0.38	0.05-0.08	1.67	0.07
B2	1500×168×180	16	30	5.32	0.38	1.0	4.50	1.0
B3	1500×168×180	16	30	5.32	0.38	0.3	2.0	0.3
B4	1500×180×180	20	30	5.32	0.47	0.3	2.38	0.35
B5	1500×186×180	22	30	5.32	0.5	0.3	1.07	0.35
B6	1500×195×180	25	30	5.32	0.57	0.3	0.98	0.31
B7	1500×188×180	16	40	7.1	0.38	0.05-0.08	1.25	0.07
B8	1500×188×180	16	40	7.1	0.38	1.0	4.05	1.05
B9	1500×188×180	16	40	7.1	0.38	0.3	2.04	0.3
B10	1500×208×180	16	50	8.87	0.38	1.0	4.11	1.1
B11	1500×208×180	16	50	8.87	0.38	0.05-0.08	0.82	0.08
B12	1500×208×180	16	50	8.87	0.38	0.3	2.65	0.34
B13	1500×228×180	16	60	10.65	0.38	0.3	4.54	0.35

[a] I is corrosion current.
[b] $w_{c,de}$ is predetermined corrosion-induced cracking width.
[c] $w_{c,ex}$ is practically measured corrosion-induced cracking width.
[d] ρ is corrosion rate of reinforcement average section.
[e] t is electro-migration time.

3. Experimental result and analysis

3.1. The feature points of corrosion-induced crack and reinforcement corrosion rate

This paper chose corrosion-induced cracks beginning to appear and crack width reaching 0.3mm and 1mm as feature point,and analyzes the influence factors and reinforcement corrosion rate.

In Table 2 B1、B7、B11 beams are used to analyze the effect of different cover thickness on corrosion rate of reinforcement at time of corrosion-induced cracking under long-term sustained loads; B3、B4、B5、B6 beams and B3、B9、B12、B13 beams are used respectively to analyze when the expected corrosion-induced crack width reaches 0.3mm,the effect of steel bar diameter and cover thickness on corrosion rate of reinforcement; While B2、B8、B10 beams are used to analyze when the expected corrosion-induced crack width reaches 1mm,the effect of cover thickness on corrosion rate of reinforcement. When constant current is adapted to accelerate corrosion, assume that reinforcement is uniform corrosion, the corrosion rate of reinforcement average cross section in Table 2 can be calculated and obtained from equation (1) - (2):

$$\Delta d_s = \frac{t_2 M_{\mathrm{Fe}} i}{\gamma_s Z_{\mathrm{Fe}} F} \quad (1)$$

$$\rho = \frac{d^2 - (d - 2\Delta d_s)^2}{d^2} \quad (2)$$

Where, Δd_s is the target corrosion depth(cm); γ_s is the density of the steel(g/cm^3); M_{Fe} is the atomic weight of the iron, taking 56g/mol; t_2 is the total galvanising time(s); Z_{Fe} is the valence of the iron, taking 2; i is the corrosion current density,taking 0.2mA/cm^2; ρ is the corrosive rate of reinforcement average cross section; d is diameter(cm) of steel bar.

The practically measured corrosion-induced crack width and the corresponding corrosion rate of reinforcement average cross section of each beam are given in Table 2 by measurement and calculation.Owing to absorbing sponges were wrapped on surface of members in the process of corrosion,crack widths can not be real-time monitored,so, under the influence of measuring point in time, the measured values and predetermined values in Table 2 exist tiny deviation,but can reflect basically the corrosion of corrosion-induced crack feature points.In addition,once cover cracks,crack width will appear,so crack width 0.05~0.08mmare taken as initial crack time in this paper.

From the analysis results of experiment and calculation for B1、B7、B11 beam it appears that under long-term sustained load,when cover thickness is between 30mm and 50mm,the corrosive rate of reinforcement at time of corrosion-induced crack is between 0.8% and 1.7%, and decreases with increasing of concrete cover thickness.The reason is that sustained load makes concrete at the bottom of beam generate longitudinal tensile stress,while steel bar corrosion makes concrete generate circular tensile stress,under the joint actions of longitudinal tensile stress and circular tensile stress,concrete is in the state of bidirectional tensile stress,the real effective cover thickness decreases,consequently the corrosive rate of reinforcement at the time of corrosion-induced crack decreases.

From the results of experiment and analysis for B3、B4、B5、B6 and B3、B9、B12、B13 beam it appears that when corrosion-induced crack width is between 0.3mm and 0.35mm,the required corrosive rate of steel bar decreases with increasing of steel bar diameter, and increases with increasing of cover thickness,the change of its corrosive rate is between 0.98% and 4.54%;When cover thickness is between 30mm and 50mm,and corrosion-induced crack width reaches 1mm,the corrosive rate of steel bar is between 4% and 4.5% (B2、B8、B10 beam).

3.2. Time-varying analysis of corrosion-induced crack width

After each dry-wet circulation time on B2、B8、B10 beam respectively,corrosion-induced crack width was measured,corresponding corrosive rate of steel bar was calculated,the relationship of corrosion-induced crack changing with corrosive rate of steel bar was obtained as shown Figure2.

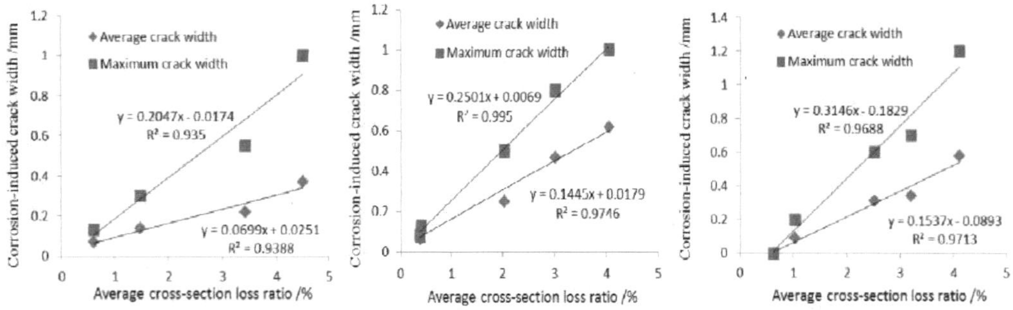

(a) B2 beam（c=30mm） (b) B8 beam（c=40mm） (c) B10 beam（c=50mm）

Figure 2.The relationships between the corrosion-induced crack width and average cross-section loss ratio

From Figure 2 when corrosive rate of steel bar was within 5%,the corrosion-induced crack maximum width and average width of each beam have a good linear relationship with the corrosive rate of steel bar.Considering the influence of cover thickness on corrosion-induced crack width on the surface of members,the development of corrosion-induced cracks can be predicted according to the following equation:

$$w_{c,max} = A\rho - B \quad \rho \leq 5\% \quad (3)$$

$$w_{c,average} = C\rho - D \quad \rho \leq 5\% \quad (4)$$

Where, $w_{c,max}$、$w_{c,average}$ are the maximum value and average value of corrosion-induced crack width respectively; ρ is the corrosive rate of reinforcement average cross section; A、B、C、D stand for coefficient,for the different cover thickness,the coefficient is given in this paper,as shown in Table 3.

Table 3.*A*、*B*、*C*、*D* coefficient table

Cover thicknes（mm）	*A*	*B*	*C*	*D*
30	0.2047	0.0174	0.0699	0.0251
40	0.2501	0.0069	0.1445	0.0179
50	0.3146	0.1829	0.1537	0.0893

The relationship of corrosion-induced crack maximum width and average width changing with time can be obtained by equation (1)~(2)formula.

4. Conclusion

1) Under long-term sustained load,when cover thickness is between 30mm and 50mm,the corrosive rate of reinforcement at time of corrosion-induced crack is between 0.8% and 1.7%, and decreases with increasing of concrete cover thickness; When corrosion-induced crack width is between 0.3mm and 0.35mm,the required corrosive rate of steel bar decreases with increasing of steel bar diameter, and increases with increasing of cover thickness,the change of its corrosive rate is between 0.98% and 4.54%;When corrosion-induced crack width reaches 1mm,the corrosive rate of steel bar is between 4% and 4.5%. When the corrosive rate of steel bar is predicted through corrosion-induced crack width,such influence factors as the diameter of steel bar and cover thickness of concrete should be considered comprehensively.

2) When corrosive rate of steel bar is within 5%,the maximum width and average width of the corrosion-induced crack have a good linear relationship with the corrosive rate of steel bar.

Acknowledgements
This work was supported by the National Natural Science Fundation of China through Projects (Project code:51678513 and 51278444).

References

[1] AndradeC, AlonsoC and MolinaFJ 1993Cover cracking as a function of bar corrosion: part 1—experimental test. *Materials and Structures***26** 453-464.

[2] OhBH，KimKH and JangBS2009Critical corrosion amount to cause cracking of reinforced concrete structures. *ACI Materials Journal***106** 333-339.

[3] VidalT, CastelA and Francois R2004Analyzing crack width to predict corrosion in reinforced concrete. *Cement and Concrete Research***34** 165-174.

[4] ZhangR，CastelA and Francois R2010Concrete cover cracking with reinforcement corrosion of RC beam during chloride-induced corrosion process. *Cement and Concrete Research***40** 415-425.

IOP Conf. Series: Materials Science and Engineering **359** (2018) 012030 doi:10.1088/1757-899X/359/1/012030

Strain Characterization and Microstructure Evolution Under Deformation in 2060 Alloy

X Jin,[1,2*] G D Zhang[1], Y F Zhao[1] and F Xue[1]

[1] Suzhou Nuclear Power Research Institute, Suzhou 215004, PR China
[2] Laboratory of Advanced Materials, School of Material Science and Engineering, Tsinghua University, Beijing 100084, PR China

*Corresponding author: Jin Xiao, PhD, jin_xiao2002@163.com

Abstract. A new method of DIC combined with EBSD is developed for the characterization of strain and microstructure evolution during bending. The traditional microhardness point and DIC methods are used to study the microstructure evolution in 2060 alloy during bending; the interested area suffers under tensile stress, the microstructure evolution is collected by SEM, EBSD, digital image correlation (DIC) method during bending. The results shows that the DIC method can both realize the strain tensor characterization of the interested area, and can also express the local strain tensor in the micro-area even more. The degree of grain division in the process of deformation is related to the strain in this region; the grains have larger strain of small angle grain boundary (SLGBs), which results in a new micro-organizational structure. The misorientation is smaller with larger strain degree while the misorientation is larger with smaller strain.

1 Introduction

2060 alloy is the third generation Al-Li production developed by Alcan Inc. in 2011, which significant improvements in elastic modulus, fatigue crack growth resistance, corrosion resistance and plane stress fracture toughness at reduced density as compared to the previous productions[1-3], used in manufacture the fuselage and lower wing skin structure[4, 5]. The key step in 2060 alloy, the process of obtaining appropriate size and shape is the most important step for working as aircraft structural component, and one commonly used deformation processing is bending, which is an environment-friendly and low waste method in the manufacture of structural parts[6].

It is regarded that grains are subdivided into cell blocks (CBs) containing ordinary dislocation cells during plastic deformation in FCC metals with high stacking fault energy[7]. During the process of bending, grains in outer side are affect by tensile stress while grains in inner side are affect by compressive stress. In general, specimen under tensile stress is easier to damage in the same stress condition[8].

In the past decades, many studies have been done to investigate the microstructure and dislocation boundaries during deformation, but it is hard to explain the relationship between strain and microstructure due to hard to characterize the relationship at the same time. In the present work, we aimed to find a feasible way to characterise simultaneous changes in strain and microstructure during bending. EBSD and DIC imaging techniques are combined as the most viable method of potential method to characterize the surface changes in specimens; thus, the statistical result of microstructure and strain information within grains is accurately collected.

2. Experimental

In the present work, the material used for this study was a commercially 2060 aluminum alloy made by Alcoa Inc., the heat treatment is T8. The standard compositional range of 2060 alloy is shown in Table 1, together with the composition of the of the studied alloy analyzed by infrared spectrometer(ICP-AES).

Table 1
Chemical composition of 2060 alloy, wt.%.

Element	Ag	Cu	Fe	Mg	Mn	Si	Ti	Zn	Zr	Li	Al
Nominal	0.05-0.5	3.4-4.5	0.07	0.6-1.1	0.1-0.5	0.07	0.1	0.3-0.5	0.05-0.15	0.6-0.9	Rest
2060	0.34	3.52	0.03	0.66	0.29	0.02	0.03	0.34	0.1	0.73	Rest

A split sample of 10×3×2 mm (along the rolling direction(RD),transverse direction (TD) and normal direction(ND),respectively) was used. Simples for characterization is surface(RD×ND) was first mechanically polished by sandpaper from 600μm to 4000μm,then electrochemically polished by means of 10% HCLO4 solution in alcohol at room temperature under applied voltage of 20V for a time of 15s-20s. The simple is bent by cold-rolled forming, and their bend radii are 10, 5, and 3 mm, respectively.

EBSD data collected by Oxford Nordly 2, Accelerating voltage of 20 kV, magnification of 300× and a working distance of 15 mm were used to scan over the same area of SEM scan, step size is 2μm.

In this study, the DIC was used to characterize the strain evolution during bending, and data are recorded during different stages of bending of a specimen and are later used as an input to a DIC system for strain calculations. According to the principle of strain calculation used in this study, a large number of precipitates (white points in Fig.1) with different sizes are defined as random speckle patterns to record the corresponding displacements, as shown in Fig.1.

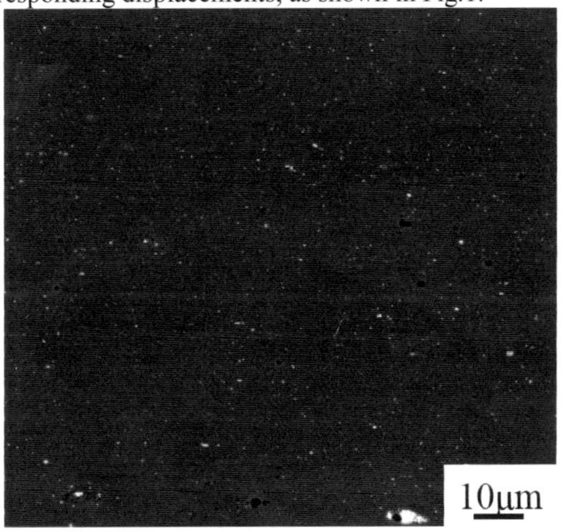

Fig.1 Microstructure of 2060 alloy with precipitates distribution in a BSE image

In Fig1, the precipitates randomly dispersed within the grains are evident; the precipitates appear as speckle patterns for strain calculations. In this study, a commercially available DIC system (VIC 2D, CSI, USA) was used. Images processed by the DIC system were collected by a BSE probe installed in a Tescan MIRA3 scanning electron microscope (Tenscan, Brno, Czech Republic).

3. Results and discussion

3.1 Traditional method of strain measurement

One of the strain characterization is micro-hardness point method, which record the displacement of point (such as micro-hardness points) and calculating the strain of the region during the deformation. It is wildly used in measurement of plane strain[4, 9].

In present work, this traditional is used to verify the error of the new method. The interested region is near the outer liner of bent simple where the tension strain is larger relatively, as the red dotted line in Fig.2. Fig.2 shows the morphology under different bend radius, four angles in rectangle is marked with micro-hardness, the relative displacement of the four points are record during bending and calculate the amount of strain in the region.

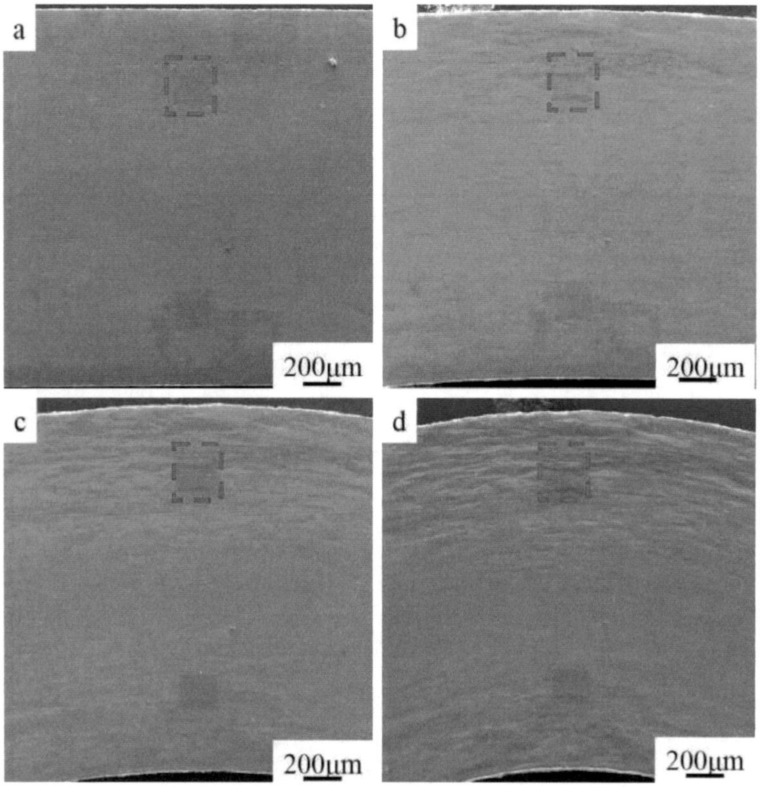

Fig.2 Morphology of simple with different bend radius: (a) unbend; (b) r=10mm; (c) r=5mm; (d) r=3mm

Fig.3 shows the strain changed in the red dotted line region during bending. From the Fig.3, the strain tensor ε_{xx} increased rapidly with the bend radius decreased, when the bend reach 3mm, the strain rise to nearly 0.14. The strain tensor ε_{yy} decreased with the bending radius decreasing, and the variation range is small. The strain tensor ε_{xy} almost unchanged during bending.

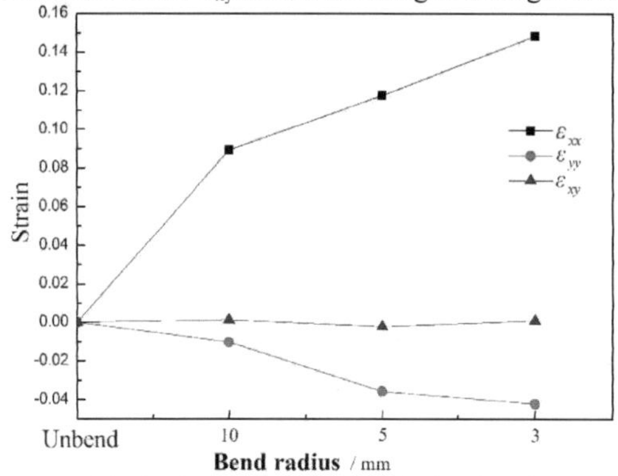

Fig.3 The strain changed during bending

From the result of changes in strain, this interested region can be believed as uniaxial tension. But one of the disadvantage of this method is not have enough accuracy, when the strain need measure, the displacement of mark points around this region should be calculated. This method appears to be insufficient when characterizing the strain in the smaller area, the reason is that there is no mark points around the region used for tracking calculation.

3.2 Strain measurement and microstructure evolution

3.2.1 DIC method of strain measurement

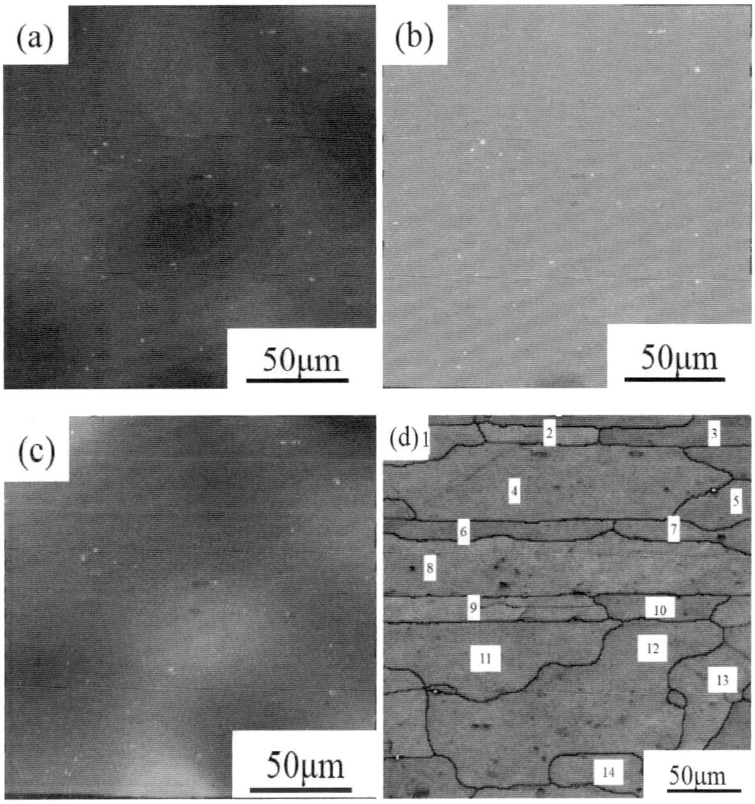

Fig.4 Strain distribution with DIC method under different bend radius:: (a) r=10mm; (b) r=5mm; (c) r=3mm; (d) EBSD map before bending

Fig.4 shows the strain distribution with DIC method under different bend radius in the region of red dotted line in Fig.2 (The darker of red colour, the larger strain which have. The darker of blue colour, the smaller strain which have.). From the Fig.4, there is a significant nonuniformity of strain distribution in the region during bending, and this distribution is uneven with the bending radius decreases. Fig.4(d) is the EBSD map from the same area before bending, contrast Fig.4(d) to other maps in Fig.4, it can be observed that there is a significant difference in the amount of strain on each of the grains in this region. There is a strain concentration on some grains, and with the decrease of the bending radius, the change of strain on different grains is also obviously different, eg. when the bend radius reach 3mm, there are large strain near the grain4, grain11 and grain 14, and strain near grain 10 and grain 12 is smaller.

From Fig.3, the main strain in this area is ε_{xx}, which means the strain in this area during bending can be approximated seemed as deformed under uniaxial stress. Fig.5 shows the Schmidt factor in this area under uniaxial tension, from the map grain11 and grain12 have the max Schmidt factor, the value near to 0.5. In general, the greater the Schmidt factor have, the more prone to deformation. From the Fig.4(c), the region where the strain distribution is greatest is not within the grains 11 and grains 12, but across the grains 12 and grains 15 and the like across the grains. So it is difficult to explain the mechanism with Schmidt factors.

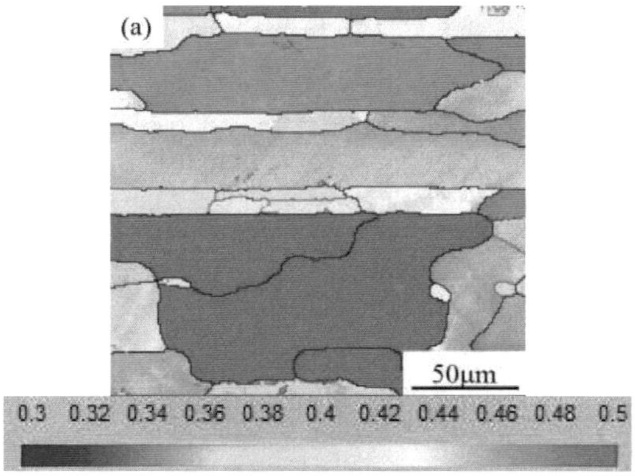

Fig.5 Schmidt factor under uniaxial tension

Further analysis of the misorientation between grains in this region, the relationship between the grains' orientation can be obtained, as shown in Table2. Combined theFig.4(d) and Table 2, the region get larger strain under tensile stress have smaller misorientation, such as: the misorientation between grain2 and grain4 is 27.4°, the misorientation between grain11 and grain12 is 29.8°, the misorientation between grain12 and grain15 is 27.9°. While the region get smaller strain have larger misorientation, such as: the misorientation between grain1 and grain2 is 56.3°, the misorientation between grain10 and grain12 is 44.6°. The similar phenomenon occurred in other region, the misorientation is smaller between grains when the strain is larger, while the misorientation is larger between grains when the strain is smaller. The reason is that the grain boundary has a blocking effect on the slip of the dislocations. The dislocation moves in the polycrystals in the process of deformation of the polycrystalline sample[10], the grain boundary has some resistance to the slip of the dislocations due to the grain orientation difference on both sides of the grain boundary, when the misorientation is large dislocation is difficult to slip from grain into the second grain. Multiple slip systems need to start at the same time in order to meet the grain boundary deformation of the coordinatio. This also leads to dislocations that are difficult to pass through the grain boundary and block at the grain boundary, which increasing the strength of the area and making the area hard to deformation[11, 12].

Table 2 The Euler angle and misorientation[13] of main grains in interested region

grain	Euler angle / °	Neighbor grain	Micorientation / °
grain1	311.02, 33.65, 69.99	grain1-grain2	56.3
grain2	31.82, 44.49, 31.41	grain1-grain4	50.7
grain3	56.99, 6.38, 86.17	grain2-grain3	24
grain4	163.23, 52.49, 47.05	grain2-grain4	27.4
grain5	285.55, 1.29, 81.34	grain3-grain4	46.7
grain6	8.12, 21.58, 80.04	grain3-grain16	31.5
grain7	30.06, 34.31, 43.82	grain4-grain5	58
grain8	290.87, 42.57, 79.77	grain4-grain6	41.3
grain9	264.54, 51.21, 50.85	grain4-grain7	27.7
grain10	61.84, 6.23, 32.03	grain6-grain8	46.8
grain11	122.69, 27.05, 59.55	grain6-grain7	23.2
grain12	40.68, 36.17, 48.5	grain7-grain8	48
grain13	152.65, 41.37, 48.16	grain8-grain9	44.8
grain14	119.96, 26.69, 63.46	grain8-grain10	44.6
grain15	266.24, 41.42, 58.07	grain9-grain11	37.5
grain16	64.65, 26.93, 12.7	grain11-grain12	41.2
		grain12-grain15	27.9
		grain12-grain10	30.7
		grain12-grain14	39.8
		grain12-grain13	29.1

3.3Microstrcture Evolution in interested area

Fig.6 shows the microstructure evolution under different bend radius. From map the number of SLGBs increased significantly with the bend radius decreased, this SLGBs result by the large number of dislocations slip in grains. The region have more number of SLGBs while the strain is large, and the region with smaller strain have less number of SLGBs. In Fig.5(c), the band structures[14] formed by SLGBs is observed in grain 4, grain8 and grain12. In the previous uniaxial tensile experiments, a similar structure has been found, and it is considered that this banded structure is a cell structure that occurs when grain is split and the strain reaches a certain degree in the process of deformation. The dislocation density is high at the cell wall, and the dislocation density is expressed as a SLGBs in the reconstructed map of EBSD, when the deformation degree is low, the strain is not enough to form a cell structure inside the grain, and there are a large number of randomly distributed dislocations in the grain, that is a large number of scattered small grain boundaries in the grain[13, 15].

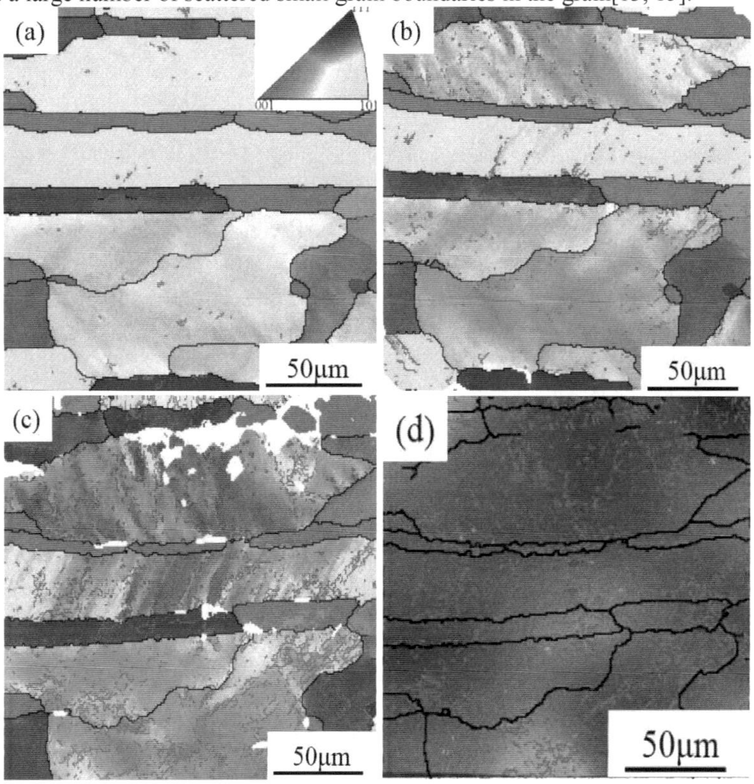

Fig.6 The microstructure of interested area under different bend radius: (a)Un bend; (b) r=10mm; (c) r=5mm; (d) Complex of microstructure and strain map when r=5mm (Black line is the large angle grain boundary, red line is SLGBs)

4. Conclusions

The traditional microhardness point and DIC methods is used to studied the microstructure evolution during deformation. A new method of strain change and microstructure evolution during bending is developed by DIC combined with EBSD, the conclusions shows as follow:

1. The microhardness point method can only characterize the average tensor tensor in the interested area, and the DIC method can both realize the strain tensor characterization of the interested area, and also can express the local strain tensor in the micro-area even more.

2. The degree of grain division in the process of deformation is related to the strain in this region, the grains have larger strain occurred a obviously split phenomenon, which resulting in a new micro-organizational structure.

3. The strain degree is relate to the misorientation between grains, and the larger misorentation have a obviously block effect on the dislocations slips. The misorientation is smaller have larger strain degree while the misorientation is larger have smaller strain.

Acknowledgements
This work was supported by National Key Research and Development Program of China (No. 2016YFC0801901).

References:

[1] *R.J. Rioja, J. Liu*, The Evolution of Al-Li Base Products for Aerospace and Space Applications, Metal. Mat. Trans. A, 43A(2012) 3325-3337.
[2] *B. Ahmed, S. Wu*. Aluminum Lithium Alloys (Al-Li-Cu-X)-New Generation Material for Aerospace Applications. In: D. Yang, T.B. Zhang, Q. Luo (Eds), Applied Mech. Mater., (2014)104-111.
[3] *R.J. Rioja*, Fabrication methods to manufacture isotropic Al-Li alloys and products for space and aerospace applications, Mate. Sci. Eng. A, 257(1998) 100-107.
[4] V.*K. Jain, K.V. Jata, R.J. Rioja, J.T. Morgan, A.K. Hopkins*, Processing of an experimental aluminum--lithium alloy for controlled microstructure, J Mater. Proc. Tech., 73(1998) 108-118.
[5] *Cheng Yuanyuan, Zheng Ziqiao, Cai Biao* et al. Initiation and Propagation Behavior of Fatigue Cracks in 2197(Al-Li)-T851 Alloy Rare Metal Mater. & Eng., ,40(2011) 1926-1930.
[6] *A. Davidkov, R. Petrov, P. Smet, B. Schepers, L. Kestens*, Microstructure controlled bending response in AA6016 Al alloys, Mate. Sci. Eng. A, 22(2011) 7068-7076.
[7] *Q. Liu, J. Wert, N. Hansen*, Location-dependent lattice rotation and shear strain in rolled aluminium single crystals of cube and Goss orientations, Acta Mater., 48(2000) 4267-4279.
[8] *W.Y. Yang and W.J. Qiang*. Mechanical Behavior of Materials, Chemical Industry Press, Beijing, 2009, p. 71.
[9] *Q. Liu, N. Hansen*, Macroscopic and microscopic subdivision of a cold-rolled aluminium single crystal of cubic orientation, P. Roy. Soc. A-Math. Phy, 454(1998) 2555-2591.
[10] *T. Zhai, A.J. Wilkinson, J.W. Martin*, A crystallographic mechanism for fatigue crack propagation through grain boundaries, Acta Mater., 48(2000) 4917-4927.
[11] *B. X Zhou*, A Study of the preyielding microstructure in 3%Si-Fe alloy, Acta Merallurgica Sin., (1983) 31-39.
[12] *R. A, H.F. J, R.G. S, H. M.* Recrystallization and related annealing phenomena, Elsevier, Netherlands, 2004.
[13] *D.I. L, K. H, H. K*, The Nucleation of Recrystallisation and the Development of Textures in Heavily Compressed Iron-Carbon Alloys, Texture, 1(1974) 151-156.
[14] *V Tvergaard and A Needleman*. Shear Band Development in Polycrystals. P. Roy. Soc. A-Math. Phy. (1993)443-547.
[15] *Pan Jinsheng*, Tong Jiangmin, and Tian Minbo. Foundamental of Materials Science[B], Tinshua Press, Beijing, 1998.

IOP Conf. Series: Materials Science and Engineering **359** (2018) 012031 doi:10.1088/1757-899X/359/1/012031

Effect of Talc on High-temperature ExothermicPeak and Properties of Polypropylene

Q Chen and R B Yu*

PolymerPhysicsLaboratory, East China University of Science and Technology, 130 Meilong Road, Shanghai 200237, China

Email: ruobingyu319@163.com

Abstract: The exothermic peak at high temperature of pure polypropylene (PP) was found in the DSC curve, which has not been reported in the literature. In this paper, Talc was added into PP to obtain talc-modified PP. When the content of talc was 20%, the exothermic peak at high temperature of the modified PP disappeared. The reason for disappearance of high-temperature exothermic peak is discussed. The conclusions are as follows: (1) With increase in talc, the enthalpy change of high-temperature exothermic peak was lower; (2) When the content of talc was up to 20%, the relaxation time increased and the mobility of molecular chain decreased;(3) The decomposition temperature of PP containing 20% talc was up to 325℃ (300℃ for pure PP), and the maximum decomposition temperature was367℃ (328℃ for pure PP).

1. Introduction

Polypropylene is a semi-crystalline polymer. It has several common disadvantages like brittleness, poor rigidity, molding shrinkage, poor high temperature creep etc.[1] The addition of talc can improve the thermal deformation temperature,[2-5]the rigidity and increase the dimensional stability of products, and reduce the molding shrinkage of modified polypropylene.

At present, there are many studies on talc-modified polypropylene,[6-8] including the rheological behaviorand the heat resistance.[9-11]In this paper, the high-temperature exothermic peak (207°C) of pure polypropylenewas found in Figure 1, which has not been reported before. As can be seen in Figure 2, the high temperature exothermic peak in the modified polypropylene disappearedwhen the talc content was 20%. The cause of disappearance of the high-temperature exothermic peaks was studied. The relationship between the structure and the properties was researched.

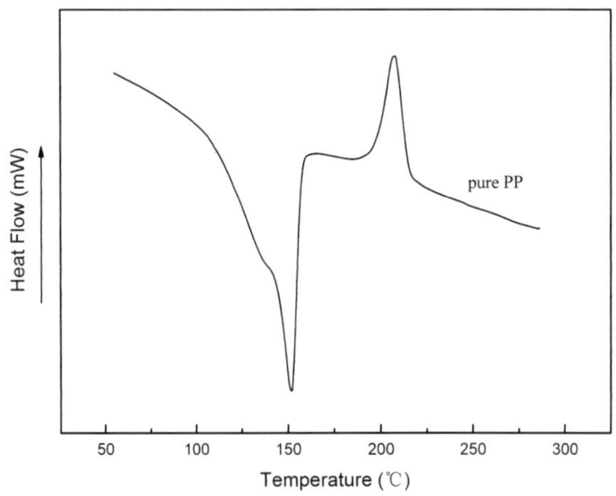

Figure.1. DSC curve of pure PP

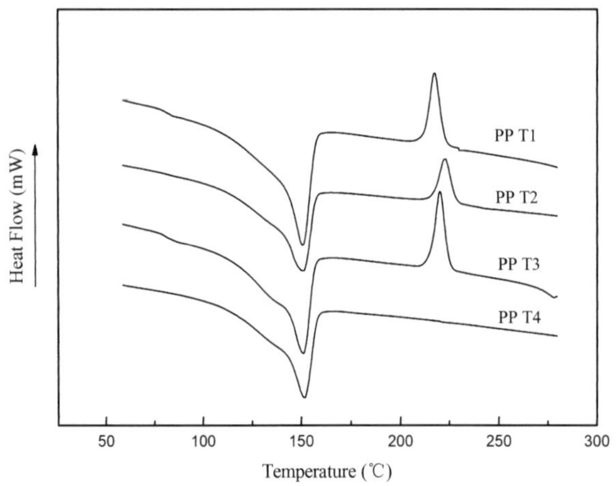

Figure.2.The DSC curve of PP modified by talc

In this paper, the 1250 mesh talc was used to modify the polypropylene. The relation between high-temperature exothermic peak and heat resistance was discussed.

2. Experimental

2.1 Materials
PP M800E was purchased from Shanghai Petrochemical Co., Ltd. Talc(1250 mesh) was purchased fromJiangyinultramicroGuangyuan Co. Ltd

2.2 Instruments
Differential scanning calorimetry (DSC) analysis was performed on Q2000differential scanning calorimetry from 40℃ to 300℃ at 10℃/min. Structural rheology analysis was scanned on HAAKE RS6000 from 0.01-100Hzat 190℃.Thermogravimetric analysis was tested on Q50 thermogravimetric analyzer (TGA) from 40℃ to 400℃ at 10℃/min

2.3Sample preparation

IOP Conf. Series: Materials Science and Engineering **359** (2018) 012031 doi:10.1088/1757-899X/359/1/012031

The composition of talc-modified polypropylene was shown in Table 1. Polypropylene and talc were weighed according to the composition ratio and then were added into the mixer for blending at 175 ℃. The talc-modified polypropylene was produced after 5 minutes.

Table 1. The Formula of PP modified by talc

Talc modified polypropylene formulations	w(PP),%	w(Talc),%
pure PP	100	0
PP T1	95	5
PP T2	90	10
PP T3	85	15
PP T4	80	20

3.Results and Discussion

The values of exothermic temperature and enthalpy change of PP modified by talccan be obtained from Figure 1 and Figure 2. The specific values are shown in Table 2.

Table 2. Exothermic temperature and enthalpy change value of PP modified by talc

Talc modified polypropylene formulations	Exothermic peak temperature/℃	Exothermic enthalpy change/(J/g)
Pure PP	207	36.0
PP T1	218	13.1
PP T2	223	11.0
PP T3	220	14.0
PP T4	/	/

The addition of a small amount of talc causes the high-temperature exothermic temperature of PP to shift to the high temperature direction. When the content of talc reaches to 20%, the exothermic peak at high temperature does not appear in therange of measurement. The addition of talc can help reduce the high temperature enthalpy change ofexothermic peak.

Figure 3 shows the relaxation time spectra of different content talc-modified polypropylene system at 190 °C.

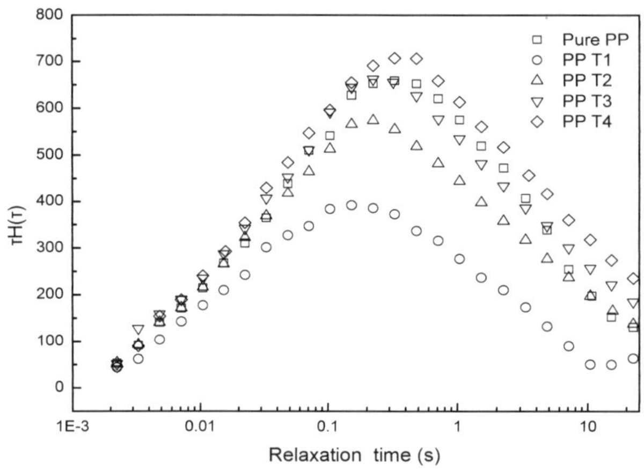

Figure 3.Weighted relaxation spectra of PP modified by talc

Table 3. Characteristic relaxation time of PP modified by talc

Talc modified polypropylene formulations	Characteristic relaxation time（s）
pure PP	0.33
PP T1	0.15
PP T2	0.23
PP T3	0.33
PP T4	0.49

As can be seen from Figure3, there is only one characteristic relaxation peak on the weighted relaxation spectra. So the corresponding time is the characteristic relaxation time. As it is shown in Table 3, the corresponding characteristic relaxation times are 0.33s, 0.15s, 0.23s, 0.33s, 0.49s, respectively. The addition of the small amount of talc can reduce the relaxation time of the system, and with the increase of talc content, the characteristic relaxation time increases gradually. The longer the relaxation time is, the more difficult the molecular chain moves.

When the temperature reaches T_d (temperature at 5% weight loss), the side chain of the polymer will break. when the temperature continues to rise, the polymer's main molecular chain will break, and then the decomposition rate increases sharply.T_p (temperature at the fastest decomposition rate) is available through the differential treatment of TGA curve. Thus the T_d and T_p are two basic parameters that characterize the heat resistance of the material. We conducted the TGA test of talc-modifiedpolypropylenein the atmosphere of air. Figure 4 is the thermogravimetric curve of talc-modified polypropylene. The T_d and T_p obtained in Figure4 are shown in Table 4.

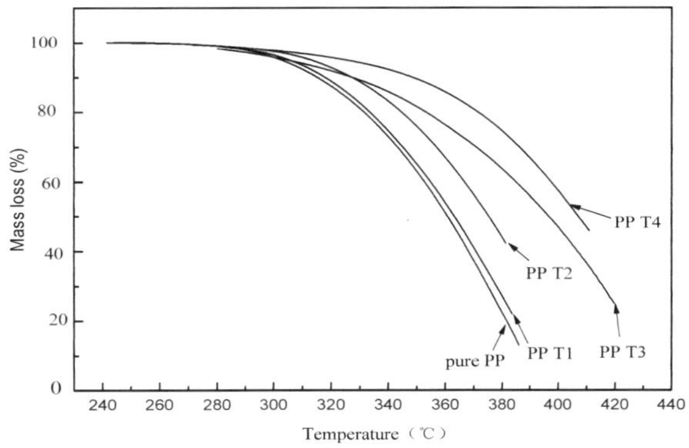

Figure 4. The TGA curve of PP modified by talc (in air)

Table 4. The T_d and $T_p$of PP modified by talc

Talc modified polypropylene formulations	T_d/℃	T_p/℃
pure PP	300	328
PP T1	305	330
PP T2	312	338
PP T3	315	356
PP T4	325	367

It can be seen from Figure 4 and Table 4 that the addition of talc in polypropylene can improve the T_d of the polypropylene system. When the content reaches 20%, T_d reaches to 325℃ and T_p is 29℃ higher than pure PP. The addition of talc helpto improve the heat resistance of polypropylene.

4. Conclusion

(1) The high-temperature exothermic peak appeared in the DSC curve of pure polypropylene, which means there are some active reaction points in the system. Adding a small amount of talc into polypropylene can significantly reduce the enthalpy change. When the talc content is 20%, theexothermic peak disappears,which means the system activity was depressed.
(2) Adding a small amount of talc into polypropylene can reduce the characteristic relaxation time of the polypropylene system. With the increase of adding amount, the characteristic relaxation time increases and the molecular chain movement becomes difficult.
(3) With the increase of talc, the stability of the system was improved gradually. Talc can effectively improve the heat resistance of polypropylene.

References

[1] Natarajan S 2015 *Mater. Manuf. Processes* **31** 1-2
[2] Wang K, Zhang Q, Wu Z, Du R and Fu Q 2001 *Eng. Plast. Appl.***12** 43-6
[3] Parija S, Nayak S K, Verma S K and Tripathy S S 2010 *Polym. Compos.***25** 646-52
[4] Zheng C, Liu J, Fan J, Luan Y, Song L 2016 *Mater. Des.***91**1-10
[5] Beltran M I, Garcia J C, MarcillaA , Hidalgo M and Mijangos C 1999 *Polym. Degrad. Stab.***65** 65-73

IOP Conf. Series: Materials Science and Engineering **359** (2018) 012031 doi:10.1088/1757-899X/359/1/012031

[6] Wong C Y, Lam F 2002 *Polym. Test.* **21** 691-6
[7] Briedis I P and Faitelson L A 1976 *Mech. Compos. Mater.* **12** 278-86
[8] Gafur M A, Nasrin R, Mina M F, Bhuiyan A H, Tamba Y and Asano T 2010 *Polym. Degrad. Stab.* **95** 1818-25
[9] Samsudin M S F, Ishak Z A M, Jikan S S, Ariff Z M and Ariffin A 2006 J. *Appl. Polym. Sci.* **102** 5421-26
[10] Alavi M, Esfandeh M, Morshedian J and Jahani Y 2013 *e-Polym.* **10** 276-91
[11] Rotzinger B 2006 *Polym. Degrad. Stab.* **91**2884-87

IOP Conf. Series: Materials Science and Engineering **359** (2018) 012032 doi:10.1088/1757-899X/359/1/012032

Surface Segregation in Ag/TiO_x 3D Nanocomposite Prepared by Physical Vapor Deposition

J Xiong[1] and L Y He[2]*

[1]Hubei Collaborative Innovation Center for High-efficiency Utilization of Solar Energy, Hubei University of Technology, Wuhan 430068, China
[2]School of chemical engineering and pharmacy, Wuhan Institute of Technology, Wuhan 430205, China

Corresponding author's e-mail address: lyhewit@163.com

Abstract. The antimicrobial activities of silver based nanocomposites are usually studied in terms of Ag content and ion release rate. Under this condition, controllable silver ions release with high antibacterial activity is the basis for silver based nanocomposite. The goal is to investigate the influence of O_2 content and titanium oxide barrier thickness on the evolution in morphology. The SEM/TEM results showed that the size of Ag nanoparticles has a clear dependence on O_2 concentration in reactive sputtering process; increased oxygen implies larger Ag nanoparticles in the matrix. In addition, a clear suppressing effect and better size distribution is obtained after the thickness of coated titanium oxide barrier is verified.

1. Introduction

3D nanocomposite thin films consisting of noble metal nanoparticles embedded in a transparent and non-toxic matrix such as titanium oxide, have received increasing attention because when these features are combined with the size-tunable optical, electronic and antibacterial properties of nanoparticles, the possibilities are wide [1]. Titanium oxide has been recognized as a well-matched biomaterial matrix, and various techniques can be applied to modify the surface properties for large range of use [2]. They show considerable promise in biomedical applications, where device surfaces are required to be biocompatible, as well as mechanically and chemically stable. An antimicrobial property of titanium oxide is also a beneficial characteristic of biomedical devices, as many clinical studies have reported that bacterial infection after implantation is a significant complication [3–5].

Silver as a noble metal is well known as one of the most interesting antibacterial materials because of its excellent broad spectrum antibacterial activates and they can come in a variety of shapes and sizes [6–9]. The antimicrobial activities of silver based nanocomposite are usually studied in terms of Ag content and ion release rate. Silver ions has been reported to interact with nucleic acids and cytoplasmic components and to bind bacterial cell membrane, in addition, due to interaction between the silver ions and DNA structure, respiratory chain enzymes of bacterial their multiplication would be prevented [10–12]. Under this condition, controllable silver ions release with high antibacterial activity is the key point of silver based nanocomposite.

Many way can be used to produce these nanocomposites thin films such as chemical synthesis, ion implantation, magnetron sputtering and others. The magnetron sputtering methods including RF sputtering and DC reactive sputtering have been widely used due to its flexibility, simplicity in the materials combination and tailoring in size distribution. RF-sputtering is one of the most feasible way because of the capability of obtaining an homogeneous surface coverage at low temperatures [13], however, Comparing with the relative lower deposition rate and simple morphology of RF magnetron sputtering, DC reactive magnetron sputtering is attracted intense interest not only provides controllable structures, sizes and morphologies flexibility and in forming the films with controllable stoichiometry, but also is straightforward to implement in industrial-scale production for the relative high deposition rate and large area coated[14-16].

In this work, varied nanocomposited of titanium oxide thin films containing Ag nanoparticles were synthesized on comercially pure Si, quatz and glass substrates by a DC reactive magnetron co-sputtering method with varied O_2 partial pressure, moreover, RF magnetron sputtering was also selected to serve as a non-oxygen model for comparing.

2. Experimental

The nanocomposite coatings were prepared in a home made deposition vacuum chamber by reactive co-sputtering of Ti target and Ag target from two independent magnetron sources. The base pressure is 10-8 mbar, and O_2/Ar were 1.6% and 16%, a pulse box also used when 16% to avoid arc and target poisoning. Ti target (99.99%, Goodfellow GmbH) was sputtered by a DC planar magnetron source (ION'X 2UHV, Thin Film Consulting) and the Ag target (99.99%, Goodfellow GmbH) by a RF magnetron source (ION'X 2UHV, Thin Film Consulting). Both continuous and independent monitoring of the deposition rates of the Ag and TiO_2 were done in situ by using the two quartz-crystal monitors. The deposition rates were monitored independently for Ag and Ti with quartz-crystal microbalances and by variation of the power applied to the magnetron sources the silver metal volume fraction (MVF) was controlled. The sample holder was rotated throughout the deposition process to achieve uniform and homogenous deposition on all the samples mounted on the sample holder. First, a titanium oxide buffer layer with a thickness of 10 nm was deposited to increased adhesion strength and relax the stress between substrate and coatings. After that, the silver containing titanium oxide nanocomposite (30 nm in thickness) was fabricated with the rotation rate of a substrate holder at a constant rate of 9 rpm. Next, A varied titanium oxide in different thickness was produced for adjusting silver ion release and morphology. Meanwhile, A set of same structure samples were prepared by co-sputtering of TiO_2 ceramic target (99.9%, Williams Advanced Material) and Ag target (99.99%, GoodfellowGmbH) as reference of producing in a pure argon atmosphere.

3. Results and Discussion

3.1. Oxygen dependence

SEM images clearly show these nanocomposite surface morphology, including the silver nanoparticles on the surface region via different oxygen content (Fig. 1(a)–(c)). Fig.1a is a typical image without oxygen influence which is consistency with TEM. However, Ag nanoparticles were irregularly sized and agglomerated after introducing oxgen, from a few nanometers to hundreds of nanometers. We can conclude that the larger Ag nanoparticles existed not only in the matrix but also on the surface. Ag islands and "nodules" were even formed in 16% O_2 because of surface energy. It was reported that the physical mechanism of Ag agglomeration was related to the surface diffusion of Ag atomswhich are driven by surface energy, resulting in the grain boundarygrooving, hillocks formation, and, finally, agglomeration.

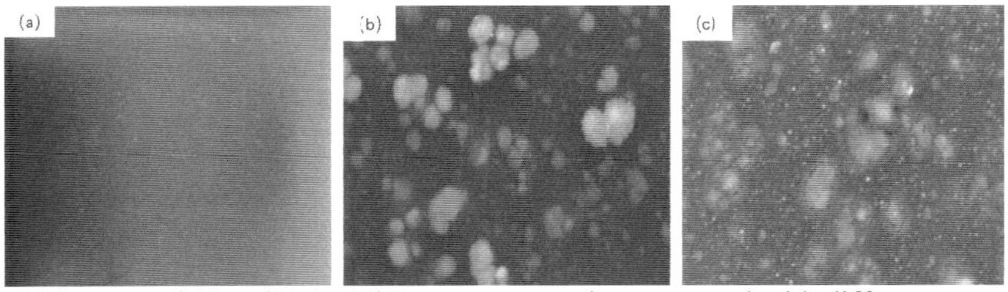

Figure 1. SEM images of 15 *vol*% Ag-TiO_2 nanocomposites prepared with different oxygen partial pressure (a)non-reactive sputtering and without oxygen(b) oxygen partial pressure 1.6% (c) oxygen partial pressure 16%

To study the size of the silver nanoparticles in the Ag/TiOx nanocomposites, TEM analysis was utilized. Fig.2 shows TEM image of the silver nanoparticles with different O_2 concentration. We can see that the Ag nanoparticles are rather small with diameters around 2 nm and distribute quite well in Fig. 2a which produced without oxygen(RF magnetron sputtering). However, the particle size of Ag Nanoparticles growing bigger after oxygen was introduced which increase to 5-10 nm. And we can see clearly from the TEM picture that more oxygen means larger Ag nanoparticles in the matrix.

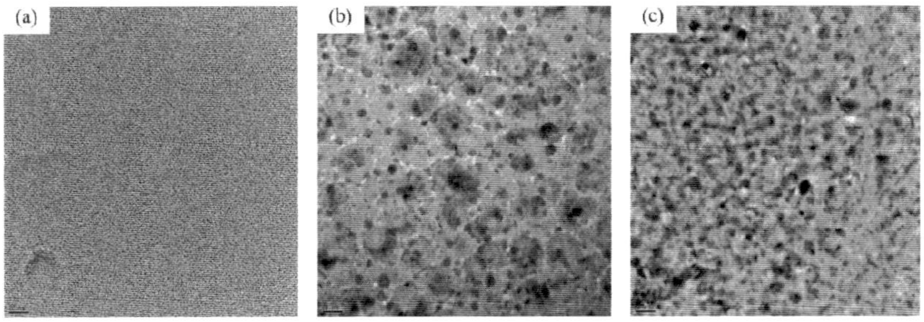

Figure 2. SEM images of 15 *vol*% Ag-TiO_2 nanocomposites prepared with different oxygen partial pressure (a)non-reactive sputtering and without oxygen(b) oxygen partial pressure 1.6% (c) oxygen partial pressure 16%

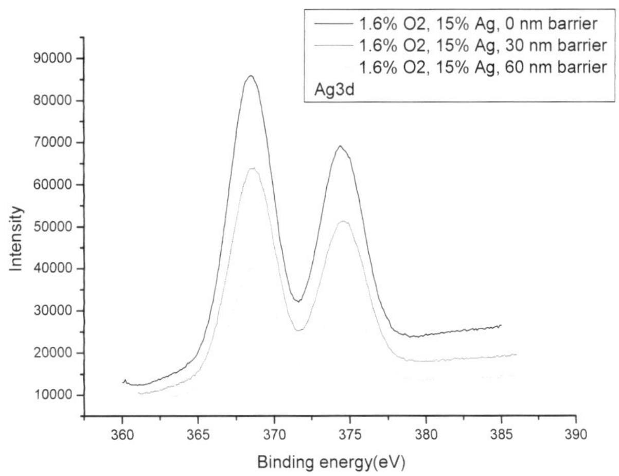

Figure 3. XPSspectra of 15 *vol*% Ag-TiO_2 nanocomposites prepared with oxygen partial pressure at1.6%

IOP Conf. Series: Materials Science and Engineering **359** (2018) 012032 doi:10.1088/1757-899X/359/1/012032

3.2. Barrier layer impact

Figure 3 shows the X-ray photoelectron spectra of the evolution of Ag 3d core levels depends on the barrier thickness during deposition at the surface (1.6% O_2). The decrease of intensity only between 1-2 times which indicates that still metallic silver appears on the surface due to the aggregation that still some big ones appears on the surface. Similar results can also be observed from the SEM images (not showed here)

4. Summary

The Ag/TiO_2 composite thin films were deposited by reactive magnetron sputtering. The morphology of Ag/TiO_2 composite thin films were studied by using X-ray diffraction, scanning electron microscopy, transmission electron microscopy, X-ray photoelectron spectra and atomic absorption spectrometry. The results show that silver nanoparticles in thin film bulk phase and surface aggregation and irregular distribution of the surface increased with the increase of oxygen flow rates.The buffer layer on the inhibitory effects of silver nanoparticles surface segregation is limited.

References

[1] Cho S H, Lee S, Ku D Y, et al. 2004*Thin Solid Films***447–448**68
[2] Geetha M, Singh A K and Asokamani R2009*Prog. Mater. Sci***54**397
[3] Ochoa RAT and Mow CS2008*Knee***15**144
[4] Padovani S, D'Acapito F, Cattaruzza E, et al. 2002*Europ. Phys. J. B***25**11
[5] Paital S R and Dahotre N B 2009 *Mat. Sci. Eng. R***66**1
[6] Morones J R, ElechiguerraJ L and CamachoA 2005*Nanotechnology***16**2346
[7] Sukdeb P, Yu K T and Song J M2007*Appl. Environ. Microb* 1712
[8] Navarro E, Piccapietra F, WagnerB, et al. 2008 *Environ. Sci. Technol***42**8959
[9] Yin L Y, Cheng Y M, Espinasse B, et al. 2011 *Environ. Sci. Technol***45**2360
[10] Lalueza P, Monzón M and Arruebo M 2011*Chem. comm***47**680
[11] Lee D, Cohen R E andRubner M F 2005*Langmuir***21**9651
[12] Song D H, Uhm S H andLee S B 2011*Thin Solid Films***509**7079
[13] Kim S, Choi JE, Choi J, et al. 2009*Toxicol. in Vitro***23**1076
[14] Kawashita M, Tsuneyama S, Miyaji F, et al. 2000*Biomaterials***21**393
[15] Kyriacou S V, Brownlow W J and Xu XH 2004 *Biochemistry***43** 140
[16] Alissawi N, Zaporojtchenko V, Strunskus T, et al. 2012*J. Nanopart. Res***14**928

IOP Conf. Series: Materials Science and Engineering **359** (2018) 012033 doi:10.1088/1757-899X/359/1/012033

High-temperature Mechanical Properties and Microstructure of ZrTiHfNbMo$_x$ (x=0.5, 1.0, 1.5) Refractory High Entropy Alloys

Y W Chen, Y K Li*,X W Cheng, C Wu and B Cheng

School of Materials Science and Engineering, Beijing Institute of Technology, Beijing 100081, China
National Key Laboratory of Science and Technology on Materials under Shock and Impact

E-mails: 2861209902@qq.com; liyunkai@bit.edu.cn

Abstract: Refractory high entropy alloys (RHEAs), with excellent properties at high temperature, have several applications. In this work, the ZrTiHfNbMo$_x$ (x=0.5, 1.0, 1.5) alloys were prepared by arc melting. All these alloys form body centered cubic (BCC) structure without other intermediate phases. The Mo element contributes to the strength of alloys at high temperature, but too much of Mo decreases the plasticity severely and enhances the strength. The ZrTiHfNbMo alloy, whose compressive stress is 1099 MPa at 800° C, is a promising material for high-temperature applications.

1.Introduction

In recent years, RHEAs were proposed due to their extraordinary properties in the high temperature application field[1-3]. Contrasted with the traditional superalloys,RHEAs containmore than five elements with near equal mole ratios (5%-35%)and form the single microstructure [4, 5]. Because the elements of RHEAsare in the group IV, V and VI with the melting point over 1650 ℃, the operating-temperature range of these alloysare more than Al-Ni superalloy [6, 7]. The special performances of RHEAs including high configural entropy, sluggish diffusion, the severe lattice distortionand unique cocktail effect [8-11] contribute to the high strength even in the extreme conditions, which become promising candidates in high temperature materials field.

2. Experimental

The $ZrTiHfNbMo_x$(x=0.5, 1.0, 2.0) alloys were prepared by arc melting under the purified argon atmosphere, and the cooling system employed the water-cooled copper crucible. The as-cast crystalline structure was investigated by X-ray diffractometer (XRD, D8 advance) with $K_\alpha$radiation. The high temperature compressive performances were performed on a Gleeble-3500 thermal simulator, each alloy was heated 800°C and 900°C in turns, the heating rate was setting as 300°C/min and the holding time was 3 minutes.A scanning electron microscope (SEM, HITACHIS4800) was adopted to analyze the microstructure with the secondary electron image.

3. Results and discussion

3.1. Crystal structure and microstructure analysis

The X-ray diffraction patterns of these as-cast $ZrTiHfNbMo_x$alloys are shown in the figure 1.According to the analysis through JADE 5.0 software, all the major peaks of these alloys are consistent withthe single phase of body center cubic without any other precipitated phase. The diffraction peaks aresharp, which indicates that these alloys have high crystallinity.With thc increase of Mo content (the radius of Mo element is apparently smaller than other element), the diffraction peak shift to the low theta due to the crystal lattice become smaller.

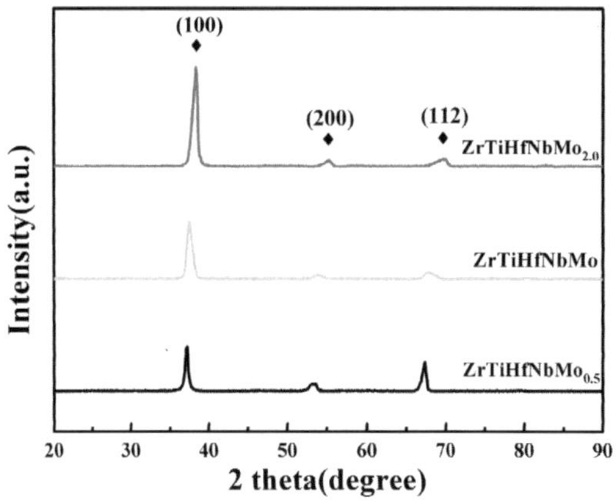

Figure 1. X-ray diffraction patterns of the RHEAs

3.2. High temperature compression performance analysis

The compressive stress-strain curves of $ZrTiHfNbMo_x$ alloys at high temperature are displayed in the figure 2.The process of stress follows that increase rapidly, reduction and remaining a stable value with the deformationincrease.The high-temperature stress of the $ZrTiHfNbMo_{0.5}$ alloy is relatively poor, which is 585 MPa at 800°C and 288 MPa at 900°C.When adjusting the content of Mo element, the stress of ZrTiHfNbMo alloy is1099MPa at 800°Cand804MPaat 900°C. The Mo element has the highest melting point and strength among these elements, which contributes to the properties of softening resistance. As the Mo content continues to increase, the stress of $ZrTiHfNbMo_{2.0}$ alloy is 1155MPa at 800°C, and 897MPa at 900°C, but the strain is less than 15% at 800°C contrasted with thestrain of ZrTiHfNbMo$_{0.5}$ and ZrTiHfNbMo alloy more than 40%, respectively. The excessive Mo element made the lattice distortion increase and the dislocation slides difficultly even at high temperature. As the high temperature could provide much more energy, then it would promote the movement of dislocation, vacancy and grain boundary, which could decrease the stress.

The compression scanning electron microscopies of ZrTiHfNbMo$_x$alloys at high temperature are shown in the figure 3. Compared the figure.3 (a) and (b) with thefigure.3 (d) and (e), the size of grains become larger and the intercrystalline crack is more serious as the temperature increases, part of the dendrites were fused. The fracture feature of ZrTiHfNbMo$_{2.0}$ alloys is the river-like appearance in the figure.3(c), which reveals that the poor plasticity at 800℃.

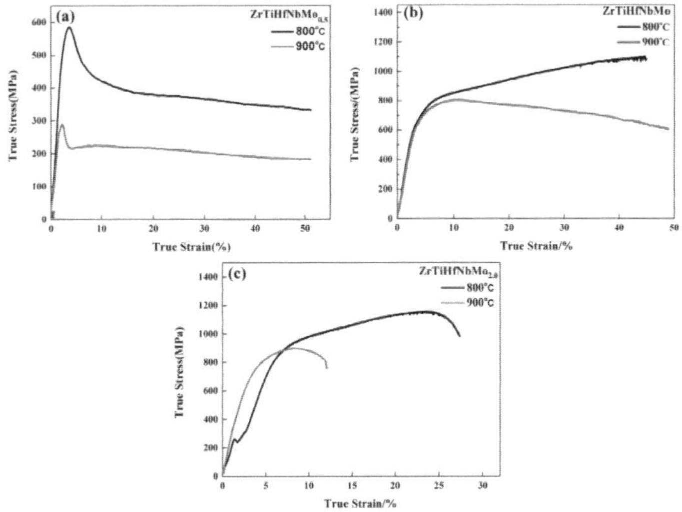

Figure 2. High-temperature compressive stress-strain curves of these alloys: (a) Mo$_{0.5}$; (b) Mo$_{1.0}$;(c) Mo$_{2.0}$

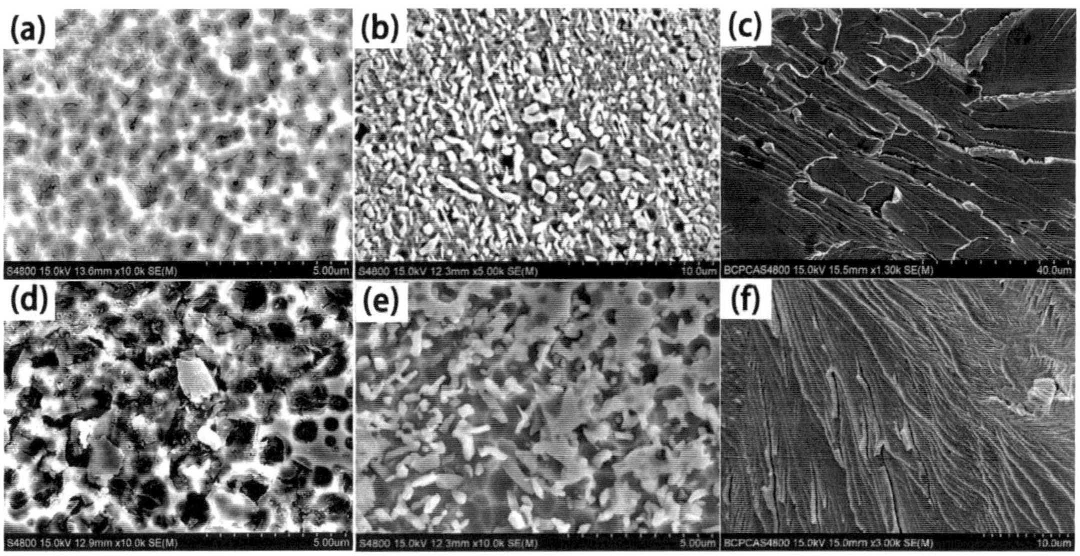

Figure 3. The high-temperature fracture scanning electron microscopies of ZrTiHfNbMo$_x$alloys. Mo$_{0.5}$: (a) at 800℃ and (d) at 900℃; Mo$_{1.0}$: (b) at 800℃ and (e) at 900℃; Mo$_{2.0}$: (c) at 800℃ and (f) at 900℃;

4. Conclusion

In this paper, the $ZrTiHfNbMo_x$ (x=0.5, 1.0, 2.0) refractoryHEAs were melted and the high-temperature properties were analyzed. These alloys formed the body centered cubic structure without other phases. The high temperature stressof these alloys is enhanced with the Mo element increase (Mo content from 0.5 to 1.0). The strength of $ZrTiHfNbMo_{2.0}$ enhances a few and their plasticity decreases rapidlyas the Mo element continues to increase, and the fracture displays the river-like appearance at 800 ℃. The reasonable content of Mo element is contributed to the high temperature mechanical properties, and the HEAs with balance between strength and plasticityat high temperature can be acquiredthrough controlling the Mo content addition.

References

[1] C.-Y. Hsu, J.-W. Yeh, S.-K. Chen, T.-T. Shun, *Wear resistance and hightemperature compression strength of FCC CuCoNiCrAl0.5Fe alloy with boron addition*, Metall. Mater. Trans. A **35A** (2004) 1465-1469.

[2] J.-W. Yeh, S.-K. Chen, J.-W. Gan, S.-J. Lin, T.-S. Chin, T.-T. Shun, C.-H. Tsau, S.Y. Chang, *Formation of simple crystal structures in Cu-Co-Ni-Cr-Al-Fe-Ti-V alloys with multiprincipal metallic elements*, Metall. Mater. Trans. A **35A** (2004) 2533-2536

[3] J.-W. Yeh, S.-K. Chen, S.-J. Lin, J.-Y. Gan, T.-S. Chin, T.-T. Shun, C.-H. Tsau, S.Y. Chang, *Nanostructured high-entropy alloys with multiple principal elements: Novel alloy design concepts and outcomes,* Adv. Eng. Mater. 6 (2004) 299-303.

[4] Yeh, J.-W., et al., *Anomalous decrease in X-ray diffraction intensities of Cu–Ni–Al–Co–Cr–Fe–Si alloy systems with multi-principal elements.* Materials Chemistry and Physics, 2007. **103**(1): p. 41-46.

[5] Cantor, B., et al., *Microstructural development in equiatomic multicomponent alloys.* Materials Science and Engineering: A, 2004. **375-377**: p. 213-218.

[6] Senkov, O.N., et al., *Development of a Refractory High Entropy Superalloy.* Entropy, 2016. **18**(3).

[7] Senkov, O.N., et al., *Microstructure and room temperature properties of a high-entropy TaNbHfZrTi alloy.* Journal of Alloys and Compounds, 2011. **509**(20): p. 6043-6048.

[8] Chuang, M.H., et al., *Microstructure and wear behavior of AlxCo1.5CrFeNi1.5Tiy high-entropy alloys.* Acta Materialia, 2011. **59**(16): p. 6308-6317.

[9] Zhou, Y.J., et al., *Solid solution alloys of AlCoCrFeNiTix with excellent room-temperature mechanical properties.* Applied Physics Letters, 2007. **90**(18): p. 181904.

[10] Yang, X. and Y. Zhang, *Prediction of high-entropy stabilized solid-solution in multi-component alloys.* Materials Chemistry and Physics, 2012. **132**(2-3): p. 233-238.

[11] JianBin Guo, YunKai Li, *Study on microstructure and mechanical properties of five-element refractory high entropy alloys*, D. Beijing Institute of Technology, 2015.

IOP Conf. Series: Materials Science and Engineering **359** (2018) 012034 doi:10.1088/1757-899X/359/1/012034

Tough and Reinforced Polypropylene/Kaolin Composites using Modified Kaolin

J L Yao[1], H X Zhu[1], Y B Qi[2],M J Guo[1], Q Hu[1]and L Gao[3]
1. School of Materials Science and Engineering, Wuhan Institute of Technology, Wuhan 430205, China;
2. National Engineering Laboratory for Fiber Optic Sensing Technology, Wuhan University of Technology, 430070, China;
3.School of Chemistry and Environmental Engineering, Jianghan University, Wuhan, 430056, China

Corresponding author e-mail: junlongyao@yahoo.com

Abstract: Polypropylene (PP)/kaolin composites have been prepared by filling modified kaolin with diethylenetriaminepentaacetic acid (DTPA) into the PP matrix. The surface modification of kaolin particles effectively improves the compatibility between kaolin and PP matrix. It is conducive for uniform dispersion of inorganic particles in the matrix, and enhances the mechanical performance of the composites. Compared with plain kaolin, the mechanical properties of the modified composites exhibit higher tensile strength, bending strength, impact strength and melt index simultaneously. The DTPA modification of kaolin overall enhances the mechanical properties of PP composites. It meets the requirements in various applications, and makes the modified experiment interesting in modern teaching.

1. Introduction

Polypropylene (PP) is one of the most important commercial thermoplastics because of relatively low cost, versatility, and recyclability good mechanical performance. However, its application is limited as an engineering thermoplastic polymer because of the relatively poor impact resistance and stiffness[1].

In order to improve the mechanical properties of polypropylene (PP),various polymer blends have been fabricated by adding different fillers [2] in PP matrix, and some toughening mechanisms have been investigated in past years. However, there is poor interfacial adhesion between the PP matrix and filler which usually results in decreased toughness. The usual incompatibility between the thermoplastic and the filler, which leads to large size of the dispersed phase inadequate for toughening, and is usually overcome by chemical modification of the filler[3-9].

In this study, PP has been mixed with kaolin modified with diethylenetriaminepentaacetic acid using conventional teaching experiment method. The phase behavior of the blends was studied by FTIR, and the mechanical behavior was studied by means of tensile, bending, impact and melt index test experiments, which make the classic experiment more interesting in modern teaching..

2.Experimental

2.1. Materials

Polypropylene（PP）, supplied by China Dushanzi Petrochemical Limited Branch
Kaolin, supplied by FoshanYufeng Powder Co., Ltd
Diethylenetriaminepentaacetic acid（DTPA）, supplied by Sinopharm Group Chemical Reagent Co., Ltd.

2.2. Preparation of modified kaolin

The kaolin, DTPA and alcohol were added into a three-neck flask agitated for 30min.Subsequently, the product was dried at room temperature.

2.3. Preparation of composites

The modified kaolin was melt-blended with PP in twin-screw extruder. And the blends were injection molded into ultimate composites standard bars for mechanical test with injection molding machine.

2.4. Mechanical test

Tensile bending impact and melt index test were conducted. For each test and type of composite, ten specimens were tested and the average values are reported.

3. Result and discussion

3.1.Tensile strength

The tensile data of the PP/kaolin composites(kaolin content are respectively 0,10,20,30,40,50 wt%) is shown in Fig.1.It can be seen that kaolin fillers increased the tensile strength of the composite in low content,but then led to decrease in the tensile strength when the fillers contents reached 20wt%.Above 20wt% there is no further increase in the tensile strength.

In comparison with composites filled by kaolin and modified kaolin, all of the composites filled by modified kaolin show higher tensile strength.

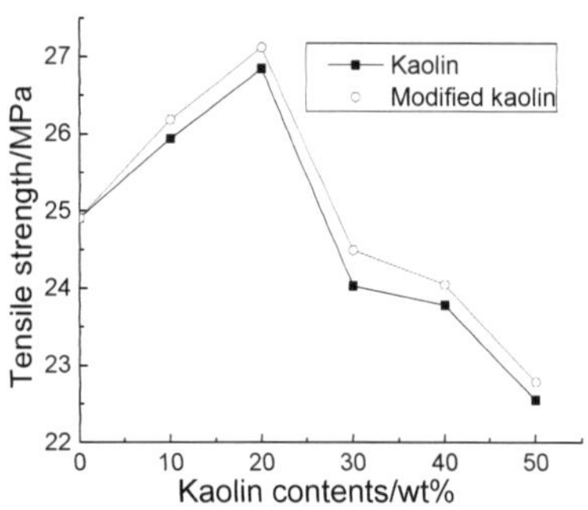

Figure1.Effect on the tensile strength of the PP/kaolin for increasing wt%kaolin

3.2. Bending property

The bending properties of PP/kaolin composites with varying amount of kaolin are shown in Fig.2a and Fig.2b. It can be seen that kaolin fillers can increased the bending strength of the composites in the lower loading region, but then led to decrease in the bending strength when the fillers contents reached 20 wt%. Above 20wt%there is no further increase of bending strength. And the bending modulus of the composites increases with the kaolin content.

The rule of composites filled by modified kaolin is similar with that of it filled by unmodified kaolin, and the bending strength of all the composites filled by modified kaolin are higher than that of the composites filled by the unmodified kaolin.

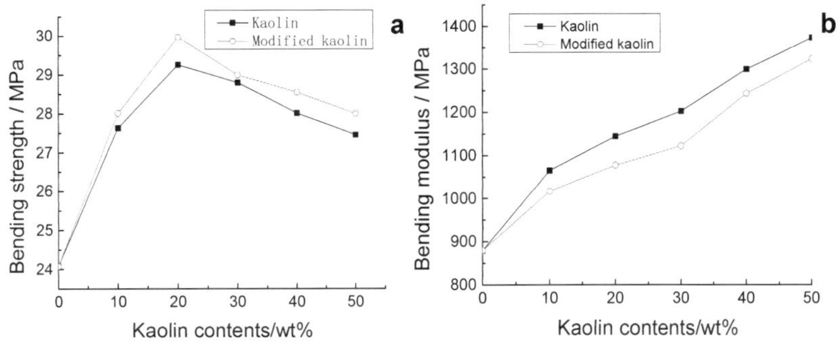

Figure 2 Effect on the bending properties of the PP/kaolin for increasing wt%kaolin

3.3. Impact strength

Fig. 3 shows the impact strength of PP/kaolin composites as a function of kaolin content.The impact strengthdecreases with addition of kaolin.The kaolin is inorganic rigid with high surface energy. When mixed with the hydrophobic PP, the agglomeration of kaolin particles became serious, and the interfacial adhesion of kaolin particles and PP matrix is small, which becomes the weak point of composite materials.The addition of rigid particles limited the movement of the PP macromolecule chains, weakening the ability to cope with external forces, resulting in an increase in the brittleness of the composites[10]. It led to the impact strength of composites decrease from 9.4kJ/m^2 to 7.0kJ/m^2.

Research on the composites filled by modified kaolin, the addition of modified kaolin led the impact strength increase at first and decreasesubsequently,which show that the impact strength of the materials has been greatly improved by surface modification.

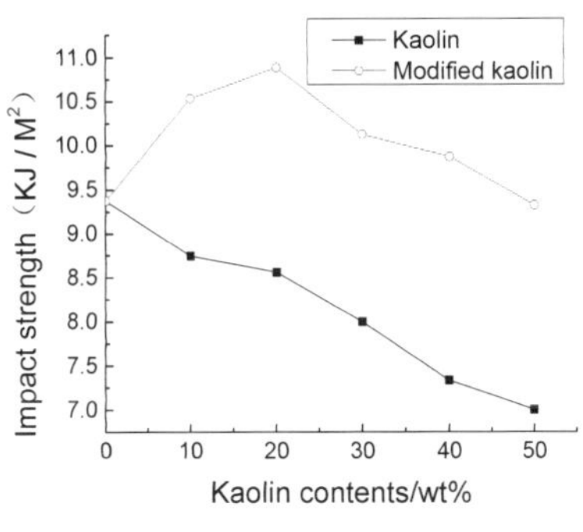

Figure 3 Effect on the impact strength of the PP/kaolin for increasing wt% kaolin

3.4. Melt index

Fig.4 shows the melt index of PP/kaolin composites with varying amount of kaolin.The melt index decrease with addition of kaolin. During the process of shear dispersion, kaolin particles tend to agglomerate and form larger particles. When the shear flow occurs, the collision and friction between particles produce flow resistance, which leads to the decrease of system mobility.

After the surface modification, the effect on melt index of composites was decrease. The melt indexes of all composites filled by modified kaolin are higher than that of composites filled by unmodified kaolin.

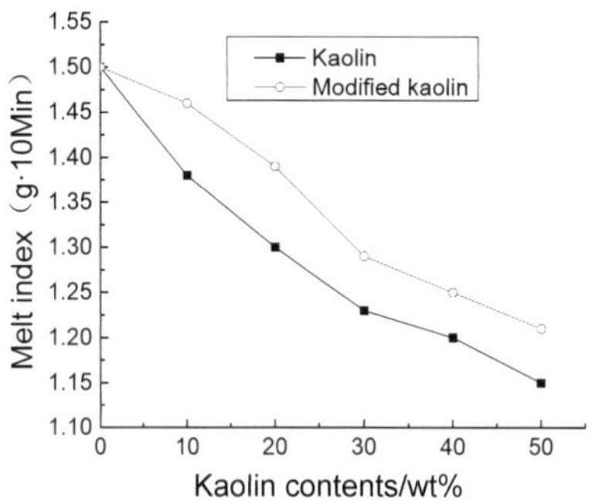

Figure 4 Effect on the melt index of the PP/kaolin for increasing wt%kaolin

3.5. FTIR

From the curves of Fig.5,in modified kaolinthe peak at 1792cm^{-1}is the absorption peak of C=O. absorption peaks at 1402cm^{-1} is smaller and deeper due to the stretching vibration of C-N bond. The absorption peak at 866cm^{-1} is the out of plane bending vibration of O-H bond in carboxylic acid, the absorption peak at 710cm^{-1} is the plane bending vibration of N-H bond,suggesting the exist of amido and carboxyl. It demonstrated that composites contains DTPA

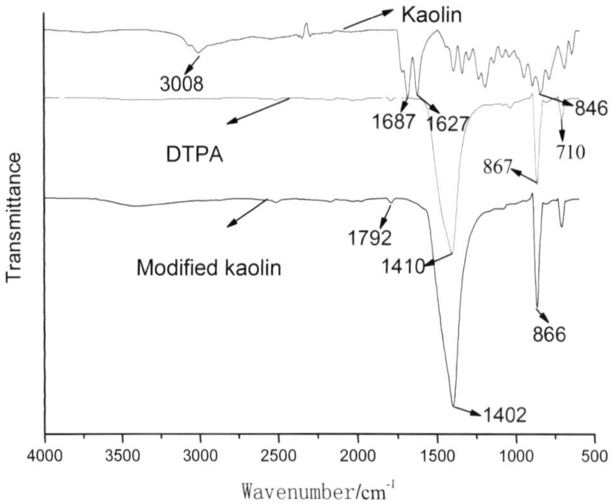

Figure 5FTIR spectra of kaolin and modified kaolin

4. Conclusion

The PP/kaolin composites showed higher tensile strength, bending strength, impact strength and melt index. When the content of modified kaolin is 20wt%,the composite has the best mechanical. Comparing the tensile strength, bending strength, impact strength and melt index of the composites filled by 20wt% modified kaolin and 20wt% unmodified kaolin, the tensile strength was increased from 26.8MPa to 27.1MPa, the bending strength was increased from 29.2MPa to 29.9MPa, the impact strength was increased from 8.5kJ/m^2 to 10.9kJ/m^2 and the melt index was increased from 1.30g/10min to 1.39g/10min. So,the conventional modification method of kaolin improved the compatibility of PP and kaolin, and tough the composites. The composite filled by DTPA modified kaolin can fully meet the demand of automotive, home appliances, construction, machinery and other areas.

Acknowledgments
The authors gratefully acknowledge the financial supports from the Teaching Project of School of Chemistry and Environmental Engineering, Jianghan University, Wuhan, China.

References

[1] Yan Z, Zhang J and Zhang H2012, *Journal of Engineering of Heilongjiang University*, **665** p 319-22.
[2] Dasari A, Zhang QX, Yu ZZ and Mai YW2010, *Macromolecules***43(13)** p 5734-9.
[3] Sivarao and Vijayaram T R2009, *T INDIAN I METALS*, **62(3)** p 245-8.
[4] Haque M, Rahman R and Islam N2010, *Composites Part A Applied Science & Manufacturing*, **40(4)** p 511-7.
[5] Herrera-Franco P J and Valadez-González A 2005,*Composites Part B Engineering*, **36(8)** p 597-608.
[6] Guan B W, Chen S F and Liu K P2013, *Applied Mechanics & Materials*, p 357-360:1019-22.
[7] Dányádi L, Móczó J andPukánszky B2010,*Composites Part A Applied Science & Manufacturing*, **41(2)** p 199-206.
[8] Mareri P, Bastide S andBinda N 1998,*Composites Science &Technology*, **58(5)** p 747-52.
[9] Dikobe D G andLuyt A S 2017,*ThermochimicaActa*, p 654.
[10] Maio E D, Iannace S andSorrentino L2004, *Polymer*, **45(26)** p 8893-900.

IOP Conf. Series: Materials Science and Engineering **359** (2018) 012035 doi:10.1088/1757-899X/359/1/012035

The VHCF experimental investigation of FV520B-I with surface roughness Ry

J L Wang, Y L Zhang, M C Ding, Q C Zhao

School of Mechanical Engineering, Dalian University of Technology, Dalian, China

E-mail: zylgzh@dlut.edu.cn

Abstract: Different surface roughness type (Ra and Ry) has different effect on the VHCF failure and life. Ra is widely employed as the quantitative expression of the surface roughness, but there are few fatigue failure mechanism analysis and experimental study under surface roughness Ry. The VHCF experiment is conducted out using the specimen with different surface roughness values. The surface roughness Ry is employed as the major research object to investigate the relationship and distribution tendency between the Ry, fatigue life and the distance between internal inclusion and surface, and a new VHCF failure character is proposed.

1. Introduction

FV520B-I has been widely applied in the manufacturing of all different kinds of large-scale mechanical equipment [1-2] and the core mechanical parts for its numerous good mechanical properties. Its fatigue life generally required to reach or over 10^7 cycles, which implies that the main factor causing the very-high cycle fatigue failure is the internal inclusion. However, the poor working condition always results in the deterioration of the surface quality during the working process. It has been adopted that the bad surface quality (that is big surface roughness value) will weaken the fatigue property to resist fatigue failure especially the fatigue strength of the material, especially the deep groove of the surface will affect the VHCF fatigue life. Different surface roughness type (Ra and Ry) has different effect on the VHCF failure and life, such as: the surface roughness can change the stress distribution of the specimen to affect internal failure sites that is the internal inclusion location.

Quite a few theories have been proposed based on the investigation of fatigue failure caused by surface roughness. Stenberg [3] investigates the influence of the surface quality to the fatigue strength. Different codes and design recommendations are provided to improve the fatigue strength of high strength steel S700 and S960. Lacerda [4] studied the effectiveness of the evolution of surface roughness to the fatigue strength under different loads and fatigue life, and the relationship between these parameters can be determined according to this study. Li [5] focuses on the fatigue life estimation with different surface roughness, S-N curves and P-S-N curves are presented in this study, and crack initiation life and crack propagation life are estimated based on the classic fatigue life model. Silva and Bagehorn [6] study the influence of surface roughness on the fatigue property of titanium alloy, it was found that the small differences of surface roughness will result in the big reduction of the fatigue strength. And the effectiveness of each treatment to the fatigue property is analyzed in detail. Zhang [4] carried out a investigation about the influence of surface roughness to VHCF life, the relationship between surface roughness and fatigue life is proposed.

Generally, Ra is widely employed as the quantitative expression of the surface roughness. But Ra means the average value of the surface quality, so it is too conservative to the quantitative analysis of

VHCF failure. Instead of Ra, Ry means the deepest groove of the surface roughness will be suitable for the quantitative analysis of VHCF failure to investigate the coupling effect of surface roughness and internal fatigue defect to VHCF life. Moreover, the Ra can be the same or in the same level for different specimens, while the value of Ry must be different. So Ry can be employed to analysis the qualitative and quantitive relationship between the surface roughness and fatigue life.

In this paper, a more detailed and targeted research is carried, the main research focuses on the study of the very-high cycle fatigue failure mechanism affected by surface roughness Ry, and the proposition of the new VHCF failure rule. Further study of effectiveness of Ry to VHCF life is carried out. A new qualitative relationship between the D_{a-s} (the distance between the internal inclusion and the surface) and the Ry is proposed based on the fatigue failure analysis, also the new qualitative relationship between D_{a-s} and fatigue life will be proposed to reflect the effectiveness of Ry to the VHCF life.

2. The very-high cycle fatigue experiment

The material studied in this research is martensitic precipitation hardening stainless steel FV520B-I. The surface roughness of the specimen is set according to the application of FV520B-I in the actual working condition and Ra is employed as the processing standard which are divided into four levels with the comprehensive consideration of actual working condition, the processing condition and the test requirement: $Ra_1 \leq 0.1\mu m$, $0.1\mu m < Ra_2 \leq 0.2\mu m$, $0.2\mu m < Ra_3 \leq 0.3\mu m$, $0.5\mu m \leq Ra_4 \leq 0.8\mu m$. The Ry can be obtained according based on the specimen manufacturing process. D_{a-s} can be measured according to the SEM observations.

Standard hourglass type test specimen is employed as the specimen in this experiment, and the details of the geometrical shape and size are designed according to the elasticity modulus and density as shown in Fig.1

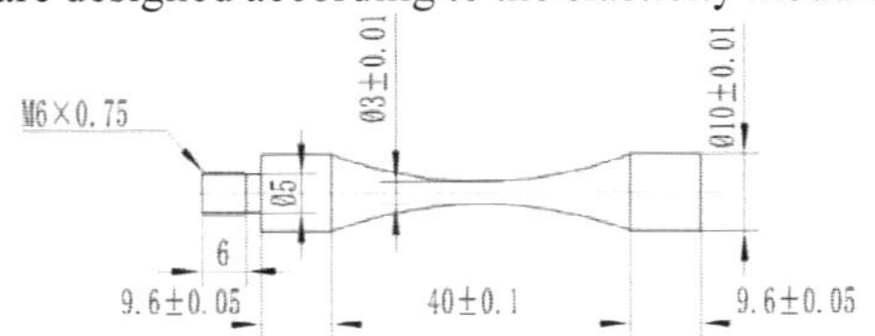

Fig.1. Specimen dimensions

USF-2000 ultrasonic fatigue test system is employed in this experiment with an operating frequency of 20 kHz. The mean stress is set to zero which presented that the stress ratio r = -1. The room temperature (20℃) w as used, which are close to the actually situation in app lication. The intermittent ratio is set to 1:2(working 250ms and stop 500ms), at same time, air cooling is employed to avoid the influence of heat. Both stress amplitude and the relevant fatigue life will be recorded.

3. The effectiveness of surface roughness Ry to VHCF life

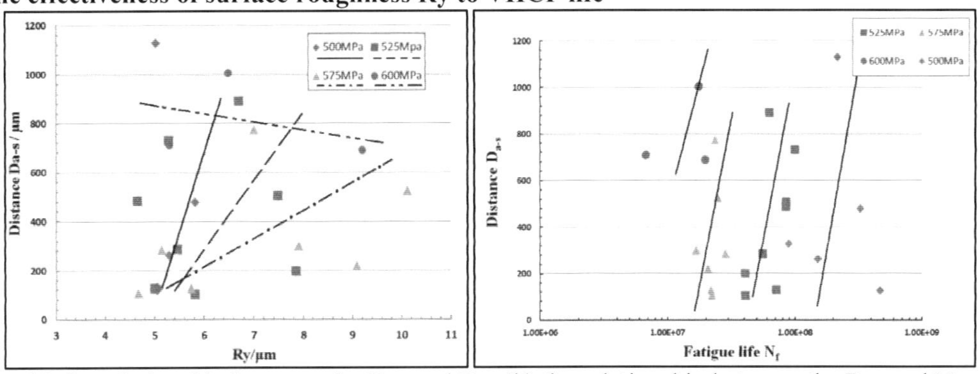

(a) the relationship between the D_{a-s} and Ry; (b) the relationship between the D_{a-s} and N_f
Fig.2. Experimental data

3.1 The effectiveness of surface roughness Ry to the location of the internal inclusion

Fig.2.(a) shows that the overall distribution character: D_{a-s} increases with the increase of the Ry under low stress amplitude situation and decreases with the increase of the Ry under high stress amplitude. A linear relationship is proposed to describe the relationship between the distance D_{a-s} and Ry. With the stress amplitude increases from 500MPa to 600MPa, the slope decreases from positive value to negative.

When the stress amplitude is less than 575MPa and the fatigue life is in VHCF regime, with the increase of Ry, the surface stress concentration is more serious and the VHCF failure becomes more possible to originate from the surface. With the comprehensive competition between surface roughness and internal inclusion, it will take more time to get surface hardening and transit the fatigue crack from surface into internal site, so the relevant VHCF failure takes more time to initiate from the internal inclusion. With the stress amplitude increases, the extra effectiveness of stress amplitude to the VHCF failure leads to the decreases of the relevant VHCF life, but the fatigue life still in the VHCF regime, so the slope of the linear relationship just decrease with the increase of stress amplitude but is still positive value. When the stress amplitude increasing to 600MPa, the relevant life decreases close to high-cycle fatigue regime, and the discretion of the internal inclusion on VHCF failure decreases and the effectiveness of Ry to fatigue failure becomes more obvious. So the fatigue failure will prefer the site close to the surface, and with the influence of big stress amplitude, the fatigue failure occurs before the fatigue crack transfer into deeper location.

According to the above phenomenon, it can be further inferred that with the stress amplitude continuous increasing over 600MPa or the fatigue life is less than 10^7 cycles, the effectiveness of surface roughness to the fatigue failure is more obvious than internal inclusion, so the fatigue initiation location moves closer to the surface and the surface failure will be obtained.

3.2 The effectiveness of distance Da-s to VHCF life

The Fig.2.(b) shows the distribution of distance D_{a-s} and VHCF life. Under the stress amplitude condition, with the change of the location of the internal inclusion, the fatigue stress distribution will also change which results in the change of the VHCF life. So the VHCF life is not only affected by the internal inclusion size, but also the location of the internal inclusion.

The analysis in chapter 3.1 indicates that the big Ry will results in the long distance D_{a-s} in the VHCF regime. Even the bad surface quality will weaken the fatigue property to resist fatigue failure especially the fatigue strength of the material, the internal inclusion is still the most serious factor to the fatigue life in the VHCF regime. So the effect of location of the internal inclusion is more obvious than the effect of the surface roughness to the VHCF life, and with the increase of the surface roughness Ry, both the distance Da-s and the relevant VHCF life increase. This conclusion is proposed under two preconditions: first, the same stress amplitude condition; second, in the VHCF regime.

Another situation is that when the stress amplitude is 600MPa, which is general bigger than the stress amplitude in VHCF regime. As the fatigue life is less than or close to 10^7 cycles, the fatigue failure prefers the surface site and the influence of big stress amplitude is more obvious than the VHCF situation. A serious stress concentration will occur on the site with big surface roughness which will promote the fatigue crack initiation and propagation process and shorten the fatigue life. Moreover, with the internal inclusion gets closer to the surface, the fatigue failure will more prefer to the HCF failure which means the fatigue life decreases. So it is proposed that Ry increases, the distance D_{a-s} decreases, and also the relevant HCF life decreases, and there is a positive correlation between the distance D_{a-s} and both of the HCF life and VHCF life as shown in the Fig.2.(b).

A linear relationship is proposed to qualitatively describe the positive correlation between the distance D_{a-s} and fatigue life N_f in Fig.2.(b). With the decrease of the stress amplitude, the linear relationship moves along the X-axis to the right, and the slopes of the linear relationship in the Fig are almost constant under different stress amplitude conditions. This means the distance D_{a-s} is affected by the coupling effect of surface roughness Ry and stress amplitude, and Ry is the dominant one.

In summary, it can be further inferred that there is a linear relationship between the Ry and VHCF life, and the more specific relationship is: in the VHCF regime, there is a positive correlation between the Ry and VHCF life; in the HCF regime or fatigue life is close to 10^7 cycles, there is a negative correlation between the Ry and fatigue life.

4. Conclusion

First, a linear relationship is proposed to describe the distribution relationship between $D_{a\text{-}s}$ and Ry: in VHCF regime, there is a positive relationship between the $D_{a\text{-}s}$ and surface roughness Ry; when the fatigue life is close to 10^7 cycles or in HCF regime, there is a neglect relationship between the distance $D_{a\text{-}s}$ and surface roughness Ry.

Second, with the stress amplitude over 600MPa or the fatigue life is less than 10^7 cycles, the effectiveness of surface roughness to the fatigue failure is more obvious than internal inclusion.

Third, the $D_{a\text{-}s}$ is employed as the intermediate variables to investigate the relationship between Ry and fatigue life and a linear relationship is proposed: in the VHCF regime, there is a positive correlation between the Ry and VHCF life; in the HCF regime or fatigue life is close to 10^7 cycles, there is a negative correlation between the Ry and fatigue life.

Acknowledgement

The authors gratefully acknowledge the research funding support from The National Natural Science Foundation of China (No.51375074) and the Fundamental Research Funds for Central Universities (No. DUT17ZD230 & No. DUT17GF214).

References

[1] Wang J L, Zhang Y L, Liu S J, Sun Q C and Lu H T Competitive giga-fatigue life analysis owing to surface defect and internal inclusion for FV520B-I. *Int. J. Fatigue*. **87** 203-209.

[2] Wang J L, Zhang Y L, Sun Q C, Liu S J, Shi B W, and Lu H T Giga-Fatigue Life Prediction of FV520B-I with Surface Roughness. *Mater. Design*. **89** 1024-1038.

[3] Stenberg T, Lindgren E, Barsoum Z and Barmicho I Fatigue assessment of cut edges in high strength steel - Influence of surface quality. *Materialwissenschaft und Werkstofftechnik*. **48(6)** 556-569

[4] de Lacerda J C, Martins G D, Signoretti V T and Teixeira R L P Evolution of the surface roughness of a low carbon steel subjected to fatigue. *International Journal of Fatigue*. **102** 143-148

[5] Li C, Dai W, Duan F, Zhang Y and He D Fatigue Life Estimation of Medium-Carbon Steel with Different Surface Roughness. *Applied Sciences*, 7**(4)**, 338.

[6] Silva P S C P D, Campanelli L C, Claros C A E, Ferreira T, Oliveira D P and Bolfarini C Prediction of the surface finishing roughness effect on the fatigue resistance of ti-6al-4v eli for implants applications. *International Journal of Fatigue*.

IOP Conf. Series: Materials Science and Engineering **359** (2018) 012036 doi:10.1088/1757-899X/359/1/012036

Preparation of Degradable Biological Carrier With LCC and its Application in Culture of Hepatocytes

H K Zhao[1], X K Chen [1], H F Wu[1] , J L Li[1] and Y M Xie[1,2,*]

[1]Research Institute of Pulp & Paper Engineering, Hubei University of Technology, Wuhan, China
[2]Hubei Provincial Key Laboratory of Green Materials for Light Industry, Hubei University of Technology, Wuhan, China

* Corresponding author: ppymxie@163.com

Abstract.The purpose of this article is to extract lignin-carbohydrate complexes (LCC) with poplar as raw material which was used to prepare bio-carrier by freeze-drying method. The, chemical properties and morphological of LCC porous biological carriers were analyzed by GPC,FT-IR, scanning electron microscopy （SEM） and optical microscopy. The FT-IR spectrum results indicated that LCC which are composed of lignin and polysaccharide, with a typical LCC structure. Galactose have a specific ability to recognize liver cells owing to the presence of receptors on hepatocytes. Cell counting results showed that the cells increases fastest while the proliferation rate of the liver cell in LCC is obviously higher than that of control group. These results indicated that poplar LCC is very biocompatible which it might be a great potential biological carrier material for human hepatocyte culture.

1. Introduction

As a kind of organic substance, Lignin has large amount in nature, just less than cellulose[1] and chitin, and mostly founded in woody plants and herbaceous plants[2-4]. The lignin-carbohydrate complex (LCC) is composed of lignin and polyglycans (polygalactose, xylose and glucomannan) hrough a firm covalent bond formed by heterogeneous polymers which is a good biological material. Many researchers [5-8] extracted LCC from plants by various methods and further studied their structure and properties. LCC has good biocompatibility and mechanical strength owing to LCC contains both hydrophobic lignin rigid blocks and hydrophilic polysaccharide flexible blocks. In recent years, Erakovic [9], Mansur [10] and Martinez [11] have studied the biocompatibility of lignin/hydroxyapatite bio-scaffolds. These results showed that the combination of lignin and polyhydroxy compounds bulk material has good biocompatibility. Previous studies have shown that the polysaccharide portion of LCC not only has good biocompatibility, but also lignin molecules dose which has aroused widespread concern about the biocompatibility of LCC-based materials.

Porous carrier have large specific surface area which increased the cell growth area, and improve cell culture density [12]. LCC is rich in galactose and mannose which has capability of recognizing by hepatocyte [13-14].In order to study the application of LCC in human hepatocytes culture, the LCC porous biological carrier materials were prepared by freeze drying method. The molecular structure, lignin and monosaccharide contents, and morphologies of porous carriers were characterized by GPC, FT-IR, scanning electron microscopy (SEM) and optical microscopy. These results indicated that LCC has good biocompatibility which can be used as carrier material for human hepatocyte culture.

IOP Conf. Series: Materials Science and Engineering **359** (2018) 012036 doi:10.1088/1757-899X/359/1/012036

2. Experimental section

2.1Materials

Poplar tree was obtained from Wuhan nursery garden. Human hepatocytes was provided by Pricells company (Wuhan, China).

2.2 Preparation of poplar LCC

Poplar wood meal extracted with benzene/ethyl alcohol (2/1,V/V) following hot water, and then dried in vacuum drying oven for 7 days. The extractives-free wood mill was further ground for 72h in a vibration ball mill with water cooling. The Poplar LCC was then purified with previous method [15],which was shown in Fig1.

2.3 Preparation of LCC based Hydrogels porous carriers.

Poplar LCC (100mg) from ball milling of poplar wood were placed in different test tubes. The NaOH solution (200μL,3.3mol/L) was then added to each tube. Subsequently, the mixtures were then stirred in an ice bath for 4h until LCC completely dissolved. To the above solutions, polyethylene glycol glycidyl ether(50μL) was added and further stirred until all polyethylene glycol glycidyl ether was dissolved. The mixture was stirred at 50 °C for 24 h to obtain hydrogel porous biological carrier. This was subsequently added to Erlemeyer flask and then washed with distilled water to until the water become clear and pH values become 7. Then the sample was freeze-drying for 24h.

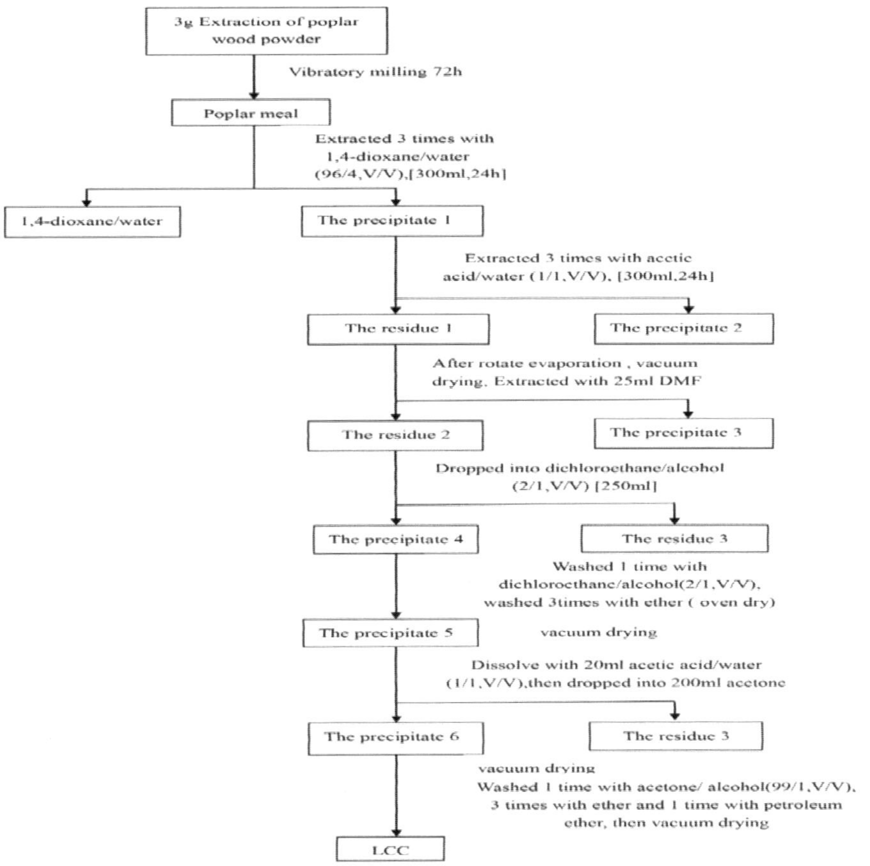

Figure 1. The isolation and purification of LCC from poplar wood.

3. Results and discussions

The peak of 3417cm^{-1} are attributed to the hydroxyl group of poplar LCC which was shown in figure 2. The absorption of alcohol based C-O at 1045cm^{-1} indicated the presence of polysaccharide. The peaks at 1509cm^{-1} and 1421cm^{-1} are assigned to the absorption of the aromatic structures in lignin moieties. The FT-IR spectrum results indicated that LCC which are composed of lignin and polysaccharide, with a typical of lignin-carbohydrate complexes structure.

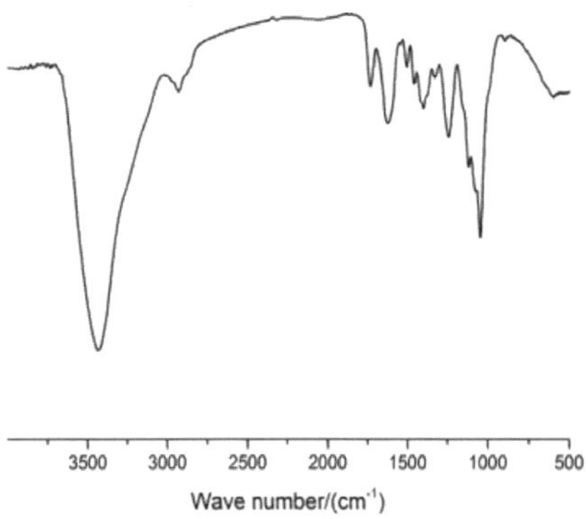

Figure 2. The FT-IR spectroscopy of the LCC.

Galactose can be recognized by the receptors on liver cells with high physiological activity to hepatocytes [16]. Galactose has been used to enhance the selective interaction between biological bio-carrier and hepatocytes. As shown in Table 1, the galactose content in LCC macromolecules is 5.67%. The contents of lignin and total sugar are 23.9% and 64.62% respectively, which will lead to good physical strength of LCC-based porous biological carrier had.

Table1. Chemical composition of poplar LCC.

Compositio n	lignin	Glucose	Xylose	Galactose	Arabinose	Mannose	Total sugar
content/%	27.49	4.83	33.15	5.67	4.25	13.52	61.42

The average molecular weight (Mw), number average (Mn) and polydispersity (Mw/Mn) of poplar LCC were shown in Table 3. The molecular weight of The LCC sample is relative low for bio-carrier. Therefore, it should be cross-linked during the preparation of bio-carrier in order to increase the strength and stability.

Table 2. Results of weight-average (Mw), number-average (Mn) molecular weights and polydispersity indexes (Mw/Mn) of LCC

Sample	Mw	Mn	Mw/Mn

LCC	9786	7153	1.370

Human hepatocytes (L-02) were stably cultured in the presence of a biological carriers. As shown in Figure 3, a large number of hepatocytes are found to adhere to the spherical biological carriers. These results showed that LCC is nonpoisonous, biocompatible and suitable for the hepatocyte culture. The cell growth experimentwith bio-carriersand the control group at 1~7th is shown by cell counting. As shown in Figure 4, the cells of cultured on the bio-carriers and the control group showed a slow increase trend in the first three days. On the 5thday, the cell proliferation reached a maximum. The proliferation rate of the experimental group was significantly higher than that of the control group.

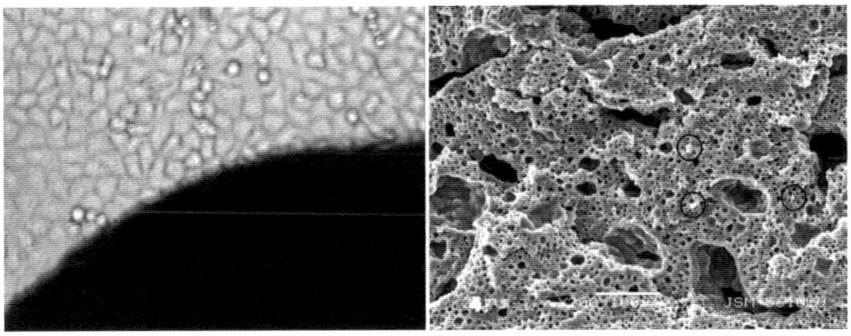

(A) (B)

Figure 3. The inverted microscope image(A) and SEM image (B) of human hepatocytes L-02 cultured on the carriers.

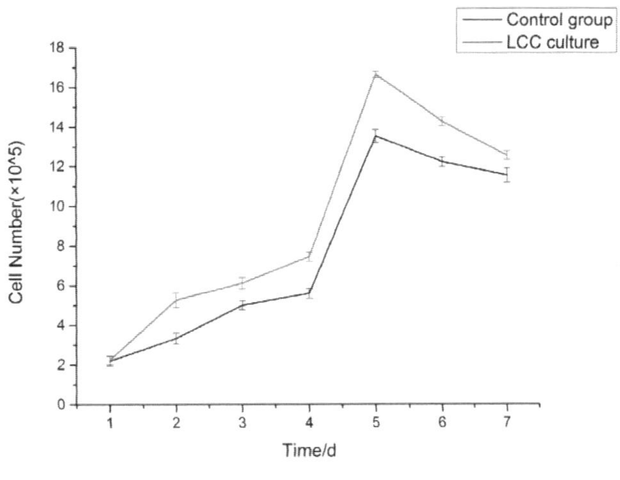

Figure 4. Hepatocytes growth curve

4. Summary

.. Lignin-carbohydrate complexes(LCC) were isolated from poplar wood by ball milling method. The FT-IR analysis showed that the LCC spherical carrier consisted of lignin and polysaccharide with a typical of LCC structure. Chemical analysis showed that the contents of galactose, lignin and total sugar were 5.67%, 27.49% and 61.42%, respectively. The spherical carrier was prepared with the LCC through cross-linking by polyethylene glycol glycidyl ether. The results indicated that the spherical biological carrier has good compatible with the liver cells owing to galactose which can be recognized by receptors of liver cells. A large number of hepatocytes were found to adhere to the spherical biological carrier. The results of cell counting show that the cell proliferation is the fastest on the 5th day.The proliferation rate of hepatocytes with LCC-based bio-carrier is significantly higher than that of control experiment. These results indicated that LCC is non-toxic, biocompatible and suitable for hepatocyte culture.

Acknowledgments
This research was financially supported by National Natural Science Foundation of China(No.31370574, 31300494).

References
[1] Wu Z, Li Y and Zhang L 2017 *Rsc. Adv.* **170** 117-123
[2] Ninan N, Muthiah M, Park I K, Elain A, and Thomas S 2013 *Carbohyd. Polym.* **98** 877-885
[3] Lin H K, Madihally S V and Palmer B 2015 Adv. Drug. Deliver. Rev. **47** 82-83
[4] Pok S, Vohra H and Wehbe C 2017 Exp. Cell. Res. **356** 48-56
[5] Shevchenko S M and Bailey G W 1996 J. mol. Struc-Theochem. **364** 197-208
[6] Sakagami H, Hashimoto K and Suzuki F 2005 Phytochemistry. **66** 2108-20
[7] Singh R, Singh S and Trimukhe K D 2005 *Carbohyd. Polym.* **62** 57-66
[8] Koshijima T, Watanabe T and Yaku F 1989 *J.Am.Chem. Soc.* **397** 11-28
[9] Erakovic S, Veljovic D and Diouf P N 2009 Int. J. *Chem. React. Eng.***7** 113-130
[10] Mansur H S, Mansur A P and Bicallho S M C M 2005 *Key. Eng. Mater.***28** 745-48
[11] Martinez M M, Pacheco B A and Vargas V M 2009 *Rev. Mvz. Cordoba.***14** 1624-32
[12] Naruse K, Tang W and Makuuchi M 2007 *World. J. Gastroentero.* **13** 1516-21
[13] Chatelet C, Damour O and Domard A 2001 *Biomaterials.* **22** 261-68
[14] Park I K, Yang J and Jeong H J 2003 *Biomaterials.* **24** 2331-37
[15] Bjorkman A 1957 *Ind. Eng. Chem. Res.* **49** 1395-98
[16] Yang J, Wang P and Fan J Y 2016 *Paper. Science. &.Technology.* **35** 1-6

IOP Conf. Series: Materials Science and Engineering **359** (2018) 012037 doi:10.1088/1757-899X/359/1/012037

Research and Application of Autodesk Fusion360 in Industrial Design

P P Song[1], Y M Qi[2]and D C Cai[3]

Tianjin University of Technology and Education, Tianjin 300222

Abstract. In 2016, Fusion 360, a productintroduced byAutodesk and integrating industrial design, structural design, mechanical simulation, and CAM, turns out a design platform supportingcollaboration and sharing both cross-platform and via the cloud. In previous products, design and manufacturing use to be isolated. In the course of design, research and development, the communication between designers and engineers used to go on through different software products, tool commands, and even industry terms. Moreover, difficulty also lies with the communication between design thoughts and machining strategies. Naturally, a difficult product design and R & D process would trigger a noticeable gap between the design model and the actual product. A complete product development process tends to cover several major areas, such as industrial design, mechanical design, rendering and animation, computer aided emulation (CAE), and computer aided manufacturing (CAM). Fusion 360, a perfect design solving the technical problems of cross-platform data exchange, realizes the effective control of cross-regional collaboration and presents an overview of collaboration and breaks the barriers between art and manufacturing, andblocks between design and processing. The "Eco-development of Fusion360 Industrial Chain" is both a significant means to and an inevitable trend forthe manufacturers and industrial designers to carry out innovation in China.

1. Background and Significance of Research

(1) Development Trend of Multi – disciplinary Integration. Fusion, literally referring to combination, does integrate a lot of correlated techniques. For example, it fuses Windows and Mac, mixes direct modeling with parametric modeling, integrates T-Splines modeling and B-Rep modeling,and absorbs desktop software into cloud computing, a term commonly pointing to 360. Fusion 360, a perfect model integrating parametric modeling and direct modeling, both preserves the flexibility of direct modeling in industrial design, and caters tothe considerationof modeling history and parametric control whichthe structural design requires. Designers can complete the entire product development process on Mac and PC and carry out data exchange between the two systems. As a result, the cooperative work among team workers could be realized at any time, in any place, and viaany device.

(2) ***Feature of 3.0-era Design.*** With the transformation of China's economy from "manufacturing" into "creation", the effect of design innovation is more and more prominent.It is changing our lives in a unique yet firm way. As is stated by Academician Lu Yongxiang, the former dean of CAS and vice president of the Standing Committee of the National People's Congress, in the third wave of industrial revolution, "innovative design" will lead the sustainable development of civilization, green, intelligent, sharable and featured byinformatizationand network. China needs to set as a significant strategy the promotion of its innovative design capability so as to enhance innovation-driving, accelerate transformative development and build an innovative country. Design by China should lead the global developmental trend and step into the design 3.0 era. "1.0 era refers to the traditional design in the agricultural society, 2.0 indicatesthe modern design in the current industrialized society", Academician Lu Yongxiang explained, "while by 3.0 is meant the innovation destined to emerge with boom of design and material innovation in the global knowledge network".

2.Functional Modules in and Analysis of Autodesk Fusion360

Fusion 360, a product introduced by Autodesk and integrating industrial design, structural design, mechanical simulation, and CAM,turns out a design platform supporting collaboration and sharing both cross-platform and via the cloud.As is shown in Fig. 1, Fusion360 comprises several working environments and modules: modeling, molding, surface patching, rendering, animation, simulation, CAM, drawings and so on.

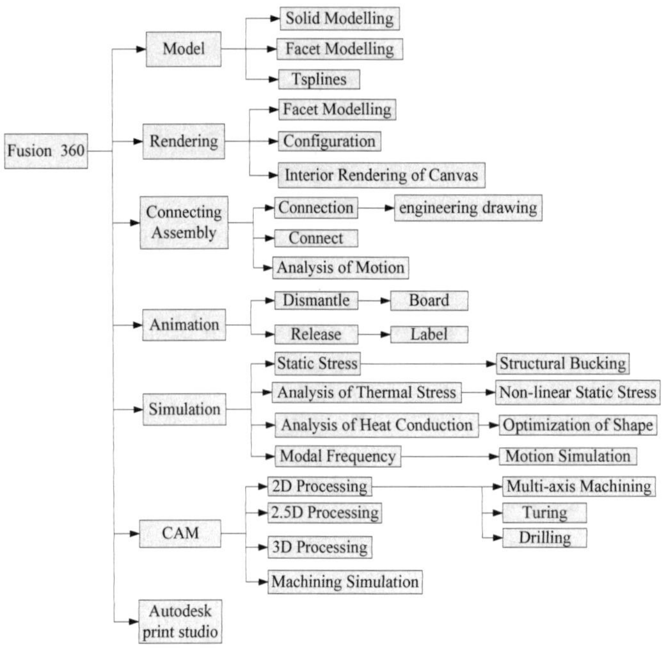

Fig. 1 Function Module of Fusion360

(1) ***Modeling.*** A salient feature of the direct modelingin Fusion360is the visibility of outcomes.The direct modeling technology enables users toperforma follow-up modeling uponmodels both featured and non-featured (such as the nonparametric models acquired by other CAD systems). In the whole process, any modification or addition makes no difference to the establishment of models.

(2) ***Rendering.*** Fusion 360 could visualize your product model basing on cloud computing.That is to

say, your models could acquire a photo-level effect via the real-time ray tracing, rendering engines or powerful cloud rendering.

*(3) **ConnectingAssembly.***Fusion 360 supports the connection-based assembly technology, a method easy to apply. Traditional practicestendto include too many meta-constraints that would affect the using effect, while the method mentioned in this paperdeals with assembly from the opposite direction, or the freedom of parts. The higher the degree of freedom is, the less constraint emerges, and vice versa. The introduction of this assembly into a large model can greatly reduce the number of meta-constraints and promotes the efficiency of design.

*(4) **Animation.***The animation module in Fusion360 is a key-frame one.Its working environment includescommands concerning with a series of subjects, such as storyboards, changesof commands, labels, views, and release. Drag the slider on the animation timeline to manipulate the model, and users could record the key-frame animation.

*(5) **2DDrawings.***When generating2D drawings associated with3D models, 2D drawings will be automatically updated when there are any updates occurring in the 3D model.

*(6) **Simulation.*** When detecting with Fusion simulation and animation, users can find the weakest place inthe model. Notable time and resources would be spared once users could fully understand or clearly learn the designed work before its delivery. Fusion360 is equipped witha kinematicpair, used to simulate theoperation of the equipment under a certain power or drive so as to verify the reasonability ofthe design.

*(7) **CAM.***Fusion 360, perfectly combining parametric design, variable design, feature modeling technology with traditional physical and surface modeling capabilities, realizes computer-aided manufacturing (CAM) and makes processing more integrated and computation more accurate. It could perform an automatic control and optimization upon a CNC machining, while providing toolsfor users to carry out a secondary development.

(8)3D Printing. Users could directly configure 3D printing strategy in Fusion360, output modelsand print them with a 3D printer. Fusion360 supports 3D printing utility tools of Spark platform, AutodeskPrintStudio, and will also integrate Ember3D printers directly.

3. Comparison and Example of Modeling

*(1) **Parametric Modeling.*** Top-down design: break the model into main components, and decompose main parts into sub-components and parts. In the process,the link between various sub-components and the assembly methods should be clarified and agrees with the overall design conception of the model. This methodadapts to the frequent modification of the product andagrees with the design concept of products.

Bottom-up design: initially determine the detailed information of the parts, gradually complete the sub-components, then assemble the main components, and eventually finish the design of the entire product. This method cannot fully reflect the design conception and increasesthe chance of design conflict or even errors, thus lacking in flexibility. This method applies to those designs not requiring frequent modification, such as the middle- and late-stage. It is a widely-employed design method.

Fusion360, incorporating both top-down and bottom-up parametric designs, supports skeletonmodels, componentsborrowing, and a distinctive segmented modeling methods.

*(2) **T-SplinesModeling.*** T-Splines modelingis a new modeling technique with the characteristics of NURBS and subdivided surface modeling techniques. as is shown in Fig. 2.

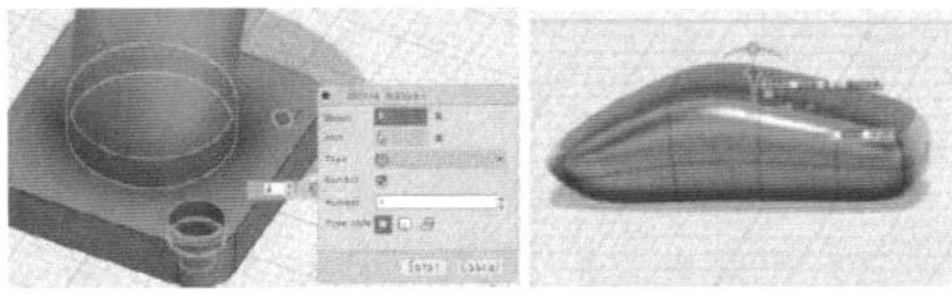

Fig. 2 Comparison of Parametric Modeling and T-Splines Modeling

*(3) **Comparison of T-Splines and NURBS Modeling.*** Fusion360 kernel is ACIS (a branch of Autodesk), but Autodesk integrates the T-Splines kernel in this kernel. If there is a better modeling technique than NURBSmodeling, that must beT-Splines. Compared with NURBS, T-Splines, greatly reducing the number of surface control points in the model, fittinglocation subdivision and raising the speed of modeling, is the very technology that can replace NURBSmodeling.

Fig. 3 respectively illustrates the chairs generated byT-Splines, NURBS, and Mesh from left to right. After acomparison, it could be readily found thatthe T-Splines parting line is sparse and even distributed, and the quality of the model surface is much better.

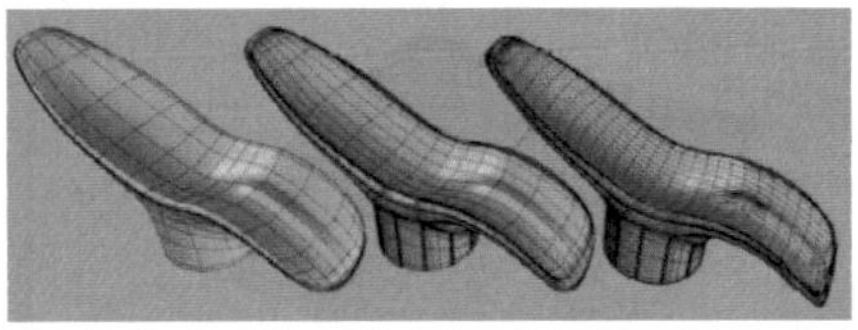

Fig. 3 A Comparison of T-Splines modeling, NURBS, and Mesh

4. Cloud Rendering andLocal Rendering

Each 3D modeling software is equipped with its own rendering tools, simple or complex, realistic or suitable for 3D and 2D rendering. Different ones cater for different needs and have their own merits. Fusion 360 is no exception.In addition to a huge rendering material database, the current version has two rendering methods, local Ray Tracing rendering, and cloud ShowCase 360 rendering.

Fusion360 also supports the inner rendering of canvas, a process referring to the real-time rendering of scenes. Fusion360 would perform continuous rendering and iteration, which gradually rendersthe rough scene delicate. As is shown in Fig. 4, what lies in the lower right corner of the view is the action bar for rendering time, progress, iteration, and quality.

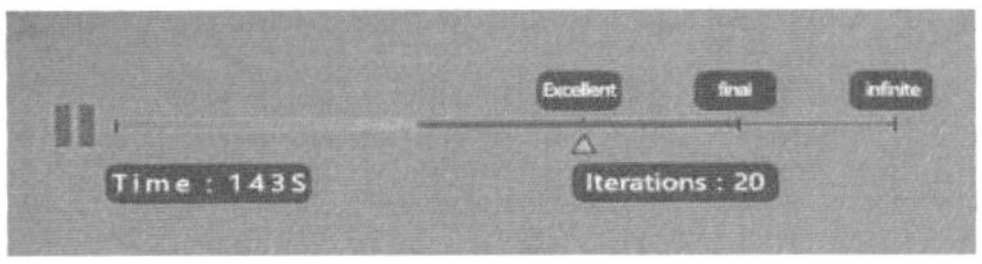

Fig. 4 Rendering Progress

Rotate the view angle, and the software shall recalculate the rendering. Some rendering settings are software-defaulted, whileothers can be manually selected by commands. Figure 5 provides a comparison of the quality of the canvas, iterations, and the actual rendering effect in different stages.

	Time	Progress	Iteration	Quality	Visual Effect

Begin	2 S	1%	0	Ordinary	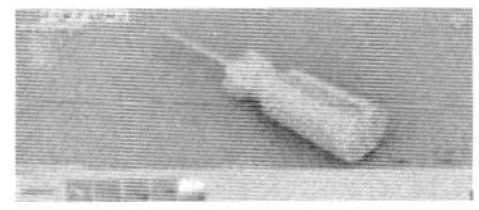
Middle	43	10%	4	Good	
Later	510	70%	41	Great	
End	1200	100%	152	Perfect	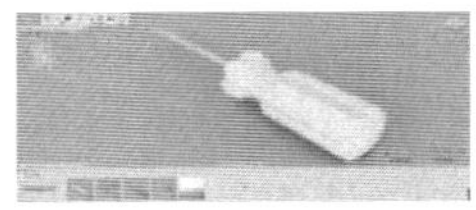

Fig. 5 RenderingComparison

5. Design and Animation of Component Assembly Based on Connection

In the process of assembly design,a project manager, once being importednecessary information in, can automaticallygenerate the list of assembly parts, and realizethe driven assembly of mechanical components under constraints. When assembling parts in the same design environment, we can assemble parts based on the same coordinate point, and set various types of joints, such as swivel, sliders, cylinders, pin troughs, and spheres. The user can also set the movement limit of the joint andmake precise configuration of the numerical settings for moving, so as to realize a real-time preview of the movement. Fusion 360 assembly commandsrender a very clear picture of the relationship between parts, and help produce assembly animation when consulting the animation modules.

The assembly system in Fusion360 provides seven kinds of contact movement: rigid, rotating, sliding, cylinder, pin slot, plane, and ball. The production of assembly animation runs as follows:

Fig. 6, illustrating theassembly design of Flash Disk, renders a clear picture of the assembly relationshipof components.

Fig. 6 Flash Disk Assembly Design

6.SimulationAnalysis

In Fusion360, a typical CAE process includes pre-processing, solving and post-processing steps. In the pre-processing phase, the designer would model the physical properties of the geometry (or expressions represented by the system), the design process and the environment by imposing loading constraints upon. Then, solve the model with appropriate mathematical formulasconcerningbasic physical field. In the post-processing phase, the results are to be presented to the designer for viewing.

The merit of Fusion 360 simulation lies with the reduced costs for product development,shorteneddevelopment time and improved product quality and durability. Moreover, users could tailortheir design decisions based on the impact of design on performance. The use of computer simulation in prototype testing can not only save time, efforts and funds, but also guarantees acorrect assessment and optimization of the design.

Fig. 7 shows a static stress analysis with a buckle model made by the author.The results can be viewed through the cloud and grid.

Static Stress Analysis Results of Buckle Model		
Displacementcloud	Stress Cloud	Grid
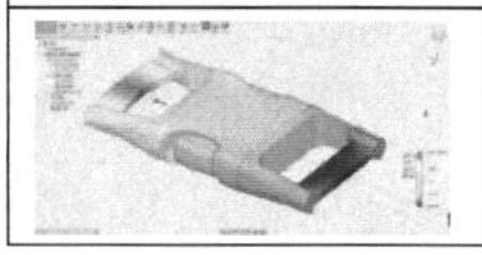	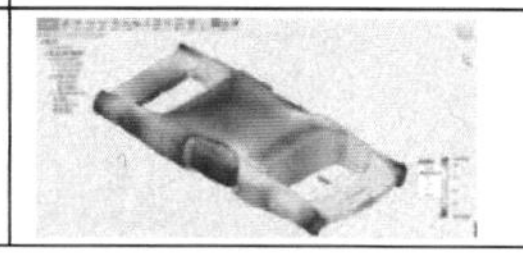	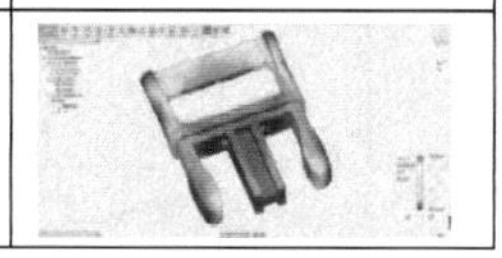

Fig. 7 Static Stress Analysis of Buckle Model

Fusion 360 simulations can show the performance in the early stage of the development processwhenthe cost of alteration and modificationminimizes.

7. CAM and 3D Print

(1) CAM. A host of processing techniques is setin the Fusion 360 CAM work environment, such as 2D processing, 3D machining, drilling, multi-axis, and turning. As Fusion 360 is still in the development phase and updatesare launched every three weeks, multi-axis machining is just been placed in the toolbarand the two orders for the side-edge processing and multi-axis contours are not perfect.

*1) 2 axis / 2.5 axis Processing.*CAM in Fusion 360 includes machining simulation, and CNC programming.

The CAM in Fusion 360 can share the same recognized CAM kernel HSMWorks and Inventor HSM ™, quickly complete, shorten the development cycle, calculate the optimized tool path, reduce the abrasion of machines and tools, and produce the best finished parts. 2D processing can perform drilling, external and interior contour processing, and surface processing.

2) 3-axis Machining. Exploiting the 3-axis machining technique, it can generate 3-axis machining paths and strategies, including rough, semi-finishing and finishing processing.

3) Processing Simulation. Possessing a good processing simulation system, it can generate a processing animation, used to observe processing.

The following figure made a 2D milling and 3D milling comparison.

Fig. 8 2D and 3D Millings

At present, there are more than 1,000 kinds of numerical control system, among whichthe ones commonly used in China are Fanuc CNC system of Japan, Siemens CNC system of Germany, DMG CNC system, Haas, CNC system, the Chinese CNC system, and so on. Fusion360 can support hundreds of current CNC systems including Farrah, Siemens, DMG, Haas, and so on.

As post-processing files are targeted, different software and CNC machine tools require their own file format and suffix. Therefore,users of Fusion360 can select and generatethe type of post-processing files according to the specific type of CNC system and view G-codes.

(2) ***3D Printing.***Fusion 360 can be easily employed in 3D printing software utilities, such as Autodesk® Print Studio, which is supported by Spark. This software preparesusers to communicate directly with Autodesk Ember ™ 3Dprinters.

As is shown in Figure 9, if the print model does not match, the entire model in the scene will show a red color, a signal indicating "error" in Print Studio.On the contrary, a blue model indicates that the model works well. Print Studio also provides the function of model analysis and repair.

Correct Model	Wrong Model
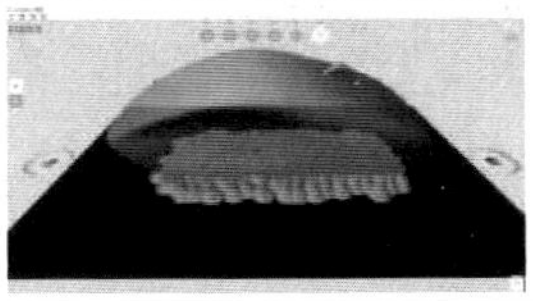	

Fig. 9 Correct Model VSWrong Model

8.Conclusion and Outlook

In summary, Fusion360,realizing the integration of design and manufacturing, rendering and analysis, cloud computing and collaborative management, and supports more than 50 document formats, making it ideal for the design and development of small products. Currently, with widely encouragement of innovation in the whole society and each field, the "Eco-development of Fusion360 Industrial Chain" is both a significant meansto and an inevitable trend forthe manufacturing and industrial designing to carry out innovation in China.

9 Acknowledgements

The authors would like to acknowledge the support of the Science Research Planning Project Tianjin University of Technology and Education, funded by the SK15-12 Program.and the key technologies R&D program of Tianjin(13ZCZDGX01500, 14ZCZDSF00022, 15ZXZNGX00260, 201510066X15).

References

[1] Tongji University News, http://news.tongji.edu.cn/classid-16-newsid-43368-t-show.html.
[2] http://www.fusion360.cn/article.html.
[3] .http://www.autodesk.com.cn/products/fusion-360/overview
[4] http://www.thearender/fusion.
[5] http://sns.zhizaoyun.com/forum.php?mod=viewthread&tid=5512
[6] http://blog.sina.com.cn/s/blog_1351daf5b0101efcv.html
[7] http://www.instructables.com/id/Fusion/
[8] https://wenku.baidu.com/view/cfda72d588eb172ded630b1c59eef8c75ebf9552.html

IOP Conf. Series: Materials Science and Engineering **359** (2018) 012038 doi:10.1088/1757-899X/359/1/012038

Ligands Exchange Process on Gold Nanoparticles in Acetone Solution

C L Hu[1], Y YMu[1], Z C Bian[1], Z H Luo[1,*],K Luo[1] and A Z Huang[2]

[1]Guangxi Key Laboratory of Universities for Clean Metallurgy Comprehensive Utilization of Nonferrous Metal Resource, Collaborative Innovation Center for Exploration of Hidden Nonferrous Metal Deposits and Development of New Materials in Guangxi, Guilin University of Technology, Guilin, 541004, PR China;

[2]Humanwell Healthcare(GROUP)CO ., LTD,Wuhan,430075, PR China

Email:luozhihong615@163.com

Abstract. The ligands exchange process on gold nanoparticles (GNPs) was proceeded by using hydrophobic group (PPh_3) and hydrophilic group (THPO) in acetone solution. The FTIR and XPS results demonstrated that part of THPO was replaced by PPh_3 which was dissolved in polar solution (acetone); the results were in accordance with the electrochemical analysis where the differential capacity decreased with increasing exchange time. After 12 h, the exchange process terminated and the final ratio of PPh_3 and THPO was about 1.4: 1.This ratio remained unchanged although the PPh_3 and THPO modified GNPs re-dispersed in the PPh_3 acetone solution demonstrating the stable adsorption of both ligands after exchanging for 12 h. The TEM images showed that the gold nanoparticles were self-assembled from scattered to arranged morphology due to the existence of hydrophilic and hydrophobic ligands and led to Janus gold nanoparticles.

1.Introduction

Surface modified gold is the most frequently used in the fields of sensors [1-4], corrosion protection [5-7], molecular electronic devices [8-9], and other nanoscale devices. Self-assembling monolayers (SAMs) to modified gold substrates is a powerful method with high orderliness, orientation, thermodynamic stability and low defect. Gold nanoparticles are obtained by using single ligands. However, Janus gold particle need the existence of hydrophilic and hydrophobic ligands on gold particle simultaneously, which possess distinct surface patchiness and exhibit at least two physicochemical properties within a single nanoparticle [10-11], and result in unique building blocks to prepare advanced ordered structures

based upon their anisotropic attributes [12-13]. Janus particles have the widespread application prospect in sensing device [14], imaging probes [15-16], catalysis [17-18], optoelectronics device [19], photoelectric biosensor [20], drug delivery [21], functional surface active agent [22], etc.
Thiol modified gold is widely applied in fundamental and practical issue due to the strong interaction between gold and sulfer [23-24]. Phosphorus ligands can also adsorb onto gold surfaces due to their electrostatic interactions although which is weaker than that of thiol ligands. However, the replacement of phosphorus ligands on gold substrate is feasible and we inferthat Janus gold particles with THPO and PPh_3 ligands can be obtained. In this paper, the ligands exchange process of THPO and $PPh_3$is investigated by using electrochemical method, FTIR and XPS. The morphology, Janus gold nanoparticles formation and the ratio of PPh_3 and THPO on the surface of gold was studied in details.

2 Experimental

2.1 Chemicals

Chloroauric acid ($HAuCl_4$, 47.8%, AR) was purchased from Shanghai Billiton Co.Ltd., China. Triphenylphosphine (PPh_3, CP) was bought from Sinopharm Chemical Reagent Co. Ltd., China. tetrakis(hydroxymethyl)phosphonium chloride (THPC, 80%, CP) was obtained from Tokyo Chemical Industry Co. Ltd., Japan. Sodium hydroxide (NaOH, 96 %,AR), sulfuric acid (H_2SO_4, AR, 95-99%) and hydroxylamine hydrochloride (96%, AR) were purchased from Shantou Xilong Chemical Co. Ltd., China.Trisphosphine oxide (THPO, 95.5%) was bought fromWuhan Jinlin Chemical Technology Co. Ltd., China.Deionized water was used in all experiments.

2.2 Preparation of gold nanoparticles (GNPs) with single ligand

NaOH (1 M, 300 μL), THPC (50 mM, 1 mL) and $HAuCl_4$ (25 mM, 2 mL) were added to a conical flask contained 47 mL deionized water under ultrasonic stirring, leading to the formation of a GNPs colloid in palm red color, which were removed in a dialysis bag (with intercept molecular weight of 3500) immersed in deionized water to remove soluble reaction products. The operation lasted for 6~7 days accompanied with the refreshment of deionized water periodically, resulted in the THPO protected nano-gold colloid (THPO-Au).

2.3 Study of exchange process on GNPs

Exchange process of the GNPs in the as-prepared colloids was carried out at the acetone/water solution by ligand exchange reaction. 5 mL as-prepared THPO-protected GNPs colloid were added into a vials contained 5 mL of 50 mM PPh3 acetone solutions, stirred for 30 min, 4 h, 12 h and 24 h, and then washed by centrifuging and dried with high pure Ar to obtain THPO- PPh_3 modified GNPs (THPO- PPh_3-Au). The THPO- PPh_3 modified GNPs (THPO- PPh_3-Au) were dispersed into DMF solution with the concentration of 0.033 mg mL^{-1}.
5 mL of THPO-Au mixed with 5 mL of PPh_3 acetone solution and stirred 24 h, and then centrifuged to remove the remained ligands, the solid was re-dispersed in 5 mL acetone to form THPO- PPh_3-Au acetone dispersion, and then 5 mL of 1 mM PPh_3 acetone solution was added into the dispersion and stirred for 24 h. The obtained samples were named as THPO- PPh_3-Au-2nd.

2.4 Characterization

UV-vis absorption spectrum (Beijing Purkinje General Instrument Co. Ltd.) was used to describe the formation of GNPs. FTIR (Thermo Nexus 470 FT-IR) and XPS (AXIS ULTRA XPS, KRATOS) were employed to determine the capping molecules of the as-prepared GNPs. TEM (JEM-2010, JEOL) was applied to measure the morphology of gold nanoparticles.

For electrochemical measurements, 5 μL of the THPO- PPh_3-Au solution was dropped onto the pre-polished grassy carbon electrode for 6 times, and dried under vacuum at 80°C for 12 h. The electrode was used as working electrode for CV test with Pt and Hg/Hg_2SO_4 as counter and reference electrode, respectively, 0.5 M H_2SO_4 was used as electrolyte. The voltage window scanned from -0.05 V- 0.05 V with 10 mV s^{-1}.

3. Results and discussion

The morphologies of the samples before and after ligands exchange are showed in Fig.1. The THPO-Au particles (the left image) are scattered with the average size of 2.1 nm. After stirring for 24 h in the 1 mM PPh_3 acetone solution, the obtained THPO- PPh_3-Au (the right image) aggregate by self-assembling.

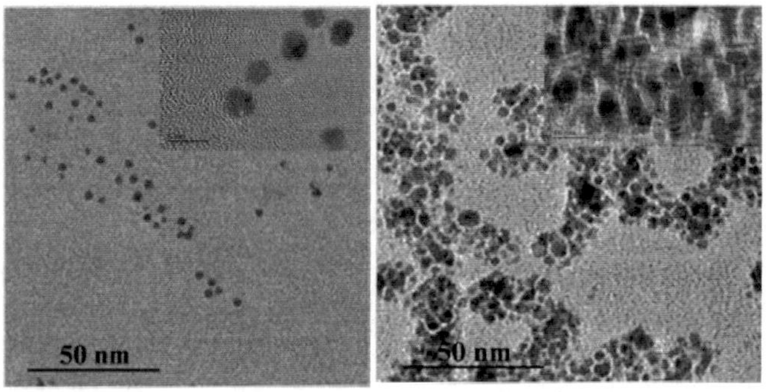

Fig.1, TEM images of THPO-Au (left) and THPO-PPh_3-Au (right)

The FTIR spectra of THPO-PPh_3-Au and THPO-PPh_3-Au-2nd are showed in Fig.2, the high intensity band at 3447 cm^{-1} belongs to the adsorbed H_2O, and the peaks at 1633 cm^{-1}, 626 cm^{-1} represent the vibration of C-H of benzene, which demonstrate the existence of PPh_3. The peaks located at 1397 cm^{-1}, 1108 cm^{-1} belong to the vibration of C-OH indicating the existence of THPO and THP.

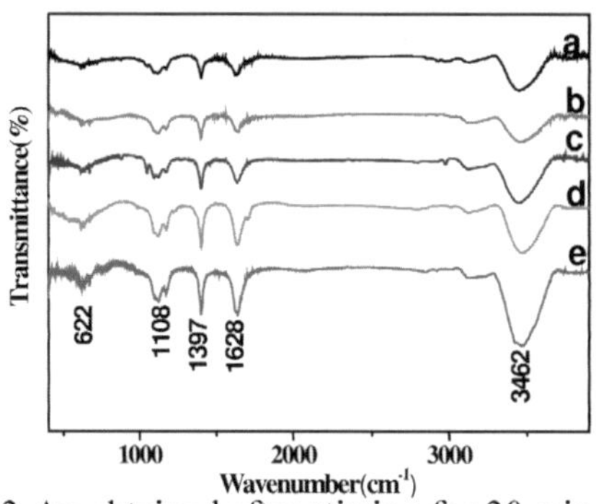

Fig.2 FTIR spectra of THPO-PPh3-Au obtained after stirring for 30 min (a), 4 h (b), 12 h (c), 24 h (d) and THPO-PPh_3-Au-2nd (e).

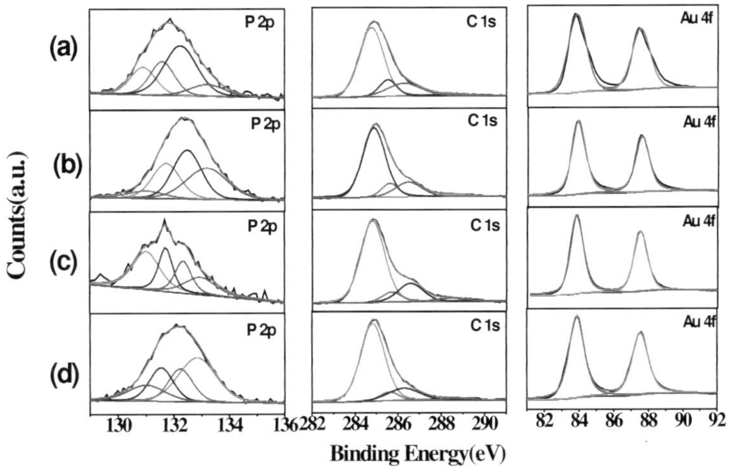

Fig.3 P 2p (left), C 1s (middle) and Au 4f (right) of THPO-PPh3-Au obtained after stirring for 30 min (a), 4 h (b), 24 h (c) and THPO-PPh_3-Au-2nd (d).

The deconvolution of the corresponding P, C and Au XPS signals are showed in Fig.3. Two fitted peaks at 83.8 eV and 87.5 eV in Fig.3 can be assigned to the Au4f5/2 and Au4f7/2[25], indicative of metallic gold. Four fitted peaks for the C1s signal at 284.8 eV, 285.3 eV, 286.4 eV and 288.4 eV were belong to –C-H [26], C-OH, C-P and O-C=O [27]groups, respectively. The fitting of the P2p curve (Fig.3) show four peaks at 131.3 eV and 132.1 eV, 131.0 eV, 131.6 eV corresponding to THP and THPO, PPh_3 and PPh_3O[28-29], respectively. The content of the above mentioned groups, according to P 2p of sample a, b, c, and d, the ratio of PPh_3, PPh_3O and THP, THPO is 0.45:1, 0.66:1, 1.38:1, 1.40:1. According to C1s, the content of C-OH came from THPO and THP and the content of C-P came from PPh_3 and PPh_3O leading to that the ratio of PPh_3, PPh_3O and THP, THPO was 0.47:1, 0.69:1, 1.37:1, 1.42:1, demonstrating of the content of hydrophobic groups of increase with increasing the mixing time. After exchanging for 12 h, the ratio of PPh_3, PPh_3O and THP, THPO almost keep constant, even the sample c re-dispersed in 1 mM PPh_3 solution, the ratio of hydrophobic groups (PPh_3, PPh_3O) and hydrophilic groups (THP, THPO) is unchanged.

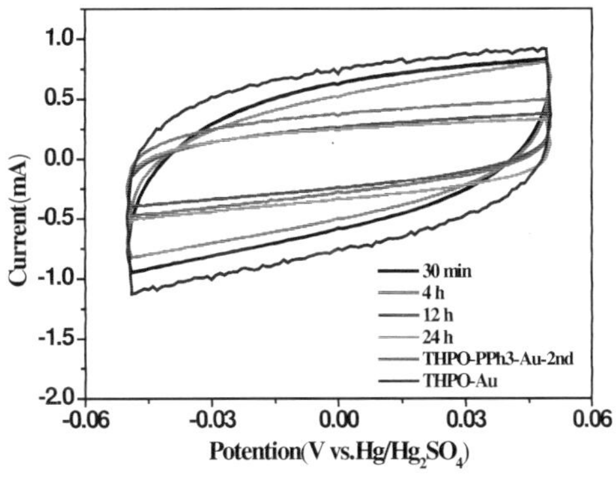

Fig. 4 CV plots of THPO-Au, THPO-PPh_3-Au obtained after stirring for 30 min, 4 h, 12 h, 24 h and THPO-PPh_3-Au-2nd.

Table 1, Differential capacity of THPO-Au, THPO-PPh_3-Au obtained after stirring for 30 min, 4 h, 12 h, 24 h and THPO-PPh_3-Au-2nd.

Echange time	30min	4h	12h	24h	THPO-PPh_3-Au-2nd
C_d (mF/mg)	61.23	51.91	28.02	28.82	29.90

The CV plots with the potential ranged from -0.05 to 0.05 V represent the differential capacity of the electrode which is calculated through the following equation: $C_d = \Delta i/2vm$, the variation of Δi indicates the adsorption of ligands. The C_d of THPO-Au is 75.45 mF/mg, Δi decrease with increasing of stirring time. The variation of C_d is showed in Table 1, with the time increasing, the value of C_d decrease, after 12 h, the C_d maintain unchanged, after exchange for the second time, the C_d increase slightly indicating that after stirring for 12 h, the ligands exchange process is completely finished that even fresh PPh_3 solution can't exchange further.

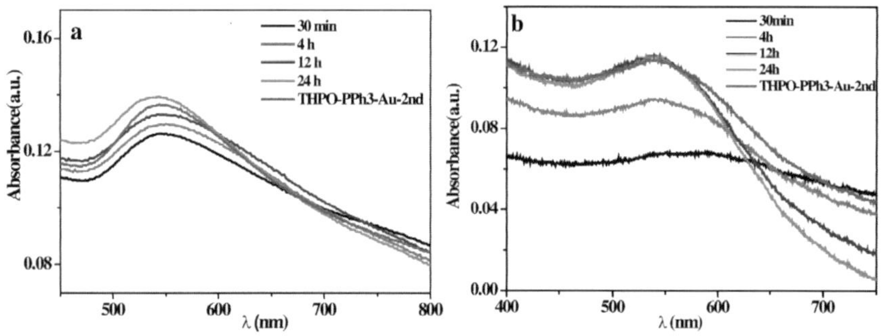

Fig.5 UV-vis spectra of THPO-PPh3-Au obtained after stirring for 30 min, 4 h, 12 h, 24 h and THPO-PPh3-Au-2nd in DMF (a) and acetone (b).

THPO-PPh_3-Au obtained after stirring for 30 min, 4 h, 12 h, 24 h and THPO-PPh_3-Au-2nd are re-dispersed in DMF and acetone, the UV-vis spectra are showed in Fig.5. The adsorption peak (Fig. 5a) at 540 nm demonstrate the existence of Au nanoparticles, with increasing the stirring time, the absorbance increase demonstrating the increase of Au content, and the variation trend of the absorbance of THPO-PPh_3-Au in acetone is the same as that in DMF (Fig.5b).

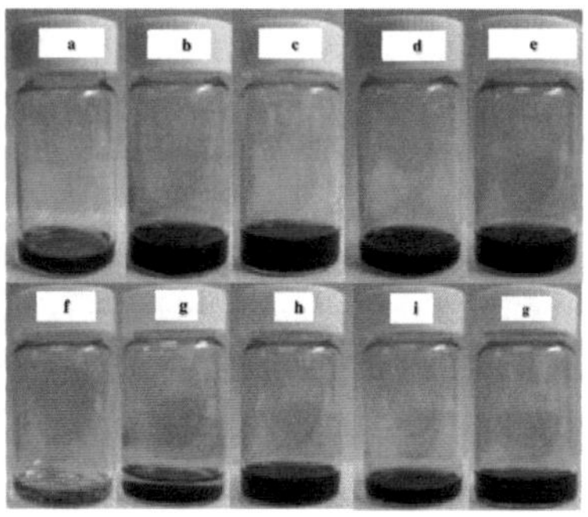

Fig.6 Images of THPO-PPh3-Au after stirring for 30 min (a, f), 4h (b, g), 12 h (c, h), 24 h (d, i) and THPO-PPh3-Au-2nd (e, g) before (a, b, c, d, e) and after (f, g, h, i, g) standing for 24 h in acetone solution.

The dispersibility and stability of the THPO-PPh_3-Au and THPO-PPh_3-Au-2nd are measured by re-dispersing in acetone and standing for 24 h. As showed in Fig.6, the color darken by increasing the stirring time, possibly due to the content of Au increases, which is in accordance with the UV-vis results. After standing for 24 h, the samples obtained by stirring 30 min and 4 h settle, and the samples obtained by stirring for more than 12 h is stable possibly due to the high content of PPh_3 which is demonstrated by XPS results.

4. Conclusions

The ligands exchange taken place by mixing the THPO modified Au in PPh_3 acetone solution and stirring for different time, with the increasing of adsorption time, the surface of bare Au was decreased as well as the differential capacity. The FTIR results demonstrated the existence of PPh_3 and THPO, and XPS results showed that the content of PPh_3 was increased with the increasing of stirring time, as well as the dispersibility and stability measured by UV-vis and standing experiments.

Acknowledgment

The work was supported Guangxi Natural Science Foundation (Grant No. 2016GXNSFAA380107 and 2015GXNSFBA139220) and the funding from the Guangxi Ministry-Province Jointly-Constructed Cultivation Base for State Key Laboratory of Processing for Nonferrous Metal and Featured Materials.

Reference

[1] S. Campuzano, M. Pedrero, Talanta, 66(2005), 1310-1319.
[2] C. Boozer, J. Ladd, Anal. Chem., 76(2004), 6967-6972.
[3] N.K. Chaki, K.Vijayamohanan, Biosens. &Bioelectron., 17(2002), 1-12.
[4] K. Motesharei, M.R. Ghadiri, J. Am. Chem. Soc., 119(1987), 11306-11312.
[5] F.P.Zamborini, J. K.Campbell, R. M.Crooks,,Langmuir, 14(1998), 640-647.
[6] G.K.Jennings, P.E. Laibinis, J. Am. Chem. Soc., 119(1997), 5208-5214.
[7] S. Ramachandran, B-L.Tsai, M. Blanco, Langmuir, 12(1996), 6419-6428.
[8] K. Kitagawa, T. Morita, S.Kimura, Langmuir, 21(2005), 10624-10631.
[9] J. Chen, M.A. Reed, A.M. Rawlett, Science, 286(1999), 1550-1552.
[10] A. Walther, A. H. EMuller, .Application. Chem. Rev., 2013, 113, 5194-5261.
[11] X. Y.Ling,I. Y.Phang, ; C.Acikgoz,J. Angew. Chem, Int. Ed., 2009, 48, 7677-7682.
[12] S. K.Smoukov, S.Gangwal,M.Marquez,Soft Matter 2009, 5, 1285–1292.
[13] S.Gangwal, O. J.Cayre,O. D.Velev,Langmuir, 2008, 24,13312– 13320.
[14] J.Lee,J. Kim,Chem. Mater., 2012, 24, 2817-2822.
[15] G. A.Sotiriou,A. M.Hirt,P. Y.Lozach,Chem.Mater., 2011, 23, 1985-1992.
[16] M.Yoshida, K. H.Roh, J.Lahann, Biomaterials, 2007, 28, 2446-2456.
[17] L.Baraban,D.Makarov,R.Steubel,ACS Nano., 2012, 6, 3383-3389.
[18] D. J.Cole-Hamilton, Science, 2010, 327, 41-42.
[19] Y. K.Takahara,S.Ikeda,S.Ishino,K.Tachi,Am. Chem. Soc., 2005, 127, 6271-6275.
[20] P. A.Suci,S.Kang,M.Yong,Am. Chem. Soc., 2009, 1, 1747 - 1754.
[21] G.Zheng,M.Chen,X.Y.Liu,Electrochim. Acta, 2014, 136, 301–309.
[22] M.Motaghedifard,M.Behpour,S. M.Ghoreishi,Sens. and Actuator B: Chem., 2014, 203, 802–811.
[23] K. Sunil, M. K.Aryan, Park, Biosens. & Bioelectron., 2014, 61, 260–265.
[24] X.Du,Z. Y.Miao,D.Zhang,Biosens. & Bioelectron., 2014, 62 , 73–78.
[25] C.J. Powell, Appl. Surf. Sci., 1995, 89: 141- 149.
[26] D Briggs, G Beamson, Anal. Chem., 1992, 64: 1729-1736.
[27] G Beamson, D Briggs, Mol. Phys., 1992,76: 919-936.
[28] J.K. Kouba, J.L. Pierce, R.A. Walton, J. Organomet. Chem., 1980, 202: C105-107.
[29] W.J .Stec, W.E. Morgan, R.G. Albridge, et al., Inorgan. Chem., 1972, 11: 219-225.

Mechanical Properties of Steel Fiber Reinforced all Lightweight Aggregate Concrete

Y M Yang, J Y Li, Y Zhen, Y N Nie and W L Dong
School of Civil Engineering,Jilin Jianzhu University,Changchun,China,130118.

E-mail:282872984@qq.com

Abstract. In order to study the basic mechanical properties and failure characteristics of alllightweight aggregateconcrete with different volume of steel fiber (0%, 1%, 2%),shale ceramsite is used as light coarse aggregate. The shale sand is made of light fine aggregate and mixed with different volume of steel fiber, and the mix proportion design of all lightweight aggregate concrete is carried out. The cubic compressive strength, axial compressive strength, flexural strength, splitting strength and modulus of elasticity of steel fiber all lightweight aggregate concrete were studied.Test results show that the incorporation of steel fiber can restrict the cracking of concrete,improve crack resistance; at the same time, it shows good plastic deformation ability and failure morphology. It lays a theoretical foundation for further research on the application of all lightweight aggregate concrete in structural systems.

1. Introduction

As a branch of lightweight aggregate concrete,all lightweight aggregate concrete is lightweight aggregate concrete made of lightweight coarse aggregate, light fine aggregate, cementitious material and cement and water[1],bulk density below 1700Kg/m^3.All lightweight aggregate concrete is a lightweight, energy-saving and environment-friendly building material,The utility model has the advantages of good structural performance, good seismic performance, heat insulation, good durability, etc.But there are some disadvantages, such as low tensile strength and poor toughness.Steel fiber is a kind of good toughness reinforced material, and has been paid much attention by many scholars[2].At present, a great deal of research has been done on fiber reinforced concrete materials,it is found that the fiber concrete member has good crack resistance, good deform ability and good energy absorption capacity,but the research on the steel fiber concrete is still very weak.With the application of fiber concrete in construction, transportation, military and other fields,it is urgent to make a thorough study of all kinds of steel fiber lightweight concrete and provide a theoretical basis for the rapid application of engineering structural members.

2. Experimental design

2.1 .Main raw material

The cement is made of ordinary 42.5 Portland cement produced by Jilin Yatai India Cements Limited;light coarse aggregate is made of shale ceramsite from light building materials factory of Tianjin Wu Tong District,the particle size is 5~30mm;Light fine aggregate used Jilin Nong'an Yixing wall materials Co. Ltd production of shale pottery,The particle size is 3~4mm;The admixture has fly ash and ordinary water reducing agent.The main properties of ceramsite and earthenware sand are shown in Table 1.Steel fiber use shear wave type steel fiber produced by Liaoning Anshan Corbett limited, The main parameters are shown in table 2.

Table 1 . main properties of ceramsite and earthenware sand

Type	Sieve analysis (mm)	Bulk density (kg/m^3)	Apparent density (kg/m^3)	Void fraction (%)	Water absorption (%)	Cylinder pressure strength (Mpa)
Shale ceramsite	5-30	765	820	43	5.6	7.2
Shale ceramic sand	3-4	480	770	38	11.2	3.6

Table 2. main parameters of steel fiber

Name	Length(mm)	Diameter(mm)	Aspect ratio	Tensile strength(Mpa)	Modulus of elasticity(Mpa)
Shear wave type steel fiber	30	0.6	50	≥770	200000

2.2. Mixture ratio design

According to JGJ51-2002 *Technical specification for lightweight aggregate concrete,*mix proportion design of lightweight concrete is carried out by using loose volume method.Through reasonable mix design,give full play to light aggregate lightweight features,at the same time, adding different volume of steel fiber can improve the toughness and shear resistance of all lightweight concrete.Steel fiber all lightweight aggregate concrete mix proportion as shown in Table 3.

Table 3. Steel fiber all lightweight aggregate concrete mix proportion

Number	Steel fiber volume frac tion	Cement/ kg	Shale ceramsite/kg	Shale ceramic sand/kg	Fly ash/kg	Water reducer/%	W/C
G0	0%	100	100	60	10	0.75	0.4
G1	1%	110	100	60	10	0.75	0.4

G2	2%	120	100	60	10	0.75	0.4

3.Experimental design

According to *Standard for test method of mechanical properties on ordinary concrete* to make standard test blocks,it is used to study the basic mechanical properties of steel fiber reinforced concrete.The design plan of each block is shown in Table 4.

Table 4. design schemes of steel fiber volume fraction (0%, 1% and 2%)

Shape of test block	Test block size	Number	Parameter index
cube	150mm×150mm×150mm	6×3	standard cube compressive strength,Splitting strength
Prism	150mm×150mm×300mm	6×3	Axial compressive strength,Modulus of elasticity
Cuboid	150mm×150mm×550mm	3×3	Rupture strength

4. Experimental study on basic mechanical properties

4.1 .Cube compressive strength and axial compressive strength

According to existing standard for test methods and standards,the cubic compressive strength and axial compressive strength of steel fiber and lightweight concrete were tested and measured.The results are shown in Table 5.

The initial stage of loading,the G0 series test block has a crisp splitting sound,no damage on the surface of the test block.When loading to approximately 50% ultimate load,a small amount of tiny cracks appear on the surface of the specimen.When loading to approximately 80% ultimate load,the number of test block surface crack increase, accelerate the development and hear concrete damage sounds.After reaching the limit load,the surface of the test block developed a main fissure from the original crack,at the same time will see the cracks in steel fiber pull-out from the distribution of clutter in concrete,the cracks continue to crack along the direction of the steel fiber until the specimen is broken.The G2 series is similar to the G1 series,the failure process of the specimen is shown in Figure 1.Judging from the test process and failure phenomenon, G0 is considered as brittle fracture,The incorporation of G1 and G2 steel fibers can obviously improve the crack resistance and energy absorption ability of the component.

Figure 1. cube compressive strength and axial compressive strength test

Table 5. cube compressive strength and axial compressive strength of test block

Test name	G0cube compressive strength/ MP	G1cube compressive strength/ MP	G1cube compressive strength/ MP	G0Axial compressive strength/MP	G1Axial compressive strength/MP	G2Axial compressive strength/MP
1	35.24	38.78	45.91	40.43	43.36	43.08
2	37.69	38.91	44.42	40.39	41.96	45.51
3	36.49	40.88	47.20	40.73	43.12	44.24
average value	36.47	39.52	45.84	40.52	42.81	44.28

4.2 .Elastic modulus and splitting strength

The magnitude of splitting tensile strength, the occurrence and development of cracks in the reaction material under Splitting Tension.The test results are shown in Table 6, and the damage pattern is shown in Figure 2 (a).Modulus of elasticity is an index to measure the degree of difficulty of elastic deformation of materials,The greater the value, the smaller the elastic deformation under a certain stress.According to the existing standard of test method, the elastic modulus of steel fiber and lightweight concrete is tested and calculated.The test results are shown in Table 7, and the damage pattern is shown in Figure 2 (b).The experimental data show that the elastic modulus of all lightweight concrete can be improved by adding steel fiber under the same loading condition,it shows that under certain stress conditions, the total lightweight concrete with volume fraction of 1% and 2% of steel fiber is more prone to plastic deformation than whole lightweight concrete without steel fiber.

Table 6. elastic modulus and splitting strength of test block

Test name	G0-Modulus of elasticity/MP	G1-Modulus of elasticity/MP	G2-Modulus of elasticity/MP	G0-Splitting strength/MP	G1-Splitting strength/MP	G2-Splitting strength/MP
1	1.94×10^4	2.25×10^4	2.33×10^4	3.42	6.88	7.95
2	1.89×10^4	2.31×10^4	2.28×10^4	3.67	6.65	7.58
3	2.04×10^4	2.18×10^4	2.39×10^4	3.76	7.23	7.64
average value	1.96×10^4	2.25×10^4	2.33×10^4	3.62	6.92	7.72

(a)Splitting failure (b)Elastic modulus test

Figure 2.test of elastic modulus and splitting strength of test block

5.Conclusion

(1) The test results show that the density of lightweight concrete is lower than that of ordinary concrete,strength did not decrease,at the same time, it has good plastic deformation ability and failure morphology.

(2) Through cube compression test and axial compression test, we can see that,the compression failure of all lightweight concrete without steel fiber starts from the tension crack of cement mortar,Then, run through the aggregate,fracture of lightweight aggregate in Failure surface.

(3) The compressive damage of all lightweight concrete mixed with steel fiber starts from the tension crack of cement mortar and then develops with the direction of steel fiber,the crack number is small, and the failure surface can be seen randomly distributed in the steel fiber,the section is not smooth and the strength is obviously higher than that of all lightweight concrete without steel fiber,it shows that steel fiber can effectively limit the amount and development of cracks, and has good crack resistance and energy absorption.

(4) Through the elastic modulus test, we can see that,illustrates under certain stress conditions,all lightweight concrete mixed with steel fiber is more prone to plastic deformation than lightweight concrete without steel fiber.

Acknowledgements

Author introduction: Yang Yan-min (1969~), female, Jilin Province, Changchun, professor.
This work was financially supported by the Jilin Science and Technology Foundation (20150203014SF, 2015267).The authors gratefully acknowledge the Natural Science Foundation of China (No.51378238).

Reference

[1] YANG yan min,Study on bearing capacity of full lightweight concrete beams[J].*Concrete*.2011(3):48-50.

[2] WANG tao,DING qing jun,HU shu guang,ZHANG feng,REN fei,Influence of polymer and fiber on properties of lightweight aggregate concrete[A].Study and application of durability of new type high performance concrete[C].Beijing: *China Building Material Industry Press*, 2003.

[3] DOU guo qin,Experimental study on dynamic performance of high strength concrete reinforced concrete beams under impact loading[D]. Beijing: *Beijing University of Technology*, 2014.
[4] China Academy of Building Research.Technical specification for lightweight aggregate concrete JGJ 51 - 2002 [S].*China Architecture and Building Press*, 2002.
[5] China Academy of Building Research.Standard Test Method for mechanical properties of ordinary concrete: GB/T50081-2002[S].*China Architecture and Building Press*, 2002.

Synthesis and Characterization of Rhodamine B-ethylenediamine-hyaluronan Acid as Potential Biological Functional Materials

Y L Li[1,2,3,4], W X Wang[1,2,3,4], Y Wang[1,2,3,4], W B Zhang[1,2,3,4], H M Gong[1,2,3,4] and M X Liu[1,2,3,4*]

[1]Key Laboratory of Fermentation Engineering (Ministry of Education),
[2]Hubei Provincial Cooperative Innovation Center of Industrial Fermentation,
[3]National "111" Center for Cellular Regulation and Molecular Pharmaceutics,
[4]Hubei Key Laboratory of Industrial Microbiology, Hubei University of Technology, Wuhan 430068 China

[*]Corresponding author: lmxing@mail.hbut.edu.cn

Abstract. The purpose of this study is to synthesize and characterize fluorescent polymers, rhodamine B-ethylenediamine-hyaluronan acid (RhB-EA-HA). RhB-EA-HA was successfully synthesized by ester ammonolysis reaction and amidation reaction. Moreover, the structural properties of RhB-EA-HA were characterized by ^{1}H-NMR spectra, UV-vis spectrometry and Fourier transform infrared spectroscopy (FT-IR). RhB-EA-HA can be grafted on the surface of silica nanomaterials, which may be potential biological functional materials for drug delivery system.

1. Introduction

The rhodamine B (RhB) is a basic dye that belongs to xanthene which has been widely used in many fields owing to their high quantum yield, high absorption coefficient and excellent light stability [1-2]. It has been used as dye due to the bright color at the early stage. In the recent years, the use of RhB for fluorescent probes, especially in single-molecule imaging [3-4] and imaging in living cells [5-7] has also been reported. HA is an unusual polysaccharide that has a simple chemical structure but excellent properties with large, negatively charged, unbranched polymer which is composed of repeating disaccharides of glucuronic acid and *N*-acetylglucosamine:[-β(1,4)-GlcUA-β(1,3)-GlcNAc-]*n*[8].It has been widely used for biomedical application owing to its excellent biocompatibility [9].Therefore, HA has attracted considerable attention as the targeting ligands of drug delivery systems owing to a specific interaction with CD44 receptors that are over expressed on various tumor cells [10].

The results showed that the number of cases diagnosed with cancer and surviving in China within 5 years was about 7.49 million (including 3.68 million male patients and 3.81 million female patients),the overall 5-year cancer prevalence was 556/10 million [11]. The traditional means of cancer treatment mainly for chemotherapy due to it can not distinguish between rapidly dividing normal cells and cancer cells therefore limit its application [12]. To overcome the limitations associated with nonselective chemotherapeutics, 'smart' drug delivery systems have been developed [13-14]. Mesoporous silica has been widely used for drug carrier and controlled release owing to its stable porous structure, tunable pore sizes and volumes, large surface area, easier chemical functionalization of the surface, and excellent biocompatibility [15-16]. In our present work, the obtained results indicated that the RhB-EA-HA were successfully synthesized. Furthermore, in the further work, we will research the combination of mesoporous silica with RhB-EA-HA and further applied to drug delivery system. In this system, HA is not only act as capping agents but also as targeting ligands and RhB is used as fluorescent probes. Therefore, the RhB-EA-HA might be a great potential modified nanomaterials as a drug delivery system in the future.

2. Experimental section

2.1. Materials

Rhodamine B, Dichloromethane, Anhydrous ethanol, Ethylenediamine, Sodium hyaluronate (MW100KDa), Dialysis bag, Sodium sulfatehydrochloric acid, Sodium hydroxide were obtained from Sinopharm Chemical Reagent Co. Ltd. 1-ethyl-3-(3-(dimethylamino) propyl) carbodiimidehydrochloride (EDC·HCl) were purchased from J&K Chemical Co. (Beijing), All the chemicals were used as received without further purification.

2.2. Synthesis of RhB-EA

RhB reacts with EA were prepared according to the published method [17]. And the synthesis route was presented in scheme 1(A). Briefly, To a 250mL round-bottom flask, RhB (2.40g,5mmol) was first dissolved in 60mL ethanol. 0.43mL of EA(6.5mmol) was then added dropwise with vigorous stirring at room temperature. After the addition, the stirred mixture was heated to reflux for 12h. The solution became clear. Then the mixture was cooled and solvent was removed under reduced pressure. 1M HCl (about 100mL) was added to the residue to remove any unreacted EA. After that, 1MNaOH (about 140mL) was added slowly with stirring until the pH of the solution reached 9–10. The resulting solution was extracted with dichloromethane (3*100mL), washed the organic layer with 50mL water, dried on Na_2SO_4 and evaporated the solvents to get the crude product.

2.3. Synthesis of RhB-EA-HA

The synthesis route was presented in scheme 1(B). HA (100mg) was dissolved in nanopure water (50ml) and then activated by using EDC (150mg) and HCL(100mg) for 1h at room temperature. Then an aqueous solution of RhB-EA as prepared (10.5mg) was added dropwise with avoiding light and stirring at room temperature for 24 h. And the mixture was dialyzed for two days then resulting an aqueous solution of RhB-EA-HA which was freeze-dried.

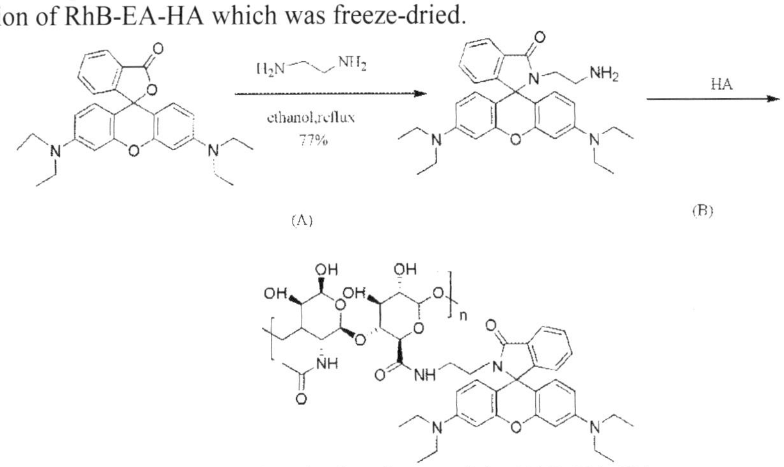

Scheme 1. Synthetic scheme of the RhB-EA-HA.

2.4. Characterizations

Fourier transform infrared (FT-IR) spectrum was noted on Nicolet Nexus 470. The UV spectrum was determined using a UV-Vis spectrophotometer (UV-2450PC,Shimadzu). ^{1}H NMR spectra were recorded on a Bruker AM-400 NMR spectrometer in CD_3OD and D_2O.

3. Results and discussions

As shown in Fig. 1(A), HA is not absorbed in the UV region. In Fig. 1(B), RhB has a strong absorption peak at 554 nm due to the presence of aromatic heterocyclic ring. In Fig. 1(C), although the peak intensity of HA by rhodamine derivative (RhB-EA) modified was reduced, the absorption peak at 554 nm was observed, these results indicating that RhB had been successfully modified to HA.

The successful incorporation of RhB and EA to HA is further confirmed by FT-IR spectra. As shown in Fig. 2, compared with RhB Fig. 2(A), a new adsorption peak appeared at1633 cm^{-1} in RhB-EA Fig. 2(B) and RhB-EA-HA Fig. 2(C) ,which is attributed to the amide bond stretching vibration of C=O peak. So, it indicated that the HA were successfully modified with the RhB and EA.

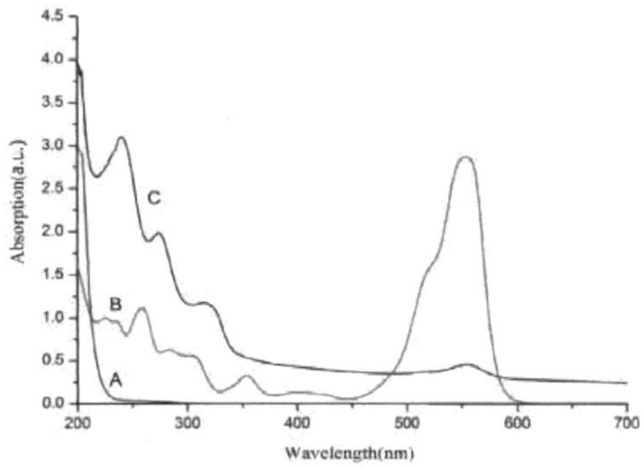

Figure 1. UV-Vis spectral of the HA(A), RhB(B) , RhB-EA-HA(C).

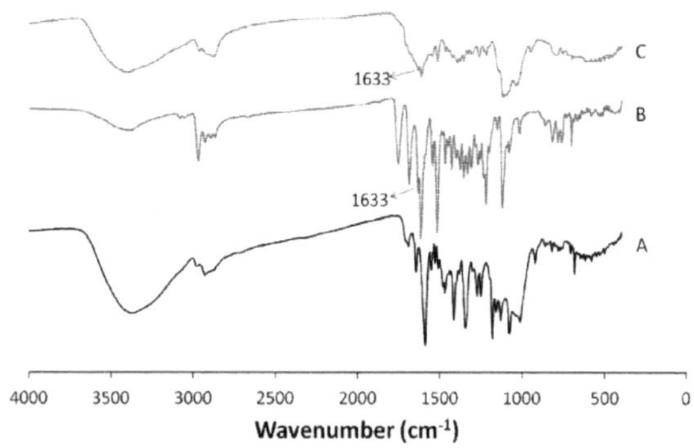

Figure 2. The FT-IR spectra of RhB (A), RhB-EA(B) and RhB-EA-HA(C).

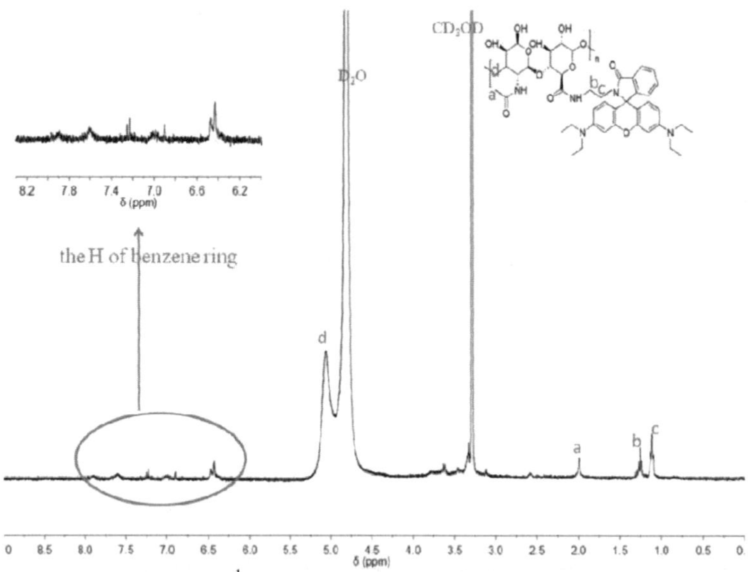

Figure 3. ^{1}H-NMR spectrum of RhB-EA-HA.

The structures of RhB-EA-HA was also determined by ^{1}H-NMR analysis (Fig. 3). As displayed in Fig. 3, the presence of peak at δ 2.00 corresponded to the protons of characteristic peak of N-acetyl methyl of HA. And two peaks appeared at δ 1.26, δ 1.12 were attributed to the protons of the methylene peak of RhB derivates(RhB- EA). The presence of peak at δ 5.06 corresponded to the protons of methylene peak of HA. The red arrows in Fig. 3 indicate the characteristic peaks of the benzene ring of the RhB. These results suggested that RhB was successfully grafted onto the main chain of HA.

4. Summary

In this paper, the RhB-EA-HA was successfully synthesized. Moreover, the RhB-EA-HA were characterized by ^{1}H-NMR, UV-Vis and FT-IR. These results would provide a potential biological functional materials for the modified mesoporous silica nanoparticles to applying in drug delivery system. So, in the next work, we will research the combination of mesoporous silica with RhB-EA-HA and further applied to drug delivery system. In this system, HA is not only act as capping agents but also as targeting ligands and RhB is used as imaging in living cells. So, the RhB-EA-HA would be a great potential modified nanomaterials as a drug delivery system in the future.

Acknowledgments
This research was financially supported by National Natural Science Foundation of China (No. 81201197).

References

[1] Beija M, Afonso C A and Martinho J M 2009 *Chem. Soc. Rev.* **38** 2410-33
[2] Gonçalves M S 2009 *Chem. Rev.* **109** 190-12
[3] Li L, Tian X, Zou G, Shi Z, Zhang X and Jin W 2008 *Anal. Chem.* **80** 3999-06
[4] Bossi M, Fölling J, Belov V N , Boyarskiy V P, Medda R and Egner A 2008 *Nano. Lett.* **8** 2463-68
[5] Zhou Z, Yu M, Yang H, Huang K, Li F, Yi T and Huang C 2008 *Chem. Commun.* **29** 3387-89
[6] Wen H, Howarth M, Greytak A B, Zheng Y, Nocera D G, Ting A Y and Bawendi M G 2008 *J. Am. Chem. Soc.* **130** 1274-84
[7] Hara C, Tateyama K, Akamatsu N , Imabayashi H, Karaki K, Nomura N, Okano H, Miyawaki A 2006 *Brain .Cell. biology.* **35** 229-37
[8] Toole B P 2004 *Nat. Rev. Cancer.* **4** 528-39
[9] Chen Z, Li Z, Lin Y, Ren J and Qu X 2013 *Chem.* **19** 1778-83
[10] Yang C, Liu Y, He Y, Du Y, Wang W, Shi X and Gao F 2013 *Biomaterials* **34** 6829-38
[11] ChenW, Zheng R, Baade P D, Zhang S, Zeng H, Bray F, Jemal A, Yu X Q and He I 2016 *Ca-Cancer. J. Clin.* **66** 115–32
[12] Vaupel P and Kallinowski F 1989 *Cancer. Res.* **49** 6449-65
[13] Li J M, Wang Y Y, Zhao M X, Tan C P, Li Y Q, Le X Y, Ji L N and Mao Z W 2012 *Biomaterials.* **33** 2780-90
[14] Shen J, Yin Q, Chen L , Zhang Z and Li Y 2012 *Biomaterials.* **33** 8613-24
[15] Wang J, Vinu A and Coppens M O 2007 *J. Mater. Chem.* **17** 4265-73
[16] Vinu A, Hossain K Z and ArigaK J 2005 *Nanosci. Nanotechnol.* **5** 347-71
[17] Soh J H, Swamy K M K, Kim S K, Kim S, Lee S H and Yoon J Y 2007 *Tetrahedron. Lett.* **48** 5966-69

IOP Conf. Series: Materials Science and Engineering **359** (2018) 012041 doi:10.1088/1757-899X/359/1/012041

Safety Performance of Exterior Wall Insulation Material Based on Large Security Concept

Q L Zuo[1,2*] ,Y J Wang[1] and J S Li[3]

[1]College of Resources and Environmental Engineering, Wuhan University of Technology, Wuhan 430070, PR. China

[2]Henan Institute of Engineering, Safety Engineering, Zhengzhou, 451191, PR China

[3]Railway Police College, Public Security Department, Henan, Zhengzhou, 450000, PR China.

E-mail:543755297@qq.com

Abstract. In order to evaluate the fire spread characteristics of building insulation materials under corner fire, an experiment is carried out with small-scale fire spread test system. The change rule of the parameters such as the average height of the flame, the average temperature of the flame and the shape of the flame are analyzed. The variations of the fire spread characteristic parameters of the building insulation materials are investigated. The results show that the average temperature of Expanded Polystyrene (EPS) board, with different thickness, decrease - rise - decrease - increase. During the combustion process, the fire of 4cm thick plate spreads faster.

1.Introduction

Hard closed-cell polystyrene foam (EPS) color steel sandwich panel is the current international, especially in the domestic market, a very common use of lightweight new building materials insulation materials[1-3].

With the increasing construction of domestic buildings in recent years, the use of polystyrene (EPS) and other cost-effective materials as decorative materials and insulation materials has been widely used in the decoration of exterior walls. This has attracted the attention of many scholars[4-5]. Edith et al. [6] have studied the critical thickness of the material during the process of fire propagation in a microgravity environment. Eskimo koala et al. [7] have experimented with the rapid propagation of PMMA flakes with different thicknesses. As the EPS fire spreads more complicated, the influence of the thickness on the fire propagation characteristics of the insulation material is relatively small, so it is necessary to study the fire spread characteristics of different thickness insulation materials in the vertical up case.

Based on this, the small-scale combustion experiment on insulation materials EPS is carried out in the laboratory. The thickness of the material is 2cm and 4cm. The size of the material is 25cm × 25cm. The thermal conductivity is 0.032 W / m • k. The compressive strength is 75 KPA and the density is 19 Kg /㎡. All of the values meet the relevant standards. The standard parameters are summarized in Table

1. In the laboratory, the propagation regularity of the flame is analyzed from three aspects of the average temperature of the flame, the average temperature of the flame and the thickness of the flame in the combustion process.

2. Experimental method and system

The experiment is carried out on a small-scale fire spread test. The experimental platform is constructed by the incombustible materials quartz. It is can be shown from the figure 1. Based on the small size experiments related to large-scale fire experiments, it is very important to predict and control the material fire at the present stage based on the parameters to predict the material fire propagation characteristics. Samples of insulation materials are from Hebei Langfang Xinda insulation plant. Before the experiment, the sample is processed. The scale is drawn every 3 cm on the surface to facilitate the high-speed camera to take pictures of the flame height at different times. And it is facilitated to measure the height of the flame height at different times. A relatively confined environment is built that there is plenty of air to support the burning of the combustion chamber. The experimental material is placed vertically to simulate the insulation material in the external wall burning scene. To avoid the interference of the outside world on the experimental system, the uniformity of the experimental environment is maintained during experiment. The sample placed vertically is kept by the use of blade system during the whole process of the experiment. The ignition device is used to ignite. The infrared thermometer is arranged 5 groups. It is layout straightly from the lowest end to the most high-end of the sample. The spacing is kept 5cm each infrared thermometer to measure the flame temperature.

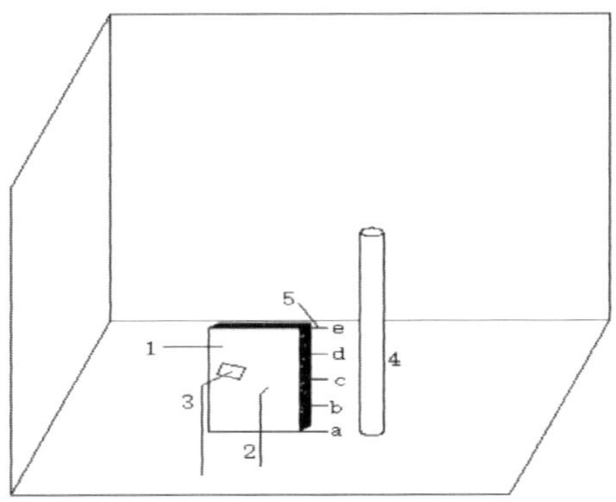

Figure 1. Schematic diagram of the experimental system

1.EPS insulation material 2. Blade system 3. High-speed camera 4. Height measuring instrument 5. Infrared thermometer

Table 1. EPS standard summary of the parameters

Project	Unit	EPS
Thermal Conductivity	W/m • k	≤0.042
Compressive strength	KPA	≥69
Water absorption	%	<2.0
Temperature (max)	℃	70
Flammability degree		Ignited spontaneous

		combustion
Burn level		B2
density	KG/㎡	18-20
Convection heat transfer		Yes
Bonding strength	MPA	>0.1

3.Results and discussions

During the experiment, the room temperature is 18.2 ℃, the humidity is 22% RH and the wind speed is 1 m/s. The combustion experiments of EPS with two kinds of thicknesses are carried out respectively. And the variation rules of mean maximum flame height, flame characteristic and flame mean temperature are obtained.

3.1 the flame characteristics of EPS under different thickness .Figure 2 is the thickness of 4cm and 2cm flame traces. They are at the beginning of combustion, burning to the material 1/4 position, burning to the material 1/2 position and the end of combustion respectively. The start time t = 0 s is defined as the moment when the flame appears before pyrolysis. It can be seen from Figure 2, the 4 cm thick average maximum flame height is greater than 2cm thick for different thickness EPS in the combustion process. And the fire spread is faster. It its mainly because the EPS board of 4 cm thick can provide more adequate fuel. It can be seen from the experiment that the combustion speed and intensity is also increasing with the increase the thickness of polystyrene (EPS) material. It can be also found that the smoke is generated during the polystyrene (EPS) combustion, and the Irritating smell is accompanied. But 2 cm thick is burnt more fully from the burning traces. The EPS board of 4 cm thick is better between them from the perspective of insulation and fire.

Figure2. Flame characteristics at different locations with different thicknesses

3.2 The average temperature of the flame of EPS under different thickness.

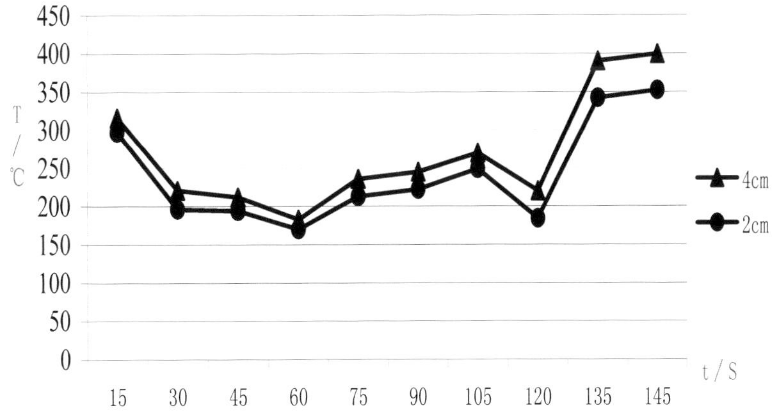

Figure3. The temperature change of the EPS board at different thicknesses

It can be seen from Figure 3, the different thickness of the EPS board material burning the average temperature exhibits the regular changes of decrease - rise - decrease – increase. It is indicated that the vertical direction of the insulation material shows a cycle of shock phenomenon in the fire during the flame. It is also consistent with the flame structure of EPS board in the combustion process which is a discontinuous oval and small phenomenon in turn from the bottom up.

3.3 The maximum flame height of EPS under different thickness.

Table 2. Flame average height and burning time of EPS

Thickness/cm	Maximum height of flame /cm	Burning time/s
4	11	152
2	6.1	162

It can be seen from Table 2, the average height of the flame changes relatively large with the increasing in thickness. It is indicated that the thickness has a greater impact on the average height of the flame.

Comparing to the burning time and the average height of the flame, it can be seen that the average height of the flame is up to 11cm in 152s for the thickness of 4cm EPS board, while maximum flame height is only 6.1cm for thickness of 2cm EPS board in 162s. This indicates that the burning of 2 cm thick polystyrene (EPS) is not sufficient. It is mainly because the EPS board can not provide sufficient fuel to maintain combustion.

4.Summary

Through the combustion experiments of EPS plates with different thicknesses, the variation rules of the mean maximum flame height, average flame temperature and flame characteristics during the vertical placement are analyzed.

(1) The fire of 4cm thick board is spread faster for different thickness EPS in the combustion process. It is mainly because the EPS board of 4 cm thick can provide more adequate fuel. The 2cm thick is burnt more fully from the burning traces of view. It is shown that the thickness of materials should be pay attention to in the selection of the external wall insulation materials. It is ensured its insulation characteristics at the same time to minimize the building fire impact.

(2) The average temperature of different thicknesses exhibits a decrease - rise - decrease - increase in the law when the EPS plates are combusted. This indicates that the flame in the vertical direction exhibits a cycle Sexual shock phenomenon.

(3) Comparing to the burning time and the average height of the flame, the maximum height of the thickness of 4cm EPS board is up to 11cm in the 152s time, while the maximum flame height of the thickness of 2cm EPS board is only 6.1cm in 162s.it is indicated that the burning of 2 cm thick polystyrene (EPS) is not sufficient.

5.Acknowledgments

This work was supported by Henan province key research projects of higher education(16B620001), Henan province key research projects of higher education(17A620001),Henan province key research projects of higher education(18B620001) and Henan Province Science and Technology Development Plans (172102310351)

References

[1] Gan Jie, Peng Nina. Polystyrene insulation material aging before and after the combustion performance experimental study [J]. Fire technology and product information, 2017,2: 16-19.
[2] YIN Ming, ZHAO Min, ZOU Weixiong. Study on Combustion Performance of Exterior Wall Insulation Materials [J]. Construction, 2016, 17: 70-73.
[3] WANG Yazhen, CHANG Haiyang, HE Zijian. Study on systematic error of thermal conductivity of thermal insulation materials by steady state method [J]. Building Science, 2016,32 (10): 33-36.
[4] Zhao Jie, Huang Xinjie, Liu Wei. Flame Propagation and Combustion Characteristics of EPS with Different Width and Density [J]. Process Engineering, 2016,16 (3): 465-469.
[5] Zhang Qinglin, Han Weiping, Wang Junsheng. Thermal Behavior and Combustion Behavior of Polymer-based Building Insulation Materials [J]. Journal of Building Materials, 2015,18 (6): 23-27.
[6] Edith. Fire safety with EPS material properties and relevance for fire safety during transport, construction and under end use conditions in external thermal insulation component systems[M]. Owned by the authors, published by EDP Sciences, 2013
[7] Eskomi koala, Tula Hacanin, Anna Matala.Fire safety of EPS insulated facades in residential multi-storey buildings[M]. Owned by the authors, published by EDP Sciences, 2013

IOP Conf. Series: Materials Science and Engineering **359** (2018) 012042 doi:10.1088/1757-899X/359/1/012042

Micro-structure and Air-tightness of Squeeze Casting Motor housing for New Energy Vehicle

Y F Jiang [1], Z Q Kang* [1], W F Jiang[1], K W Wang[1], D L Sha[1], M L Li[2] and J Sun[2]
[1] School of mechanicalengineering, JiangSu University, Zhenjiang, China, 212013
[2]SuZhou SanJi Foundry Equipment CO.,LT, SuZhou, China, 215105

Corresponding author: Z Q KangEmail:1085581593@qq.com

Abstract: In order to improve the performance of automobile parts, the influence of squeeze casting process parameters on casting defects, material structure and air-tightness of aluminum alloy motor housing for new energy vehicle was studied. The results show that the density of the castings increases with the increase in pressure and mold temperature. With increase in pouring temperature, it increases first and then decreases. Pressure has the greatest influence on the density of the castings. Under a certain pressure, with moderate increase in casting temperature and mold temperature, the grain growth begins to increase; the dendrites become less, the new α - Al grains are spherical and granular, the micro-structure is uniform. Also, with increase in pressure, this effect is more pronounced, the air-tightness of castings improve. In conclusion, when the pressure is 110MPa, pouring temperature is 680° C, mold temperature is 280° C, pressure holding for 30s, and punch speed of 0.1m/s, there is no clear shrinkage in the casting, the structure is uniform, the qualified rate of air-tightness of production reaches 86%, and the performance is excellent.

1. Introduction

Relevant data shows that the number of private cars in China have reached 137 million[1]. As the key component of the motor module, excellent air-tightness is necessary for the safe and efficient operation of the motor housing.

The mainly material of motor housing is aluminum alloy, which not only improving the self cooling effect, but also reducing the weight of the automobile. The literature pointed out that the car's weight reduce 100kg, the fuel consumption will reduce 0.3-0.5L per 100 kilometers[2]. The aluminum alloy motor housing manufacturing main processes are low pressure casting and pressure casting, etc. As a kind of near net forming technology, squeeze casting is getting more attention[3]. This process is characterized in that a high mechanical pressure is applied to the metal liquid which is poured into the

IOP Conf. Series: Materials Science and Engineering **359** (2018) 012042 doi:10.1088/1757-899X/359/1/012042

mold cavity with low speed and high flow rate, and the metal liquid is solidified under high pressure to obtain the casting with complicated shape, high density and good air-tightness. This technology is widely used in automobile, motorcycle, mechanical, aerospace and other fields[4-6]. However, there are many casting defects in the castings, and low air-tightness, which is difficult to meet the needs of customers.

This paper aims at the existing problems of automobile motor housing. Through orthogonal test, the optimum extrusion casting process parameters of aluminum alloy motor housing are obtained. Under the optimal process parameters, obtained the motor housing products with excellent air-tightness.

2. Experiment procedure

2.1 Design of casting gate and overflow system for motor housing

The structure of motor housing is very complicated, and it is difficult to produce the casting by direct extrusion casting. As the channel of the metal liquid flowing into the mold, the design of the channel should ensure that the metal liquid flow into cavity smoothly and continuously. The design of gating system has a direct influence on the quality of the production[7-8]. Unreasonable shape and size of the channel will cause metal liquid collision, forming a vortex, roll gas, which will result in low air-tightness after forming. The position, shape and size of the inner gate which are directly connected with the flow rate and flow direction of the molten metal into the mold cavity, which influences the forming quality of the product. According to the structure characteristics of the motor housing, the inner gate is opened in the center hole position, and the inner gate diameter is set to 60mm. From the gate to the material handle, the sectional area of layers increases, the longitudinal angle is 8 degrees. In order to ensure that the castings are fully filled, the thickness of the handle is controlled between 25-40mm. The design of the drainage system should contribute to the smooth discharge of gas and the collection of oxide in the filling process. The overflow grooves of the motor housing is provided on the parting surface, and metal liquid reachs the position finally. The concrete pouring system modeled by Pro-E is shown in Figure 1.

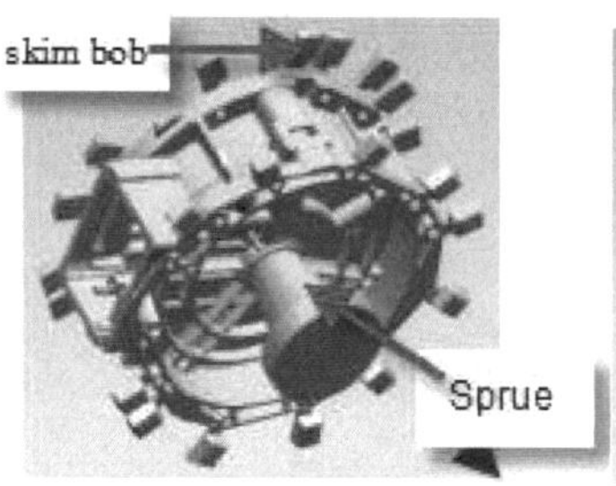

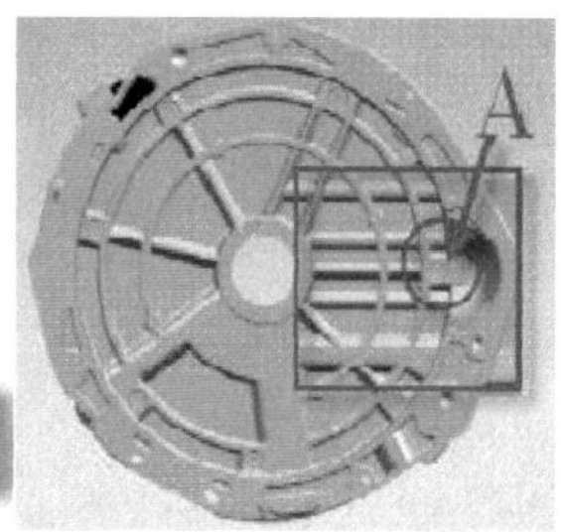

Fig. 1 motor housing casting model Fig. 2 detection position of motor housing

2.2 Test method

The test material is ADC12 aluminum alloy. The alloy melting process is that when the temperature of aluminum liquid is 720-730 degree, the nitrogen gas is used to remove the gas which into liquid. Then, a certain amount of Al-Ti-B and Al-10Sr are added to make aluminum liquid more refined. In the end, deslagging agent is used to remove impurities. Squeeze casting equipment consists of SCV-2000T vertical squeeze casting machine, motor housing mold and ancillary equipment.The orthogonal test with three factors and three levels is adopted in squeeze casting experiment. According to the principle of orthogonal test, The specific combination of process parameters are shown in Table 1, the pressure holding time of 30s and punch velocity of 0.1m/s are remained unchanged.

Table 1 Orthogonal test plan

Number	Pressure/Mpa A	Casting temperature/℃ B	Mold preheating temperature/℃ C
G1	1（70）	1（660）	1（220）
G2	1（70）	2（680）	2（250）
G3	1（70）	3（700）	3（280）
G4	2（90）	1（660）	2（250）
G5	2（90）	2（680）	3（280）
G6	2（90）	3（700）	1（220）
G7	3（110）	1（660）	3（280）
G8	3（110）	2（680）	1（220）
G9	3（110）	3（700）	2（250）

After the test, A position (Figure 2, which is modeled by Pro-E) which should have good tightness requirement. For this, by use XG-1604L/C type X-ray real-time imaging equipment for internal detection, then calculate the area density and analysis microstructure by OLYMPUS metallographic microscope, perform the air-tightness test finally.

3. Results and discussion

3.1 Analysis of Process Parameters on filling

The effect of process parameters on the filling of the casting is shoot by the camera, as shown in Figure 3. In figure 3-(a), 70MPa/660℃/220℃, respectively expressed as the ratio of pressure, casting temperature and mold temperature, others empathy. As shown in Figure 3, G1 casting has obvious appearance defects. The lower pressure, pouring temperature and mold temperature, result in cold liquid alloy reduce liquidity, casting will appear cold lap, flow marks, and even less casting defects. With the pouring temperature and mold temperature increasing, the motor housing casting cooling slowly. Keep pressure at the same time, casting can not completely solidified. When taking out the casting, it is easy to have a casting injury. As shown in figure 3- (b).

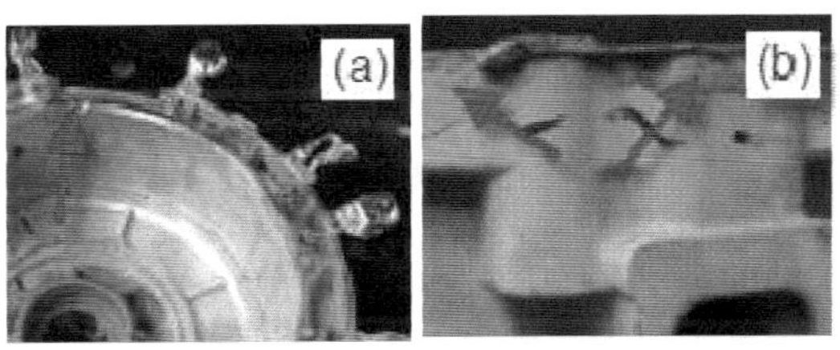

G1(70MPa 660℃ 220℃) G3(70MPa 700℃280℃)

Fig. 3 Effect of process parameters on filling pattern

Because the G1 and G3 castings have obvious defects, X-ray test was detected at A group only for the rest of the castings. Figure 4 is obtained by an X-ray inspection apparatus. Test results show that the shrinkages of G2, G4, G6 groups are in the pipeline with gas tight requirements, as shown in figure 4- (a). It can be concluded that the air-tightness of these castings is not qualified. The shrinkages of G7 group is far from the pipe, and has little effect on the air-tightness. For the rest of the castings, at the appropriate temperature, with the increasing of the pressure, castings don't have shrinkage porosity and other defects, meeting the requirements of air tightness, as shown in figure 4-(b).

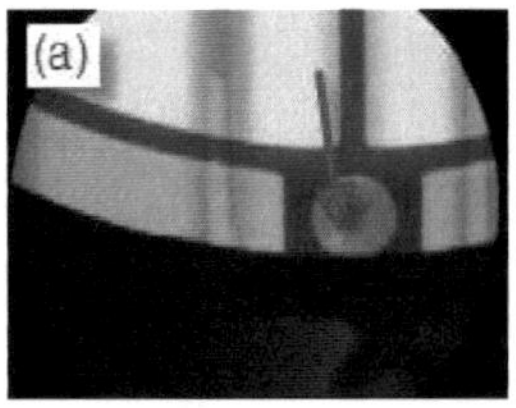

Defective castings

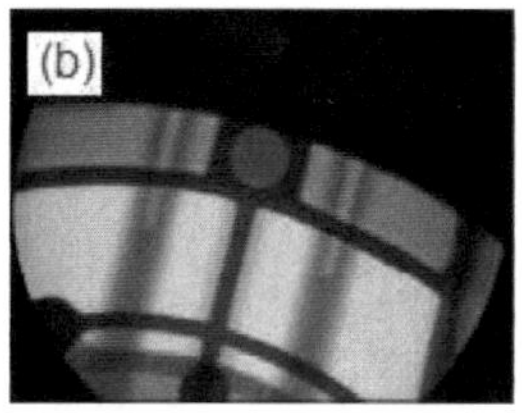

Defect free castings

Fig. 4 Effect of process parameters on the inner ofthe

3.2 Effect of process parameters on density

To a certain extent, the density of the casting can reflect the air tightness of the casting. In order to reflect all influence of the selected parameters on the density, the total density of the 9 groups of castings was measured. For different process parameters, the test results of the motor housing density are shown in table 2.

Table 2 The density of motor housing casting

Number	Pressure/MPa	Casting temperature/℃	Mold preheating temperatur/℃	Density/g·cm^{-3}
	A	B	C	
G 1	70	660	220	2.7293
G 2	70	680	250	2.7380
G 3	70	700	280	2.7358
G 4	90	660	250	2.7342
G 5	90	680	280	2.7432
G 6	90	700	220	2.7372
G 7	110	660	280	2.7403
G 8	110	680	220	2.7386
G 9	110	700	250	2.7436
I	8.2031	8.2038	8.2051	
II	8.2146	8.2198	8.2158	
III	8.2225	8.2166	8.2193	
K1	2.7344	2.7346	2.7350	
K2	2.7382	2.7399	2.7386	
K3	2.7408	2.7389	2.7398	

R	0.0065	0.0053	0.0047
Primaryfactor → secondary	A B C		
Optimal scheme	A_3 B_2 C_3		

Among them:

I =The sum of the results on the level number 1 on the J column;

II =The sum of the results on the level number 2 on the J column;

III=The sum of the results on the level number 3 on the J column;

K1= I /3，K2= II /3，K3=III/3；K1, K2 and K3 respectively represent the average of the sum.

R=The disparity between the maximum and minimum of K1, K2 and K3on the J column.

The results presented in Table 3 , groups 5 and 9 correspond to the high density, which correspond to the process parameters A2B2C3 and A3B3C2. The greater the density, the better the air-tightness. Corresponding parameters combination are more reasonable. In Table 3, the effect of the factors on the density of the R value can be with descending order: A>B>C. The effect of pressure on the density of squeeze casting is maximum. Through the analysis of the experimental results, it can be concluded that the casting density increase with the increase of the specific pressure and the temperature of the mold. With the increase of the casting temperature, the density of the motor housing increase firstly and then decrease. Overall, the optimal process parameters are A3B2C3, the pressure is 110MPa, pouring temperature is 680℃, the mold temperature is 280℃.

3.3 Effect of process parameters on Microstructure

Microstructures of the A region in the casting under different process parameters are shown in figure 5. Figure 5 is obtained by an optical microscope. At low pressure, pouring temperature and mold temperature, the majority of $\alpha - Al$ microstructures present dendrite, only a handful is ball and granular, distribution is very loose, irregular grain shapes and sizes, relatively small, poor mechanical properties, as shown in figure 5- (a). Keep the same pressure, increasing pouring temperature and mold temperature, the fine grain structures disappeared obviously, grain begin to grow, the number of dendrite morphology begin to decrease. Enhance the pouring temperature to 700℃and mold temperature to 280℃, the average cooling rate of metal solidification and casting solidification rate slowed down, the solidification time became longer. It can be seen that the $\alpha - Al$ grain is further grown up, the dendrite coarsening, the dendrite space decreases, as shown in figure 5- (b).In figure 5- (c), The microstructure of the casting is almost without dendrites, $\alpha - Al$ grains present spherical and granular. The microstructure is uniform, and its shape and size are roughly equal. Because of at a suitable pouring temperature and mold temperature, increasing the pressure, the casting and the mold cavity fit more closely, and the heat transfer coefficient between the housing and the wall increases. The increase effective area between the two make the alloy cooling rate accelerated and the nucleation rate increase. In addition, the pressure inhibits the diffusion of atoms and the growth of the crystal nucleus, so that the grain is refined, and the microstructure become uniform[9-11].

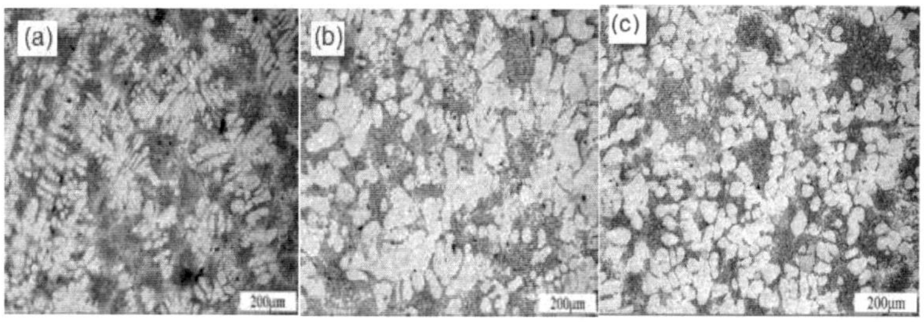

G1(70MPa /660℃ /220℃) G3(70MPa /700℃ /280℃)G5(90MPa /680℃/ 280℃)

Fig. 5 microstructure of A

It can be drawn from the figure 3 and table 5 that the microstructure of the casting is mainly present dendrite, and the gap is larger, the density is the smallest, so the air-tightness is poor in the G1 group; Keeping the pressure, increasing the casting temperature and mold temperature, the dendrite structure of the G3 group reduced significantly, grain growth, casting density increase, air-tightness is also good; At appropriate pouring temperature and mold temperature, improve the pressure, microstructure of G5 basically presents spherical shape, uniform round, the shape and size of roughly equal. So the G5 group have larger density and better air-tightness. In general, at proper pouring temperature, increasing the pressure of squeeze casting, the microstructure of production become denser and the grains uniform distribution, so the air-tightness is bound to increase.

3.4 Air-tightness test

The air-tightness of the casting is not only determined by the macroscopic filling of the casting, but also the density and microstructure of the casting. By analyzing the filling result, density and microstructure of the casting, it is found that at proper pouring temperature and mold temperature, improve pressure, the casting will get better air-tightness. Thus, the best process parameters are A3B2C3. The pressure is 110MPa, pouring temperature is 680℃, mold temperature is 280℃, holding pressure time is 30s, the injection rate is 0.1m/s.

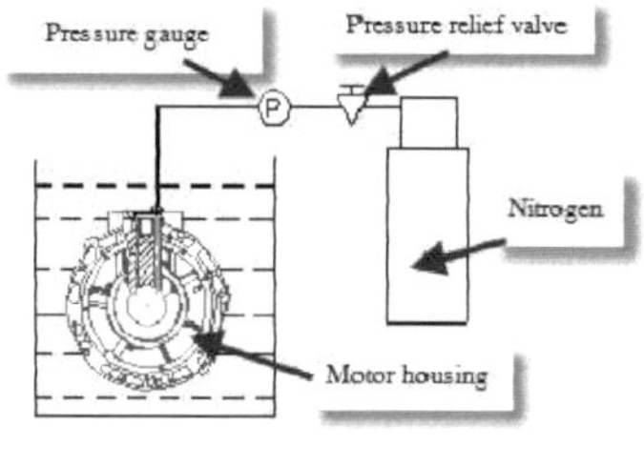

Fig.6 Air-tightness testingdevice

Producing castings under optimum process parameters, then the air-tightness test is carried out, and the experimental device as shown in Figure 6. Place the casting into the water, then fill the casting with nitrogen gas, when the pressure gauge reach 1.8MPa, close the gas source , and keep 3min. To observe whether the bubble is formed and the pressure gauge is reduced. Finally, the experimental results show that the qualified rate of air-tightness reached 86%.

4. Conclusion

In the case of low specific pressure, pouring temperature and mold temperature, the fluidity of the aluminum alloy liquid is reduced, the casting will appear cold interval, flow marks and other defects; Improved the casting temperature and mold temperature, while increasing the pressure, casting defects are significantly reduced, and the air-tightness of castings increases.

The results show that the pressure has the greatest influence on the density of the castings, and further analysis shows that the density of the castings increases with the increase of the pressure and the mold temperature, while increasing the pouring temperature, the density of the motor housing increases first and then decreases ;

Under the same condition of the pressure, as the casting temperature and mold temperature increases, the fine grain structures disappeared obviously, grain begin to grow, the number of dendrite morphology begin to decrease and α - Al grains present spherical , granular, and the microstructure is uniform. With the pressure increases, this effect is more obvious, the airtight of castings improve inevitably. Therefore, at a proper casting temperature and mold temperature to increase squeeze punch pressure, the air-tightness of castings will be improved.

The optimum process parameters of new energy vehicle motor housing produced by squeeze casting is: the pressure is 110MPa, the pouring temperature is 680℃, the mold temperature is 280℃, the pressure holding time is 30s, and the punch speed is 0.1m/s. Under the above conditions, there is no obvious shrinkage into the motor housings, it has excellent mechanical properties, and the air-tightness rate exceeds 86%.

References

[1] Xu, C.M.. 2014. Passenger car market will continue to maintain a stable and rapid development in [J]. Vehicle Engineers. 2014 (01)

[2] Fan, Z.J.; Gui, L.J.; Su, R.Y.. 2014. Research and development of automobile lightweight technology [J]. Journal of automobile safety and energy saving, 5 (1): 1-16.

[3] HuangH,TianY,YuanG.Y.,et al. 1994. Formation Mechanism of Quasicrystals at Nanoscale during Hot Compression of Mg Alloys[J]. Scripta Materialia .

[4] Morton, J.R.; Barlow, J.. 1994. Squeeze casting: from a theory to profit and a future[J]. The Foundryman, 87(1): 23-28.

[5] Wang, S.Z.; Ji, Z.S.; Zhou, J.M.; et al. 2014. Formation and Growth Mechanism of Globular Crystal of ADC12 Aluminum Alloy by Near-Liquidus Squeeze Casting[J]. Advances in Mechanical Engineering . 2014

[6] Dai, W.; Wu, S.S.; Lü, S.L.; et al. 2012. Effects of rheo-squeeze casting parameters on microstructure and mechanical properties of AlCuMnTi alloy[J]. Materials Science & Engineering A .

[7] Jiang, Y.F.; Gu, W.X.. 2009. Die casting die engineer manual [M].Beijing: Mechanical Industry Press.

[8] Liao, H.M.; Long, S.Y.; Li, F.G.; et al. 2008. Numerical simulation of casting magnesium alloy wheel casting system [J]. Special casting and nonferrous alloys, 28 (4): 271-273.

[9] Zhang, M.;Zhang, W.W.; Zhao, H.D.; et al. 2007. Effect of Pressure on Microstructures and Mechanical Properties of Al-Cu-Based Alloy Prepared by Squeeze Casting[J]. Trans. Nonferrous Met. Soc. China. 17: 496-501.

[10] Wang, K.; Liu, C.M.; Zhai, Y.B.; et al.2010. Microstructural characteristics of near-liquidus cast AZ91D alloy during semi-solid die casting[J]. Transactions of Nonferrous Metals Society of China. (02)

[11] Lin, B.; Zhang, W.W.; Cheng, P.; et al. 2013. Microstructure and mechanical properties of Al-5.0Cu-0.6Mn-0.5Fe alloy by squeeze casting [J]. Chinese Journal of nonferrous metals (05)

IOP Conf. Series: Materials Science and Engineering **359** (2018) 012043 doi:10.1088/1757-899X/359/1/012043

Upconversion enhancement in $NaYF_4$:Yb^{3+}, Er^{3+} crystal via Bi^{3+} doping

S X Xiao, Z S Chen*, H Q Huang, J G Zheng and J P Xu

Jiangxi Key Laboratory for Mass Spectrometry and Instrumentation, East China University of Technology, Nanchang, Jiangxi 330013, P.R. China

Email: zhshcheng@ecit.cn

Abstract. In this research, $NaY_{0.8-x}F_4$:18%Yb^{3+}, 2%Er^{3+}, x% Bi^{3+} (x = 0, 1, 2, 4, 7, 10mol%) crystals were preparedby a simple hydrothermal method. Their purity process and upconversion (UC) intensity were controlled by different amounts of Bi^{3+} doping.The intensity of UC red and green emissionwere enhanced by 5.5 and 4.9foldsrespectively at 4 mol% Bi^{3+} doping.

1.Introduction

Rare earth-dopedupconversion (UC) luminescence is based on two-photon or multi photon mechanism of the long-wave radiation into fluorescent material shortwave radiation, which contains the anti-Stokes luminescence properties[1]. UC materials are widely used in the fields of biological imaging[2], 3D printing[3], sensors[4] and solar cell[5] because of their good chemical stability, optical stability and less toxicity. However, the low efficiency of conversion hinders its further development[6].

From the perspective of the crystal, the UC luminescence of $NaYF_4$:Yb^{3+}, Er^{3+}crystal is mainly depend on the transition of Er^{3+}energy levels to produce different UC emission, but the luminescence properties of Er^{3+} are strongly affected by the surrounding environment of crystal[7]. In theory, the asymmetry of the crystal environment is easy to make Er^{3+} f-f resonance transitions[8], so the change amount of some ions in $NaYF_4$:Yb^{3+}, Er^{3+} crystals can achieve this purpose[9]. The regulation of heterogeneous ion doping on the luminescence properties of $NaYF_4$:Yb^{3+}, Er^{3+}crystal plays a very important role. Different kinds of heterogeneous ion doping enhances UC luminescence intensity by changing the distance of interplanar spacing and the electron cloud density of the crystal which affect the distance between the luminescence center and the energy transfer center[10].

Various heterogeneous ions, such as Li^+,Mn^{2+}, Ca^{2+}, Ba^{2+}, Zn^{2+}, have been doped to enhance UC intensity[11]. To the best of our knowledge, however, the effect of Bi^{3+} doping on the UC properties has not been reported yet. Herein, a series of Bi^{3+} doped $NaYF_4$:Yb^{3+}, Er^{3+} were fabricated via a simple hydrothermal method. The influence of Bi^{3+} doping has been discussed in detail.

2.Experiment

1 mmol nitrates contained Y^{3+}, Yb^{3+}, Er^{3+} and Bi^{3+} in a ratio of (80-x):18:2:x (x =0~10), 1.2g NaOH, 10 mL OA, and 10 mL ethanol were added into a 40 ml of Telfon-line autoclave under vigorous stirring for 30 min, 4 mmol NH_4F was dissolved in distilled water and added in the former liquid dropwise and heated at 200°C for 24 h in muffle furnace. The samples were cooled down to room temperature and concentrated by a centrifuge with deionized water and ethanol to wash for three times, respectively. The final products were drying at 60 °C for 24 h to collect the powders.

The crystalline structure of as-obtained samples was determined by X-ray diffraction (XRD, D/ruax2550PC, Tokyo, Japan) using CuK_α radiation (λ = 0.154056 nm) over a 2θ range from 10° to 80° at a scanning rate of 4 °/min. The morphology of the particles was performed by the scanning election microscope (SEM, Nova Nano SEM 450, FEI) operating at an acceleration voltage of 200 kV after a gold spray process. The UC emission spectra were recorded by a spectrophotometer (Hitachi F-4500, Tokyo, Japan) equipped with a power-controllable 980 nm diode laser (LOS-BLD-0980-C Hite photoelectric, China).

3.Results and discussion

Figure.1(a) shows when the concentration of Bi^{3+} is free and 1 mol%,there is a small amount of cubic phase, and when the concentration of Bi^{3+} is more than 1 mol%, the pure hexagonal is produced. It can be inferred that the doping of Bi^{3+} is helpful to produce pure hexagonal $NaYF_4:Yb^{3+}$, Er^{3+} crystal. However, it shifted from an obvious peak to a smaller angle that can be observed by amplified an angle of 43° the peak of (201) in Figure.1(b)

According to Bragg's law[12]:

$2d\sin\theta = n\lambda$.

Combined with the angle shift in Figure.1(a), when the diffraction angle is lower, the space between the ions in the lattice is larger. That is to say, the lattice plane spacing becomes larger with the doping of Bi^{3+}. This is because the Bi^{3+} radius (r_{Bi} = 0.103 nm) is larger than that of the Y^{3+} (r_Y = 0.089 nm)[13] ,so the introduction of Bi^{3+} will lead to the volume expansion of the $NaYF_4:Yb^{3+}$, Er^{3+} crystal, when the crystal plane spacing becomes larger and the peak value shifts to 42°.

It is shown that Bi^{3+} doping $NaYF_4:Yb^{3+}$, Er^{3+} crystal, only played a role in the substitution because of no indication that the peak is shifted to 42° and then move reversely in Figure.1(b).

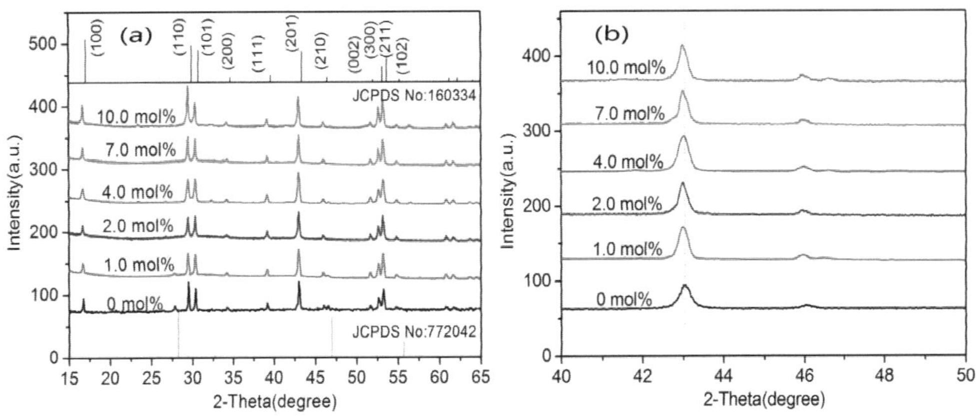

Figure. 1XRD patterns of (a): $NaY_{0.8-x}F_4$:18%Yb^{3+}, 2%Er^{3+}, x% Bi^{3+}(x = 0, 1, 2, 4, 7,

IOP Conf. Series: Materials Science and Engineering **359** (2018) 012043 doi:10.1088/1757-899X/359/1/012043

10mol%)products, standard XRD pattern of hexagonal $NaYF_4$(JCPDS No.16-0334), and (b): amplified of the main diffraction peaks (201) at 43° from Figure.1(a).

The SEM images showed the different amount of Bi^{3+} doped in$NaYF_4$:Yb^{3+}, Er^{3+} crystals in the Figure.2. In Figure 2(a) and (b) showed the presence of micro-rods and nanoparticles, the length of the micro-rods in the range of 600 ~ 800 nm, while the morphology of nanoparticles is not very uniform. According to the results of XRD, the Bi^{3+}-free and 1 mol% Bi^{3+} sample are cubic and hexagonal phase, the other samples are almost pure hexagonal phase. It can be considered that the nanoparticles are cubic phase and the micro-rods particles are hexagonal. And in Figure 2(c)~(f), most of nanoparticles have been converted into micro-rods and their length is ranged from 600 to 800 nm. The results showed that the substitution of low concentration (Bi^{3+}=1~10mol%) does not change the crystal morphology drastically[14].

Figure.2 The SEM images of $NaY_{0.8-x}F_4$:18%Yb^{3+}, 2%Er^{3+}, x% Bi^{3+} (x = 0, 1, 2, 4, 7, 10mol%) : (a) 0 mol%, (b) 1.0 mol%, (c) 3.0 mol%, (d) 4.0 mol%, (e) 7.0 mol% and (f) 10.0 mol%.

As shown in Figure. 3(a), the UC spectra of $NaY_{0.8\text{-}x}F_4$:18%Yb^{3+}, 2%Er^{3+}, x%Bi^{3+} (x = 0, 1, 2, 4, 7, 10mol%) exhibited three emission bands at peak 522nm, 543nm (green) and 655 nm (red), which derived from the transitions between the energy levels ($^2H_{11/2}$, $^4S_{3/2}$)→$^4I_{15/2}$ and $^4F_{9/2}$→$^4I_{15/2}$ of Er^{3+}, respectively[15]. It can be found with the increase of the doping ion concentration, the intensity of green emission and red emission on $NaYF_4$ have changed greatly. When Bi^{3+} doping concentration is 4 mol%, the red and green emissions reached the maximum value. After that, the UC intensity decreases and even quenches. Connection with Figure.1, The effect of Bi^{3+} ions on the lattice is mainly substitutional doping, which increases the distance between the crystal planes of the ions and leads to the increase of the distance between the energy centers and the central ions of the Er^{3+} ions. In comparison with the increase of the crystal spacing caused by the ionic radius and the decrease of energy transfer, the 4f resonance transition has a greater enhance on the UC luminescence of $NaYF_4$[16]. Therefore, the intensity of UC emission has increased. As indicated in Figure. 3(b), when 4 mol% Bi^{3+} is doped, the intensity of red and green emission is enhanced by 5.5 and 4.9 times, respectively. In addition, the intensity ratio of red to green emissionis found to be approximately equal to 1 at different level of Bi^{3+} ions. Hence, it can be concluded that the influence of Bi^{3+} doping on the UC materials intensity of Er^{3+} is relatively constant[17].

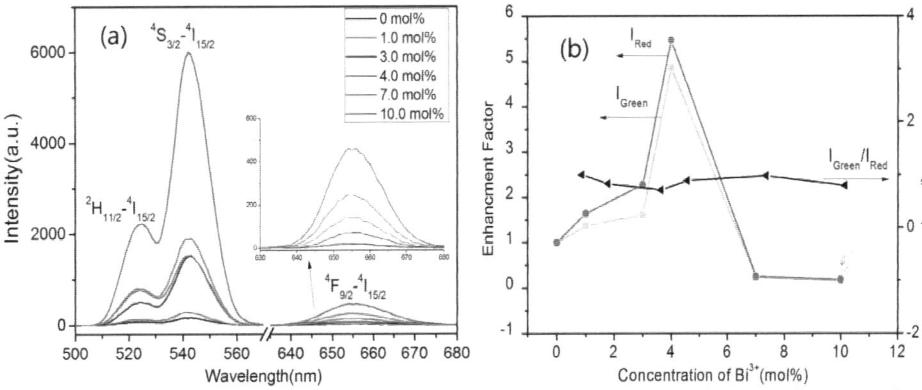

Figure. 3(a): UC luminescence spectra of the samples $NaY_{0.8\text{-}x}F_4$:18%Yb^{3+}, 2%Er^{3+}, x% Bi^{3+} (x = 0, 1, 2, 4, 7, 10mol%) under 980 nm excitation and Figure. 3(b): Enhancement factors of the green, red and total up-conversion emissions of $NaY_{0.8\text{-}x}F_4$:18%Yb^{3+}, 2%Er^{3+}, x% Bi^{3+} (x = 0, 1, 2, 4, 7, 10mol%) crystals as a function of Bi^{3+}concentrations (The integrated intensity of Bi^{3+}-free sample was scaled as unit, excitation at 980 nm).

4.Conclusion

In conclusion, we have prepared a series of $NaY_{0.8\text{-}x}F_4$:18%Yb^{3+},2%Er^{3+},xBi^{3+} (x = 0 ~ 10%) crystals through a simple hydrothermal method. The phase purity and UC intensity of $NaYF_4$:Yb^{3+}, Er^{3+} crystals were controlled by doping different concentrations of Bi^{3+}. Compared to Bi^{3+}-free samples, when incorporating 4.0 mol% Bi^{3+} ion, the UC red and green emission intensity was increased by a factor of 5.5 and 4.9, respectively, which was mainly due to modification of the crystal symmetry of Er^{3+} bythe Bi^{3+} doping.

5.Acknowledgments

This work is supported by NSFC (51362002), Natural Science Foundation of Jiangxi Province

(20161BAB203098, 20171BAB206016), Key Research and Development Projects of Jiangxi Province (20171BBH80021), Education Department of Jiangxi Province (GJJ150599, GJJ150583), and the financial supports from Jiangxi Key Laboratory for Mass Spectrometry and Instrumentation (JXMS201506) are also greatly appreciated.

References

[1] Wang, M.; Wu, J.; Chen, Z., *Key Engineering Materials*, (2016) **703**, 242.

[2] Tan, G. R.; Wang, M.; Hsu, C. Y.; Chen, N.; Zhang, Y., *Advanced Optical Materials*, (2016) **4**, 984.

[3] Méndez-Ramos, J.; Ruiz-Morales, J. C.; Acosta-Mora, P.; Khaidukov, N., *Journal of Materials Chemistry C*, (2016) **4**, 801.

[4] Tian, X.; Wei, X.; Chen, Y.; Duan, C.; Yin, M., *Opt Express*, (2014) **22**, 30333.

[5] Shan, G. B.; Demopoulos, G. P., *Advanced materials*, (2010) **22**, 4373.

[6] Trindade, T.; O'Brien, P.; Pickett, N. L., *Chem Mater*, (2001) **13**, 3843.

[7] Chen, D.; Wang, Y., *Nanoscale*, (2013) **5**, 4621.

[8] Boyer, J. C.; Cuccia, L. A.; Capobianco, J. A., *Nano Letters*, (2007) **7**, 847.

[9] He, E.; Zheng, H.; Gao, W.; Tu, Y.; Lu, Y.; Tian, H.; Li, G., *Journal of Nanoscience & Nanotechnology*, (2014) **14**, 4139.

[10] Lei, L.; Chen, D.; Xu, J.; Zhang, R.; Wang, Y., *Chemistry – An Asian Journal*, (2014) **9**, 728.

[11] Kumar, D.; Verma, K.; Verma, S.; Chaudhary, B.; Som, S.; Sharma, V.; Kumar, V.; Swart, H. C., *Physica B Condensed Matter*, (2017),

[12] Liang, Z.; Cui, Y.; Zhao, S.; Tian, L.; Zhang, J.; Xu, Z., *Journal of Alloys & Compounds*, (2014) **610**, 432.

[13] Shannon, R. D., *Acta Crystallogr A*, (1976) **A32**, 751.

[14] Wang, C.; Cheng, X., *Journal of Alloys & Compounds*, (2015) **649**, 196.

[15] Rai, M.; Singh, S. K.; Singh, A. K.; Prasad, R.; Koch, B.; Mishra, K.; Rai, S. B., *ACS Applied Materials & Interfaces*, (2015) **7**, 15339.

[16] Liang, Z.-Q.; Zhao, S.-L.; Cui, Y.; Tian, L.-J.; Zhang, J.-J.; Xu, Z., *Chinese Physics B*, (2015) **24**, 037801.

[17] Chen, Z.; Wang, M.; Wang, H.; Le, Z.; Huang, G.; Zou, L.; Liu, Z.; Wang, D.; Wang, Q.; Gong, W., *Journal of Alloys & Compounds*, (2014) **608**, 165.

IOP Conf. Series: Materials Science and Engineering **359** (2018) 012044 doi:10.1088/1757-899X/359/1/012044

Synthesis and Characterization of Magnetic Carriers Based on Immobilized Enzyme

F H Li [1,2],N Tang [1,2], Y Q Wang [1,2], L Zhang [1,2,*], W Du [1,2], J Xiang [1,2]and P G Cheng [1,2]

[1]College of Chemical Engineering and Materials, Tianjin University of Science & Technology

[2]Tianjin Key Laboratory of Marine Resources and Chemistry,Tianjin300457, China

Email: leizhang@tust.edu.cn

Abstract: Several new types of carriers and technologies have been implemented to improve traditional enzyme immobilization in industrial biotechnology. The magnetic immobilized enzyme is a kind of new method of enzyme immobilization developed in recent years. An external magnetic field can be used to control the motion mode and direction of immobilized enzyme, and to improve the catalytic efficiency of immobilized enzyme. In this paper, Fe_3O_4-$CaCO_3$-PDA complex and $CaCO_3/Fe_3O_4$ composite modified by PEI were prepared. The results show that the morphology of Fe_3O_4-$CaCO_3$-PDA complex formation is irregular, while the morphology of $CaCO_3/Fe_3O_4$ composite modified by PEI is regular and has a porous structure.

1 Introduction

Immobilized enzyme is a kind of technology that the enzyme is bound or confined in a certain area by a solid material[1,2], which combines the water soluble enzyme and the water insoluble carrier into a water-insoluble immobilized enzyme. In order to obtain the ideal immobilized enzyme, the efficient immobilization method and ideal immobilization carrier are indispensable. Carrier properties have great influence on the activity of the immobilized enzyme, the ideal carrier should have appropriate structure and excellent physical and chemical properties, immobilized carrier are mainly inorganic carrier, polymer carrier and composite carrier[3], magnetic carrier material in recent years is the development of the new material for enzyme immobilization[4][5][6]. Compared with non-magnetic materials, magnetic immobilized enzyme can be used to separate and recover the immobilized enzyme from the reaction system.Magnetic composite materials have been widely used in the field of immobilized enzyme[7],it has a variety of elemental compounds or can be used as magnetic materials, such as Fe_3O_4, Fe_2O_3[8,9], Fe, Co[10,11], Mn, Fe_2O_4, Co Fe_2O_4[12], Co Pt_3 and Fe Pt_9[13, 14], as the magnetic material for the preparation of carrier Fe_3O_4. The synthetic materials were characterized by SEM, EDS,FT-IR and so on. The results show that the morphology of Fe_3O_4-$CaCO_3$-PDA complex is not regular, and the $CaCO_3$/Fe_3O_4 composite modified by PEI form a porous surface, and the morphology is regular.

2.Materials and methods

2.1 Materials. Dopamine(pure analysis,WuHan); sodium 4-styrenesulfonate (PSS, average Mw z 70000 g/mol, powder);polyethylenimine(PEI) were purchased fromAladdin,Calcium carbonate.

2.2 Preparation of Fe_3O_4-$CaCO_3$-PDA complex.$CaCO_3$-PDA were prepared: the fresh and dry $CaCO_3$ and0.05g of dopamine were dispersedin10 mL of Tris-HCl buffer(pH=8.5), covering with plastic wrap and paying attention to the air permeability,The mixed solution was stirred for 16h, then, the prepared $CaCO_3$-PDA microspheres were centrifuged and washed 4 times with deionized water, and dispersed in 10 mL ofTris-HClbuffer (pH=8.5), with the addition of 0.08g Fe_3O_4. Covering with plastic wrap and paying attention to the hole permeability. The mixed solution was stirred for 8 h. The magnets washed conspicuously colorless, namely Fe_3O_4-PDA-$CaCO_3$ complex.

2.3 Preparation of $CaCO_3$/Fe_3O_4 magnetic carrier. The 2mg/mL polyethyleneimine (PEI) solution was prepared by mixing the NaCl solution of 0.1mol/L with 50% PEI solution.Fe_3O_4suspension and PEI solution according to the volume ratio of 1:1 mixture, placed 15 min, washing magnets 5 times, its full suspension to 5 mL Na_2CO_3 solution. The solution was quickly added to a solution of 5 mg $CaCl_2$ dissolved 15mg PSS,for a quick stirring of 30 s and a static of about 20. After 10 times of washing, Dry, that is, $CaCO_3$/Fe_3O_4 solid.

3. Results and discussion

3.1Fe_3O_4-$CaCO_3$-PDA complex.As can be seen from Figure1, $CaCO_3$-PDA microspheres are spherical particles with regular morphology and particle size of 2-5 μm. PDA forms a complete membrane on the surface of almost all $CaCO_3$microspheres.

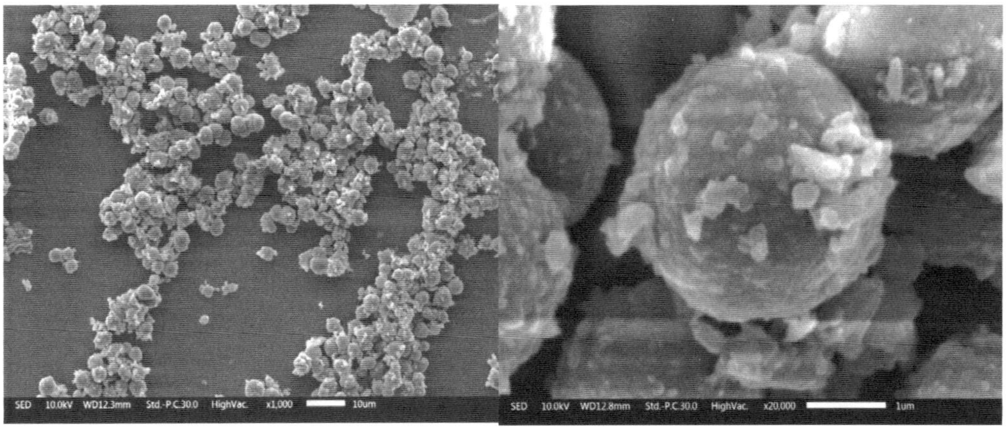

Fig.1 SEM images of $CaCO_3$-PDA

As shown in Figure 2, the Fe_3O_4-$CaCO_3$-PDA complex forms a reunion.This is because the Fe_3O_4 particles are nanometer level, so there is a very high surface tension, coupled with its magnetism, leads to the agglomeration of Fe_3O_4-$CaCO_3$-PDA particles, Fe_3O_4-$CaCO_3$-PDA particles are irregular, uneven particle size.

Fig. 2 SEM images of Fe_3O_4-$CaCO_3$-PDA composite

3.2. $CaCO_3/Fe_3O_4$ complex by PEI modified Fe_3O_4. Figure 3 is a scanning electron microscope photograph of $CaCO_3/Fe_3O_4$ particles obtained by PEI treatment of 3 mL, indicating the binding force between Fe_3O_4 and $CaCO_3$ greatly enhanced. This is because Fe_3O_4 surface modified by the PEI will have more positive charge[15]. Due to the increase of electrostatic force, it can not only strengthen the binding force between Fe_3O_4 and $CaCO_3$, but also increase the number of $CaCO_3$ particles combined with Fe_3O_4. According to figure 3, $CaCO_3/Fe_3O_4$ particles have a porous surface. Figure 4, the system contains C, O, Ca, Fe elements.

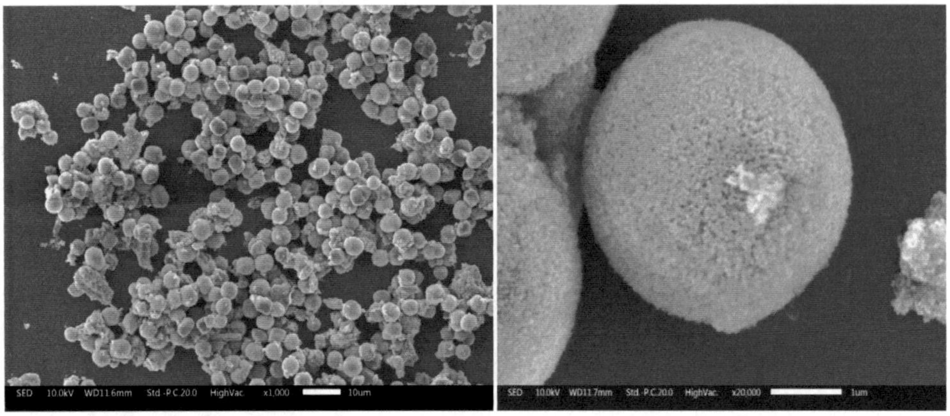

Fig.3 SEM images of $CaCO_3/Fe_3O_4$ which was made from Fe_3O_4 coating with PEI

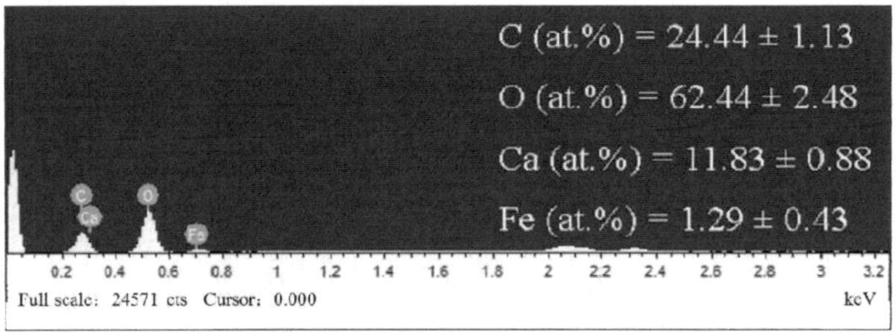

Fig. 4 EDS spectrum of $CaCO_3/Fe_3O_4$ which was made from Fe_3O_4 coating with PEI

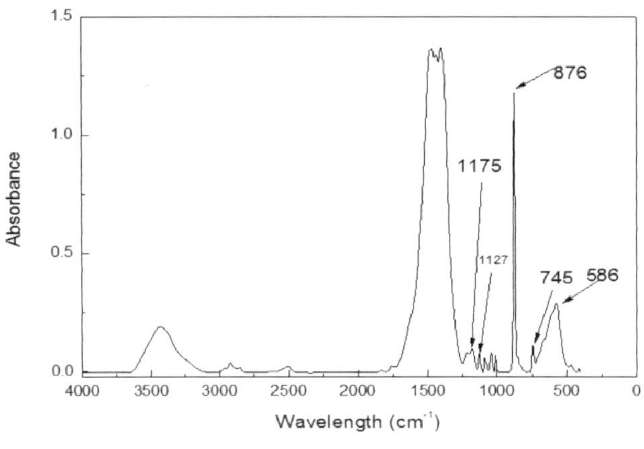

Fig.5 FT-IR spectrum of $CaCO_3/Fe_3O_4$ composite

In the FT-IR spectrum of $CaCO_3/Fe_3O_4$ composite(Figure 5), 876 cm^{-1}, 745cm^{-1} is the characteristic absorption peak of CO_3^{2-}, and the characteristic peaks of PSS are 1175 cm^{-1}, 1127 cm^{-1}, and 1046 cm^{-1}. 586cm^{-1} is the Fe-O stretching vibration absorption peak. The results show that the product contains both the characteristic peaks of Fe_3O_4 and the characteristic peaks of $CaCO_3$, so it can be concluded that Fe_3O_4 and $CaCO_3$ have been successfully combined to form Fe_3O_4- $CaCO_3$ complex.

4. Conduction

In the field of immobilized enzymes, the study of magnetic carriers is a booming topic. The magnetic immobilized enzyme is easier to separate and recover from the reaction system, and it is easy to operate. For this purpose, two magnetic carriers, Fe_3O_4-$CaCO_3$-PDA and $CaCO_3/Fe_3O_4$, were prepared. The results show that Fe_3O_4-$CaCO_3$-PDA composite is easy to agglomerate, and $Fe_3O_4/CaCO_3$ is a regular spherical particle with a diameter of between 3~5μmand has a porous structure.

5. Acknowledgements

The authors thank for the financial support of National Nature Science Foundation (21406169),Tianjin Research Program of Application Foundation and Advanced Technology (15JCYBJC20200), Training Program for Changjiang Scholars and Innovative Research Team in University ([2013]373) and the Innovative Research Team of Tianjin Municipral Education Commission (TD12-5004).

References

[1]Zhao X B, QiF, YuanC L and et al 2015 Lipase-catalyzed process for biodiesel production: Enzyme immobilization, process simulation and optimization. *Renewable and Sustainable Energy Reviews*, 44: 182-197.

[2]Surinenaĭte B R, Bendikene V G and Iuodka B A 1996 Immobilization of enzymes on carriers with magnetic properties: the search for optimum conditions for immobilization of alkaline phosphatase from the chicken gut. *PrikladnaiaBiokhimiia I Mikrobiologiia*, 32(6):609-614.

[3]Niu Y, Hou H 2011 Research Progress in Carriers of Immobilized Enzymes. *Liquor-Making Science & Technology*.

[4]Koneracka M, Kopcansky P, Antalik M and et al 1999 *Immobilization of proteins and enzymes to fine magnetic particles ournal of Magnetism and Magnetic Materials*, 201: 427-430.

[5]Guo Z, Bai S and Sun Y 2003 Preparation and characterization of immobilized lipase on magnetic hydrophobic microspheres. *Enzyme and Microbial Technology*,32(7):776-782.

[6]Varlan A, Roxana, Sansen and et al 1996 Covalent enzyme immobilization onparamagnetic polyacrolein beads. *Biosensors & Bioelectronics*, 11(4):443-448.

[7]OznurKandHakan K 2016 A simple way to obtain high saturation magnetization for superparamagnetic iron oxide nanoparticles synthesized in air atmosphere: Optimization by experimental design. *Journal of Magnetism and Magnetic Materials*, 409: 116-123.

[8]Neveu S, Bee A, Robineau M and et al 2002 Size-Selective Chemical Synthesis of TartrateStabilized Cobalt Ferrite Ionic Magnetic Fluid. *Journal of Colloid and Interface Science*,255(2):293-298.

[9]Grasset F, Labhsetwar N, Park L and et al 2002 Synthesis and Magnetic Characterization of Zinc Ferrite Nanoparticles with Different Environments: *Powder, Colloidal Solution, and Zinc Ferrite-Silica Core-Shell Nanoparticles. Langmuir*, 18(21): 8209-8216.

[10]Sun S H and Zeng H 2002 Size-Controlled Synthesis of Magnetite Nanoparticles. *Journal of the American Chemical Society*,124(28): 8204-8205.

[11]Park S J, Kim S, Lee S andetal. 2000 Synthesis and magnetic studies of uniform iron nanorods and nanospheres. *Journal of the American Chemical Society*, 122(35): 8581-8582.

[12]Puntes V F, Krishnan K M and Alivisatos A P 2001 Colloidal nanocrystal shape and size control: *The case of cobalt. Science*,291(5511):2115-2117.

[13]Park J, An K, Hwang Y and et al 2004 Ultra-large-scale syntheses of monodispersenanocrystals.*Nature Materials*, 3(12): 891-895.

[14]Sun S H and Murray C B 2000 Monodisperse Fe Pt nanoparticles and ferromagnetic6 Fe Pt nanocrystal superlattices. *Science*, 287(5460): 1989-1992.

IOP Conf. Series: Materials Science and Engineering **359** (2018) 012044 doi:10.1088/1757-899X/359/1/012044

[15]Diaz I L and Perez L D 2015 Synthesis and micellization properties of triblock copolymers PDMAEMA-b-PCL-b-PDMAEMA and their applications in the fabrication of amphotericin B-loaded nanocontainers. *Colloid and Polymer Science*, 293(3):913-923.

IOP Conf. Series: Materials Science and Engineering **359** (2018) 012045 doi:10.1088/1757-899X/359/1/012045

Effects of Activated Carbon Surface Property on Structure and Activity of Ru/AC Catalysts

S K Xu[a, b], L M Li[b] and N N Guo[b]

[a] College of Material Engineering, Henan University of Technology, Zhengzhou 450001, China

[b]Department of Chemistry, Zhengzhou University, Zhengzhou 450001, China

E-mail: sankui_xu@haut.edu.cn

Abstract. The activated carbon (AC) was modified by supercritical (SC) methanol, HNO_3 oxidation, or HNO_3 oxidation plus SC methanol, respectively. Then, the original and the modified AC were used as supports for Ru/AC catalysts prepared via the impregnation method. The results showed that the SC methanol modification decreased the content of surface acidic groups of AC. While HNO_3 oxidation displayed the opposite behavior.Furthermore, the dispersion of ruthenium and the activity of catalysts were highly dependent on the content of surface acidic groups, and the SC methanol modified sample exhibited the highest activity for hydrogenation of glucose.

1. Introduction

Activated carbon (AC) is a material that has all the required characteristics to be used as a catalyst support. Additionally, the unique properties such as the stability in both acidic and basic media, the possibility of tailoring both its textural and surface chemical properties according to the targeted aims of the catalyst producers also make AC an ideal candidate as catalyst support [1-3].It was known that catalytic properties are determined both by the texture and surface chemistry of support. Despite a lot of literatures have been published on it, the effect of AC texture and surface groups on the metal dispersion and activity is still under controversy. It has been well established in the literature that the surface groups, which form anchoring sites for metallic precursors, dominantly determine the properties of AC as a catalyst support [1-3]. The acidic groups on the surface decrease the hydrophobicity of AC, leading to the accessibility of the surface to aqueous metal precursors, while the less acidic groups increase the interaction of the metal precursor or the metal particle with the support and consequently minimize the sintering propensity of metal on carbon [3-4]. On the other hand, the nature and concentration of surface functional groups may be modified by suitable thermal or chemical post-treatments such as gas oxidation, liquidoxidation and microwave treatment [5-10].Thesupercritical fluid(SCF)is being explored as potential replacement for conventional solvents in various extractive processes, chemical reaction and the preparation of new materials[11].However, the literature on SCF modified AC was rarely reported.On the other hand, the hydrogenation of glucose to sorbitol is of great industrial importance.The frequently used catalyst for this process is Raney-Ni owning to its excellent settling properties and low cost.However, dissolution of Ni occurs during the hydrogenation of glucose.As a consequence, purification of sorbitol is necessary and renders this process economically less attractive. Thus, great efforts have been made to develop the new efficient and environment-friendly catalysts. Recently, the use of ruthenium-based catalysts could overcome this problem since ruthenium is stable against leaching and has exceptional activity [12-13].

2. Experimental

2.1 Chemicals. Coconut AC was purchased from Fuzhou Carbon Material Company. Glucose, HNO_3 and methanol were supplied by Tianjin Fine Chemical Incorporated Company. $RuCl_3$ (37wt.%) was provided by Guiyan Rare Metal Material Company. All of the chemicals were used as received.

2.2 Activated carbon modification. The typical procedure for the modification of the original AC (size 0.25-0.38 mm) was as follows:

(i) For SC CH_3OH treated sample (marked as ACS): Firstly, 56mLof CH_3OH was placed into a stainless-steel autoclaveof 150 mL. Secondly, 2.0 gof AC was placed in a cagefixed at the upper part of the autoclave. The autoclave temperature was adjusted up to 573 K and maintained for 4h. Then, the autoclave was cooled to below 353 K. Finally, the AC was removed and dried at 383 K.

(ii) For HNO_3 oxidized sample (marked as ACH): 2.0gof AC sample was oxidized in 30wt.% HNO_3 solution at 333 K for 4 h and washed with distilled water till the pH of the rinsed solution reached 7.0. The ACH sample was then dried at 383 K.

(iii) For the sample of HNO_3 oxidation plus SC CH_3OH treatment (marked as ACHS): 2.0g of ACH sample was modified with the method described in section 2.2(i).

2.3 Ru catalysts preparation. Ru/AC catalysts were prepared by impregnation of aqueous solution of $RuCl_3$ with AC, ACS, ACH and ACSH as supports. In a typical experiment, 0.6 gof AC was added to 6.0 ml $RuCl_3$ solutionandimpregnated for 6 h. The mixture was then dried at 393 K for 8 h. The solid products were reducedwith H_2 at 573 K for 3 h.

2.4 Characterizations. The contents of surface acidic groups onAC were titrated by the Boehm method [14]. The chemical constituent difference between methanol mother solution which was used to treat AC and pure methanol was analyzed by GC-MS (DSQ II).The textural properties of AC samples were measured by N_2 absorption at 77 K. The adsorption capacitiesof Ru^{3+} on ACs wereanalyzed by ICAT6000SERIES.The surface morphologies of Ru catalysts were determined by TEM (HRTEM, JEM-2010).

2.5 Activity test. The activity was performed in a 150ml stainless-autoclave containing 0.3 g of Ru-based catalyst and 50 ml of glucose aqueous solution. The H_2 pressure was in the range of 4.0-2.0 MPa with a reaction temperature at 393 K and a stirring rate at 1000rpm.The activity of the hydrogenation was obtained by recording the decrease of H_2 pressure with time, which was then converted to the hydrogen uptake rate per gram ofRu according to the ideal gas equation.

3. Results and discussion

3.1 Textural properties of AC. The textural properties of AC samples are given in Table1. Although the ACS sample displayed larger pore volume than the othersamples, all samples have quite similar total surface area and pore volume, which indicates that the AC has a relatively stable physical structure during treatment. We think that the increase in pore volume after SC methanol treatment can be attributed to the extraction of SCF. Because the experimental material is coconut AC, it surely contains some organic ingredients. Thus the extraction process must take place in SC CH_3OH. Some organic ingredients on the surface of AC were leached.

3.2 Bohem titration. The acidic properties of AC may be resulted from the presence of carboxyl groups, lactones or lactols, and hydroxyl groups of phenolic character. These groups differ in their acidities and can be differentiated by neutralization with solutions of $NaHCO_3$, Na_2CO_3 and NaOH, respectively. Table2 shows the concentration of the surface acidic groups of AC samples. It can be seen that SC CH_3OH treatment can decrease the content of acidic groups on AC. It can be explained that the SCF extracts some surface acidic groups of AC.Meanwhile, some surface acidic groups, for example carboxylic etc., may react with CH_3OH in supercritical state.To further prove the SC extraction and reaction, the different chemical composition of methanol mother solutions and pure methanol was analyzed by GC-MS. The results show that the methanol mother solution contains certain amount of organic chemical compositions such as dimethyl ether and methyl propionate etc.

Table 1 The textural properties of AC samples

Sample	Surface area (m^2/g)	Pore diameter (nm)	Pore volmue (cm^3/g)
AC	1430.7	1.040	1.440
ACS	1413.1	1.038	1.4689
ACH	1430.7	1.030	1.440
ACHS	1489.2	1.075	1.444

Table 2 Concentration of surface groups on AC samples

Sample	Total acidic (μmol/g)	Carboxylic (μmol/g)	Lactones (μmol/g)	Phenol (μmol/g)
AC	225.1	11.6	14.5	199.1
ACS	33.9	—	—	—
ACH	1012.5	458.0	80.3	474.2
ACHS	358.0	68.1	87.2	202.7

AC was oxidized by HNO_3 in order to examine the surface chemical characterization. The results indicate that the HNO_3 oxidation significantly increases the content of acidicgroups, evidenced by forty times higher of carboxylic group content. It is proposed that the HNO_3 oxidation could lead to the formation of surface acidic groups. And the ACHS sample has lower contentof acidic groups, obviously one seventh of carboxylic group content in comparison with ACH sample. The results suggest that the SC methanol has the potential of extraction and/or chemical reaction with acidic surface groups, which would result in a decrease in the content of acidic surface groups.

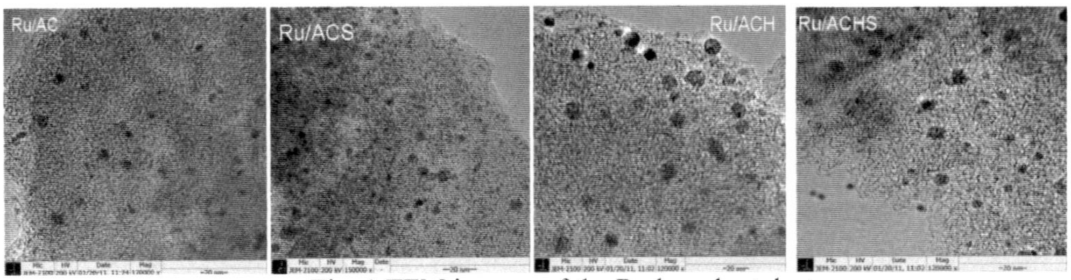

Fig. 1 TEM images of the Ru based catalysts

3.3 TEM of catalysts. Fig.1 shows the TEM morphologies of these supported Ru catalysts.It can be seen from Fig.1 that the order of Ru particle size is ACH >ACHS >AC>ACS.The ACS supported catalyst exhibits uniformly dispersed Ru nano-particles at a narrow particle size distribution (1.2-3.7 nm). And the ACH supported sample had the largest particle Ru size(2.6-25.8nm) with the broadest particle size distribution. The resultis consistent with reference [4], which reported that metal dispersion was highly dependent on the degree of carbon support oxidation, being lower for the support containing higher amount of surface acidic groups. This is attributed to the decomposition of surface complexes, which act as anchoring centers for the metal precursor upon the catalysts reduction. It can be concluded that SC CH_3OH modification could increase the Ru dispersion as it extracts the surface acidic groups of AC and reduces Ru gathering chance. HNO_3 oxidation greatly increases the surface acidic groups, and enhanced the adsorption of Ru^{3+}. However, some of these groups have a limited thermal stability at temperatures which the catalyst is treated for achieving the reduction of the metal precursor. The decomposition of the less stable surface complexes could result in metal gathering on the surface of support, thus would affect the final dispersion of the metal.

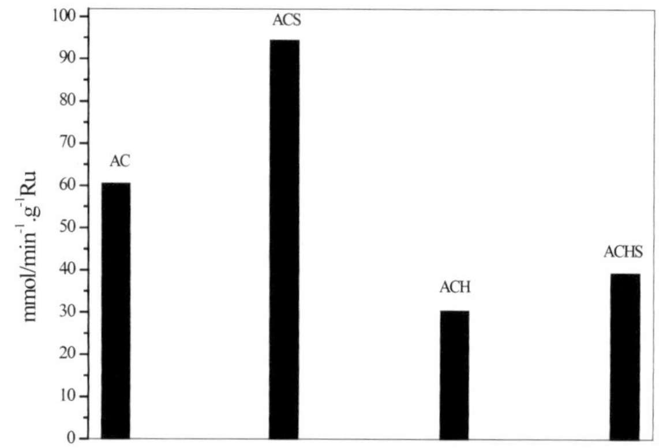

Fig. 2 Activity of Ru based catalysts

3.4 Activity. The catalytic behaviors of the Ru catalystsduring the glucose hydrogenation are summarized in Fig.2. It can be seen from Fig.2that the activity ofcatalysts follows the order of Ru/ACS >Ru/AC >Ru/ACHS >Ru/ACH, which is the same order as Ru dispersion. The Ru/ACS catalyst shows the highest activity, which is 1.56 times as large as that of Ru/AC catalyst. It shows that SC methanol treatment can increase the activity, whereas HNO_3 oxidation would greatly decrease the activity. The extraction and some chemical reaction may take place onthesurface of AC in SC CH_3OH as mentioned above, and therefore result in the leaching of some surface groups on AC. This could be confirmed by the previous Bohem titration and GC-MS results. It was known that the surface groups on the support has profound influence during the catalyst preparation step, acting as anchoring centers for the metal precursor [4]. However, some of these groups have low thermal stability when the catalyst is reduced. The decomposition of the less stable surface complexes could result in metal gathering on the surface of support, thus affect the final dispersion of the metal phase. The SC methanol treatment can decrease the surface acidic groups of AC, reduce metal gathering and finally increase Ru dispersion. The higher metal dispersion could increase the electron interaction between Ru and support.And the interaction results in Ru lacking electron benefiting chemisorb hydrogen.Therefore,the higher Ru dispersion increases the activity of catalysts. HNO_3 oxidation can increase the content of surface acidic groups of AC.Butsome of these groups have a limited thermal stability when the catalyst is reduced.This resultsin lower Ru dispersion and decreases the activity of catalysts.This is consistent with reference [15] which reported HNO_3 treatment could increase the surface acidic groups of AC, leading to lower Ru dispersion and decrease the activity of Ru/AC ammonia synthesis catalysts.

4 Conclusions

Three modifications of AC would have great effects on the content of surface acidic groups and the adsorption capacity of AC, whereas their effects on textural properties are rather limited. The dispersion of Ru and activity of catalysts are found to be highly dependent on the content of surface groups of AC, which is higher for the ACsupportcontaining lower amount of surface acidic groups. HNO_3 oxidation could lead to the formation of surface groups that decomposes at a relatively higher temperature or reduction which results in metal gathering on the surface of AC support. SC methanol modification could greatly decrease the surface acidic groups, obviously carboxylic group.This minimizes the gathering propensity of metal, leads to higher dispersion of Ru.Thus obviously increased the activity of the corresponding Ru/ACS catalyst.

Acknowledgments

This work was supported by the School high-level talent fund (No. 2013BS065).

References

[1] Rodriguez R F 1998 *Carbon*. 36 159-175.
[2] Auer E, Freund A, Pietsch J and Tacke T 1998 Appl. *Catal. A Gen.*173259-271.
[3] HagenS, BarfodR and FehrmannR 2003 *Catal.* 214 (2) 327-335.
[4] ColomaF, Sepulveda-EscribanoA, FierroJ LG 1997 Appl. *Catal. A Gen.*150 165-183.
[5] AksoyluA E, FreitasM A, FigueiredoJ L 2000 Appl. *Catal. A Gen.* 192 29-42.
[6] SongW, LiY, GuoXH, LiJ, HuangXM and ShenWJ 2010 *Mol. Catal. A Chem.* 328 53-59.
[7] ZhengXL, ZhangSJ, XuJ X and WeiK M 2002 *Carbon.* 40 2597-2603.
[8] HoweJY, DadmunMD, BrittP F 2007 Carbon. 45 1072-1080.
[9] HanW F, LiuH Z andZhuH 2007 Caal. *Commun.* 8 351-354.
[10] TakaokaM, YokokawaH and TakedaN 2007 Appl. *Catal. B. Environ.* 74179-186.
[11] UshikiI, TakahashiN 2017 *J. Supercritical Fluids*, *120*, 240-248.
[12] HofferB W, CrezeeE, MooijmanP R M, LangeveldA D, KapteijnF and MoulijnJ A 2003 *Catal. Today* 79-80:35-41.
[13] KusserowB, SchimpfS and ClausP 2003 Adv. Synth. Catal. 345:289-299.
[14] BoehmH P 1994 *Carbon*. 32:759-769.
[15] HanWF, ZhaoB, HuoC and LiuH Z 2004 *Chinese J. Catal.* 25194-198.

IOP Conf. Series: Materials Science and Engineering **359** (2018) 012046 doi:10.1088/1757-899X/359/1/012046

Lignin-based Biochar/graphene Oxide Composites as Supercapacitor Electrode Materials

Z Cai, C Jiang, X F Xiao, Y S Zhang and L Liang

School of Materials Science and Engineering, Wuhan Institute of Technology

E-mail:mc_cjiang@wit.edu.cn

Abstract. The lignin-based biochar/graphene composites were effectively obtained via an easy and rapid co-precipitation method. The chemical structure, microstructure, electrochemical properties of lignin/graphene oxide composites before and after carbonization were investigated by Fourier transformation infrared spectrum (FTIR), Scanning electron microscope (SEM), x-ray diffraction (XRD) and cyclic voltammetry (CV). FTIR results confirmed that the oxygen-containing groups of lignin, GO and their composites were partly removed after 800 ℃ carbonization and GO had a positive impact on the formation of graphitic structure for lignin. XRD results showed that lignin could completely block the restacking of GO sheets. The electrochemical test presented that lignin/graphene oxide composites exhibited a typical CV curve and the specific capacitance reached ~103F/g at a scan rate of 20mv/s.

1. Introduction

Recently, graphene has attracted many attentions in such application area as energy storage, gas molecular sensors, fuel battery, etc., due to its superb characteristics of chemical stability, high electrical conductivity (~10^6 S/m), and large specific surface area (~2630 m^2/g). More importantly, its theoretical capacitance is up to 550 F/g, showing a huge potential in supercapacitors as electrode materials.[1] However, it is extremely difficult to realize such high capacitance because single-layer graphene is prone to restacking with each other via strong Van der Waals interactions. Currently, an effective way to improve the capacitance is building 3D carbon-based structure via the combination of 2D graphene with 0D or 1D carbon materials to prevent its self-aggregation.[2] This can maintain high specific surface area and porous constructions, thereby providing more high-speed electronic and ionic channel of transmission for capacitors. It has been reported that the incorporation of carbon-based materials such as carbon nanotube (CNT), carbon black, biomass-derived carbon materials, into graphene can easily form a stable three-dimentional network and effectively increase the interlayer spacing of graphene. Among these carbonaceous materials, biochar from the carbonization of biomass resources exhibits more advantages in cost, environmental protection and renewability. Therefore, biochar/graphene composites have received many interests from researchers.[3] Lignin, as the second most renewable resource after cellulose and by-products of pulping and paper industry, is regarded as one of the most promising biopolymers. But the complex chemical and physical structure of lignin seriously hinders its progress of industrial applications. Lignin is natural polyphenolic material which contains a high content of carbon elements (40-60 wt %). Hence, many efforts have been focused to convert lignin into a high value carbonaceous material such as carbon fiber, carbon black and active carbon.[4] It is noted that lignin can be mixed with graphene oxide (GO) at any ratio in alkaline solution, and stabilize reduced GO via π-π interactions.[5] This makes it possible to prevent GO from restacking during co-precipitation and carbonization process. In return, GO may render lignin a special morphology and electro chemical properties. There is no doubt that lignin is a low-cost, non-toxic and high efficiency modifier for graphene as supercapacitor electrode materials.

2. Materials and methods

2.1. Preparation of lignin/graphene oxide (Lignin-GO) composites

80 g alkaline lignin was dispersed into 320 ml deionized water. Then a small amount of NaOH was added into the lignin solution to adjust pH to 12.5. After that, 80 ml GO solution with GO concentration of ~ 5mg/ml was mixed in aforesaid solution and stirred under room temperature for 1h, followed by ultrasonic treatment for 20 min. The pH of mixture solution was adjusted to 2.7 by 5wt % H_2SO_4, and then the mixture solution was heated to 90℃ for 1 h of flocculation. Finally the hot mixture solution was filtered and washed by deionized water for five times, followed by drying in an oven at 40℃ for 48 h.

2.2. Carbonation

Lignin, dried GO powder and Lignin-GO composites were placed into a chamber furnace and heated to 800℃ with the protection of nitrogen at heating rate of 10 ℃/min. Those carbonized samples are named as C-Lignin for lignin, C-GO for GO and C-Lignin-GO for lignin-GO composites, respectively.

2.3. Methods

FTIR spectra of the samples were analysed using a Nicolet-6000 spectrometer (America). XRD analysis was carried out on a Bruker-D8 ADVANCE and DAVINCI DESIGN diffractometer (Germany). SEM images were determined on a Philips-FEI sirion 200 microscope (Netherland). Electrochemical measurements were tested on a CHI900C electrochemical working station (Shanghai).

3. Results and Discussion

Figure 1 shows the FTIR spectra of graphite, GO, C-GO, lignin, C-Lignin, Lignin-GO composites and C-Lignin-GO. As shown in Figure 1B, graphite showed no absorption peaks in all observed range. However, after oxidation of graphite, a series of oxygen-containing absorption peaks appeared in the sample of GO, such as –OH (~3400 cm^{-1}), C=O (1720 cm^{-1}) and C-O (1176 cm^{-1}, 1070 $cm^{-1}$and 1010 cm^{-1}). After carbonization of GO, an obvious decrease in the peak intensity of oxygen-containing groups was observed in the sample of C-GO. This indicated the loss of oxygen-containing groups after 800 ℃ carbonization. But, the absorption peaks of –OH and C-O groups were existed in C-GO, suggesting the incomplete reduction of GO. Lignin showed abundant absorption peaks of various groups, such as C=O of carboxylic and carbonyl groups (1710 cm^{-1}), C=C of benzene ring skeleton (1604 cm^{-1}) and C-O of aromatic and aliphatic ether (1218 cm^{-1}, 1117 cm^{-1}), as shown in Figure 1B. Similarly, after 800 ℃ carbonization, the absorption peak of C=O almost disappeared, and an overlapped absorption band of C-O bonds at 1000-1250 cm^{-1} was observed in C-lignin. Regarding Lignin-GO composites, the FTIR spectra showed a very similar profile as pure lignin, due to the high lignin content in Lignin-GO composites. It is interesting that C-Lignin-GO exhibited different spectra from C-lignin and C-GO. Especially, the vibration of benzene ring skeleton for C-Lignin-GO composites located at 1556 cm^{-1}, while it was 1623 cm^{-1} for C-Lignin and 1564 cm^{-1} for C-GO. This suggested that GO facilitated the carbonization of lignin to form more sp^2 C atoms.

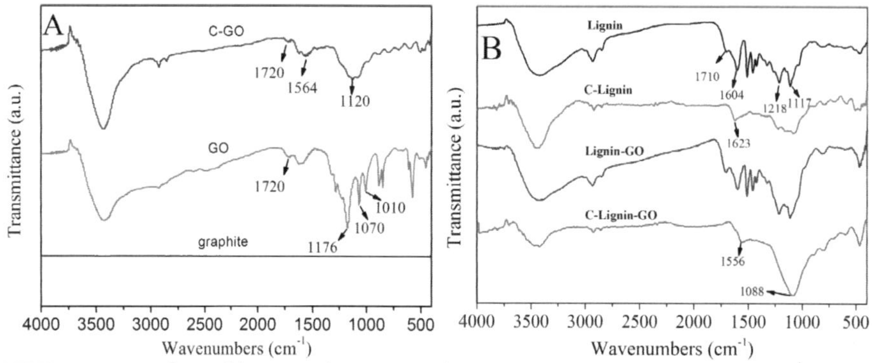

Figure 1. FTIR spectra of graphite powder, GO and C-GO (A); FTIR spectra of lignin and lignin-GO composites before and after carbonization (B).

Figure 2 shows SEM images and corresponding digital photos of lignin and Lignin-GO composites before and after carbonization. As shown in Figure 2 A1, lignin exhibited a blurry and irregular morphology due to its amorphous characteristic of chemical structure. After carbonization (Figure 2 A2), a bigger micro-meter sized carbon particle with numerous porous structures was found. In fact, these carbon particles were fused with each other to form a free standing bulk with high bonding strength (Figure 2 A3), due to the condensation reaction of lignin during carbonization process. As for Lignin-GO composites, the profile became more clear and some layered structures were observed because of the addition of GO. It was obvious that many lamellar structures appeared in C-Lignin-GO (Figure 2 B2) after carbonization, which very differed from the morphology of C-Lignin. This indicated that GO affected the carbonization behaviour of lignin. Furthermore, the carbon particles of C-Lignin-GO composites were loosely accumulated together, as shown in Figure 2 B3.

Figure 2. SEM images of lignin before (A1) and after carbonization (A2), Lignin-GO composites before (B1) and after (B2) carbonization, and the corresponding digital photos of C-Lignin (A3) and C-Lignin-GO (B3) composites.

Figure 3 presents the XRD patterns of graphite, C-Lignin and C-Lignin-GO. Graphite showed a sharp diffraction peak at $2\theta = 26.5\ ^{o}$, corresponding to the interlayer spacing of 0.336 nm. This was well consistent with the theoretical interlayer spacing of graphite with perfect graphitic structure. The XRD pattern of C-Lignin showed two wide diffraction peaks at $2\theta = 26\ ^{o}$ and $44\ ^{o}$. They were assigned to the diffractions from the (002) plane and (100/101) plane of graphited carbon structure, respectively. This indicated that lignin could form a small amount of graphitic framework to enhance its conductivity. Interestingly, C-Lignin-GO showed an almost duplicated XRD pattern as C-Lignin did.

Furthermore, no intense diffraction peaks were found in the profile. This sufficiently confirmed that lignin could prevent GO from restacking during co-precipitation and carbonization process.

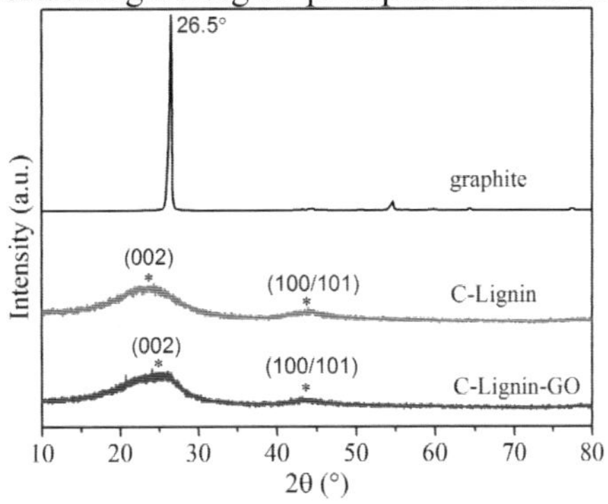

Figure 3. XRD patterns of graphite, C-Lignin and C-Lignin-GO.

Figure 4 shows the CV curves of C-Lignin and C-Lignin-GO at scan rates of 5 mV/s, 10 mV/s, 20 mV/s, 50 mV/s, 100 mV/s, 150 mV/s, 200 mV/s. The almost rectangular CV curves can be observed, revealing a typical electrical behaviour of double layer capacitor. With the increase of scan rate, the shape of CV curves become less rectangular, which may be owing to the complete infiltration of electrolyte ions into porous structures at low scan rate. As the incorporation of GO, the integral area of CV curve become bigger due to the high specific surface area and excellent electrical conductivity. The specific capacitance of C-Lignin-GO was calculated to be 103 F/g at a scan rate of 20 mV/s.

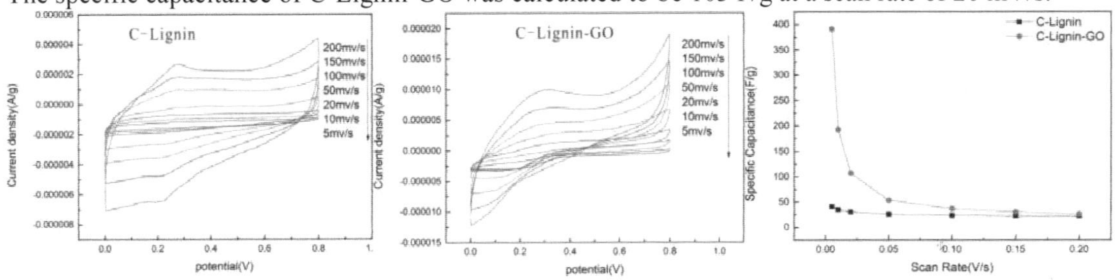

Figure 4. CV curves of C-Lignin and C-Lignin-GO and specific capacitance at different scan rate.

4. Conclusions

Lignin-GO composites were successfully prepared by a simple co-precipitation method. The Carbonized lignin-GO composites showed a high specific capacitance as electrode materials, due to the strong lignin-GO interactions to prevent GO from restacking.

Acknowledgement

This work was financially supported by National Natural Science Foundation of China (Grant No. 21606173) and the scientific research found of Wuhan Institute of Technology (k201627).

References:

[1] Xu Y., Lin Z., Zhong X., Huang X., Weiss N. O., Huang Y. and Duan X., 2014 *Nat. Commun.* **5**, 4554
[2] Wang Y., Wu Y., Huang Y., Zhang F., Yang X., Ma Y. and Chen Y., 2011 *The Journal of Physical Chemistry C* **115**, 23192-23197
[3] Abioye A. M. and Ani F. N., 2015 *Renewable and Sustainable Energy Reviews* **52**, 1282-1293
[4] Chatterjee S., Jones E. B., Clingenpeel A. C., McKenna A. M., Rios O., McNutt N. W., Keffer D. J. and Johs A., 2014 *Acs Sustain. Chem. Eng.* **2**, 2002-2010
[5] Liu W., Zhou R., Zhou D., Ding G., Soah J. M., Yue C. Y. and Lu X., 2015 *Carbon* **83**, 188-197

Preparation and Corrosion Resistance of Trivalent Chromium-Zirconium Composite Coating

J Z Huang

Foshan Supervision and Testing Centre of Quality and Metrology, Foshan, Guangdong 528000, China

Email: 350857423@qq.com

Abstract. Aluminum alloy are widely used in the industries because of its superior advantages. However there will be a thin oxide layer on the surface of the pure aluminum to inhibit corrosion, when adding some other elements, the obtained aluminum alloy is easy to be corroded. Surface protection is an important means to improve the corrosion resistance of aluminum alloy. The formal research had already confirmed that the trivalent chromium conversion coating can significantly improve the corrosion resistance, and the usage of the zirconium solution can also protect the aluminum alloy from corrosion. In this study, we constructed the binary conversion coating with the $Cr_2(SO_4)_3$ and the K_2ZrF_6. The optimum reaction conditions are as follows: 10g/L H_3PO_4, 2g/L K_2ZrF_6, 28g/L $Cr_2(SO_4)_3$, pH=2.5~3.5, temperature 40°C, and reaction time 10 min. Copper sulfate titration experiment confirmed that the corrosion resistance was significantly improved.

1. Introduction

However there is a thin surface oxide layer on the pure aluminum, which makes it stable in the external environment, the lacking of the high specific modulus of elasticity, specific strength and hardness limits the wide range of applications. When adding some other elements, such mechanical and physical properties will be significantly improved. But the new formed aluminum alloy will be easily corroded. there will be hundreds of millions economic loss each year because of the metal corrosion, so it is necessary to development a new simple and efficient method to improve the corrosion resistance of the alloy. The using of the chromate conversion coatings to protect the alloy from corroding has been confirmed to be a most effective method. Apart from the the excellent corrosion resistance ability, hexavalent chromium conversion coating also has brilliant wear resistance and self-repair ability, which makes it widely used in many industries. However, because of the toxicity of the hexavalent chromate, it can lead to the digestive tract, respiratory tract, skin and mucosal damage, more serious even cancer. Thus it is prohibited in electrical and electronic products according to the the Restriction of Hazardous Substances (RoHS) [1-3].

The researchers have developed new hexavalent chromium conversion coating alternatives to improve the corrosion resistance of aluminum alloys, while avoiding the toxicity. Among them trivalent chromium has been considered as most promising substitutes of the hexavalent chromium coating, because of the similar characteristic and low toxicity. Yu et al. used the $KCr(SO_4)_2$ to make the green trivalent chromium coating on the surface of the 6063 aluminum alloy, they adjusted the coating temperature, deposition time, the concentrations of the $KCr(SO_4)_2$ and H_3PO_4 and the pH to formed the trivalent chromium conversion coating, which showed excellent corrosion resistance ability, also they found that the quality and formation of the coating were most sensitive to the pH.[2,3] Although the researchers have devoted their effort to explore the optimum reaction conditions, the additives [4,5] and the mechanism of the formation of the coating [6,7], the anti-corrosion mechanism is still not clear. Apart from this, the Zr-based conversion coating has also been widely studied and has showed good application prospects [8-10]. Treatment of aluminum alloy with zirconium solution can increase the adhesion of the coating to the substrate and improve the corrosion resistance. However, the treated metal can not be used directly, it is necessary to paint the paint, and the widely use is limited. So in this study, the $Cr_2(SO_4)_3$ and the K_2ZrF_6 were used to constructed the binary conversion coating. We designed the experiment to confirmed the optimum reaction conditions, such as the concentration of the $Cr_2(SO_4)_3$, the K_2ZrF_6 and the H_3PO_4, pH, temperature and the reaction time. Copper sulfate titration experiment was used to evaluate the corrosion resistance of the conversion coating and the morphology and electrochemical properties were studied with SEM.

2. Experimental

2.1. Materials

The 6063 aluminum alloy was utilized to obtained the binary conversion coating in this study. The major elements of aluminum alloy are listed in Table 1. All chemicals of this work were AR grade. And deionized water was used to prepare all solutions.

Table 1. Chemical composition of 6063 aluminum alloy

Element	Mg	Si	Fe	Cu	Zn	Mn	Cr	Ti	Al
W/%	0.54	0.38	0.22	0.02	0.03	0.03	0.01	0.02	Balance

2.2. Preparation of trivalent chromium conversion coating

The 6063 aluminum was polished with 400#, 600#, 800#, 1000#, 1200# abrasive paper. Then the aluminum was washed with acetone and deionized water. After that the aluminum was reacted with 50g/L NaOH at 60 °C for two minutes and 20%HNO_3 at room temperature for two minutes. After the surface oxide film is removed, the aluminum was washed with deionized water.

After the water evaporates completely, the aluminum was immersed in the solution. And the reaction conditions for each experiments were described in Table 2.

Table 2. The reaction conditions for each experiments

	T/°C	pH	$Cr_2(SO_4)_3$/(g/L)	H_3PO_4/(g/L)	t/min
Fig. 1.	40	2.5	4,10,16,22,28,34	20	10
Fig. 2.	40	2.5	10,28	5, 10, 20, 30	10
Fig. 3.	40	2, 2.5, 3, 3.5	28	10	10
Fig. 4.	30, 40, 50	2.5	28	10	10

2.3. Preparation of binary conversion coating

When the reaction conditions conditions were determined in the one-component conversion coating, the K_2ZrF_6 was added in to the solution. The reaction conditions for the experiment were described in Table3.

Table 3. The reaction conditions for binary conversion coating

	T/°C	pH	$Cr_2(SO_4)_3$/(g/L)	H_3PO_4/(g/L)	t/min	K_2ZrF_6/(g/L)
Fig. 5.	40	2.5	10	20	10	1,2,3,4

After that five-factor four-level orthogonal test were designed to find the best conditions and the influence of each factor in the preparation of binary conversion coating. The reaction conditions for the orthogonal test were described in Table4.

Table 4. The reaction conditions for the orthogonal test

	T/°C	pH	$Cr_2(SO_4)_3$/(g/L)	H_3PO_4/(g/L)	K_2ZrF_6/(g/L)
1	20	2	2	5	1
2	30	2.5	6	10	2
3	40	3	10	20	3
4	50	3.5	14	30	4

2.4. Copper sulfate titration experiment

6.2g $CuSO_4 \cdot 5H_2O$ and 2mL Concentrated hydrochloric acid were dissolved into 150mL Deionized water. The obtained solution was dropped on the surface of the aluminum. the time from which the color changed from blue to red was recorded.

3. Results and discussion

3.1. Effect of $Cr_2(SO_4)_3$ concentration

Chromium ion is the main element of the film-forming reaction. According to the literature, we set six equal gradient parameters. 6063 was subjected to pretreatment for 10 min at a solution of different $Cr_2(SO_4)_3$ containing 20 g / L H_3PO_4 at pH 2.5, 40 ° C. From the results, 10g / L and 28g / L group of film effect and anti-corrosion effect is very prominent, anti-corrosion effect compared to other groups and control group are better, can reach 60 seconds. At the same time, the bubbling time of the 10 g / L group and the time of starting the replacement of copper were significantly later, so that 10 g / L and 28 g / L of the two preferred groups were considered to be Under these conditions the best reaction of $Cr_2(SO_4)_3$ concentration.

Table 5. The concentration gradient of $Cr_2(SO_4)_3$ and the corrosion resistance of conversion coating

ph2.5, 40℃, 10min, 20g/L H_3PO_4	Bubbling time t1（s）	Starting turning red t2(s)	Finishing turning red t3(s)	Δt1（t3-t2）	Δt2（t3-t1）
4	32	43.5	75	31.5	43
10	40.67	55.67	98	42.33	57.33
16	41.67	51.67	79.33	27.66	37.66
22	34.33	41.33	65	23.67	30.67
28	24	42.33	87	44.67	63
34	34.5	55	74.5	19.5	40
control	32	50	76	26	44

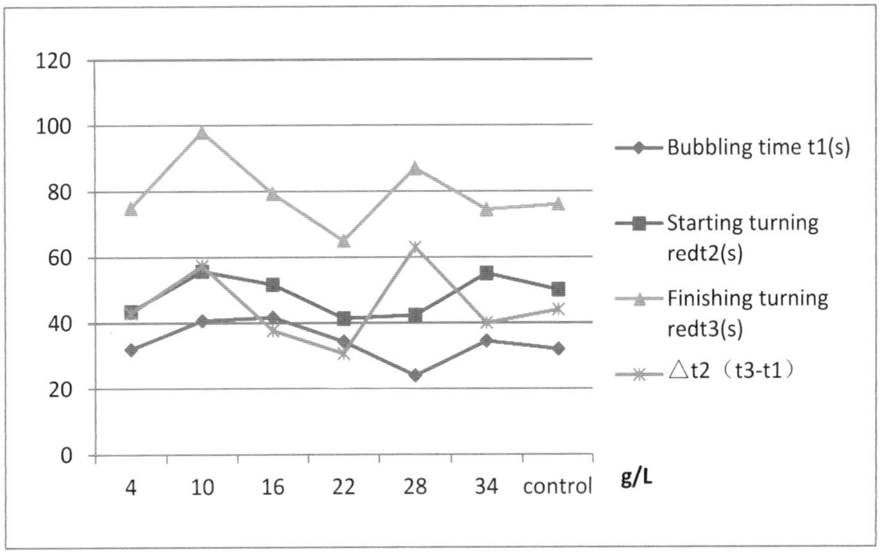

Figure 1. Effect of $Cr_2(SO_4)_3$ concentration on the corrosion resistance of conversion coating

3.2. Effect of H_3PO_4 concentration

Considering the phosphoric acid in the formation of the conversion film also has a great role, due to the presence of non-metallic non-conductive isolation layer, hindering the formation of micro-cell surface of the metal process, so the concentration of phosphoric acid will also affect the film quality. The concentration of $Cr_2(SO_4)_3$ in the two groups was 10g / L and 28g / L, respectively. Therefore, the gradient of 5-30g / L was set up in the literature. It can be seen from the experimental results that the optimum concentration of $Cr_2(SO_4)_3$ is 10g / L when the phosphoric acid concentration is 20g / L. However, when the concentration of phosphoric acid is 10g / L, the optimum concentration of $Cr_2(SO_4)_3$ 28g / L. The foaming time of the 10 g / L group and the time of starting copper replacement were significantly later, but the anti-corrosion effect was better in the 28 g / L group than in the 10 g / L group for less than 70 seconds , 28g / L of the overall red time up to 81 seconds. Therefore, a combination of a number of conditions, the concentration of phosphoric acid 10g / L, $Cr_2(SO_4)_3$ concentration 28g / L can be considered the reaction of the best phosphoric acid and chromium sulfate concentration.

Table 6. The concentration gradient of H_3PO_4 and the corrosion resistance of conversion coating

c（Cr3+）、c（H_3PO_4）	Bubbling time t1（s）	Starting turning red t2(s)	Finishing turning red t3(s)	Δt1（t3-t2）	Δt2（t3-t1）
10.5	11.67	30.33	54.67	24.34	43
10.10	14	35.67	62.33	26.66	48.33
10.20	16	37	51.33	38.33	67.33
10.30	9.33	34	63.67	29.67	54.34
28.5	15	32.33	72.67	40.34	57.67
28.10	13.33	50	94.67	44.67	81.34
28.20	13.67	38.67	77.33	38.66	63.66
28.30	17.67	46	82.67	36.67	65
Control	15.5	30.75	66.33	35.58	50.83

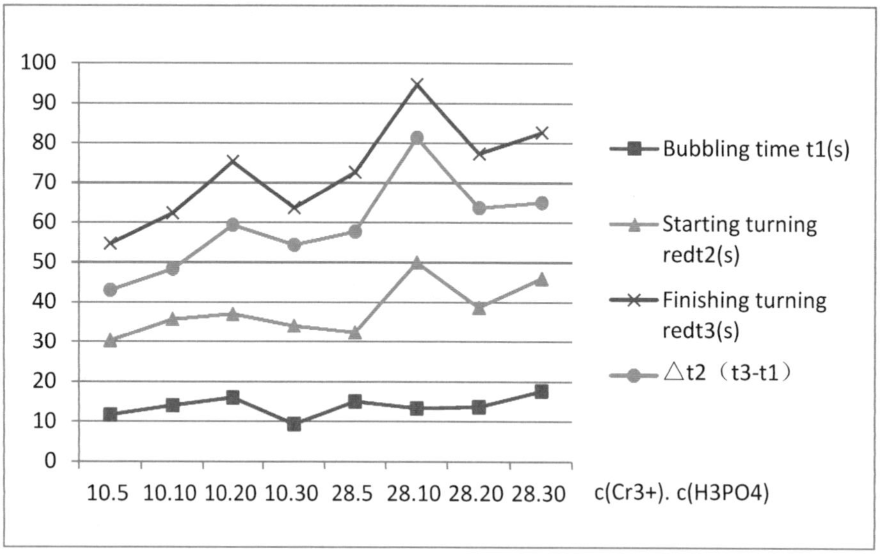

Figure 2. Effect of H_3PO_4 concentration on the corrosion resistance of conversion coating

3.3. Effect of pH

The pH of the reaction has an important effect on the formation of the conversion film. For 10 g / L phosphoric acid and 28 g / L $Cr_2(SO_4)_3$ solution at different pH values, the coating reaction was carried out at the same temperature. The time of bubbling was delayed with increasing pH, while the aluminum plate was replaced with copper at pH 3-3.5, and the corrosion resistance was better between pH 2.5-3.5. From this result, the best pH is between 2.5 and 3.5.

Table 7. The concentration gradient of pH and the corrosion resistance of conversion coating

28g/L,40℃, 10min, 10g/L H_3PO_4	Bubbling time t1（s）	Starting turning red t2(s)	Finishing turning red t3(s)	Δt1（t3-t2）	Δt2（t3-t1）
ph2	8.67	40.33	94.67	54.34	86
ph2.5	17.67	33.33	95.67	62.34	78
ph3	24.67	60	101.67	41.67	77
ph3.5	27	57.33	105.33	48	78.33
ph4	80	133.33	∞		
control	25	44	90	46	65

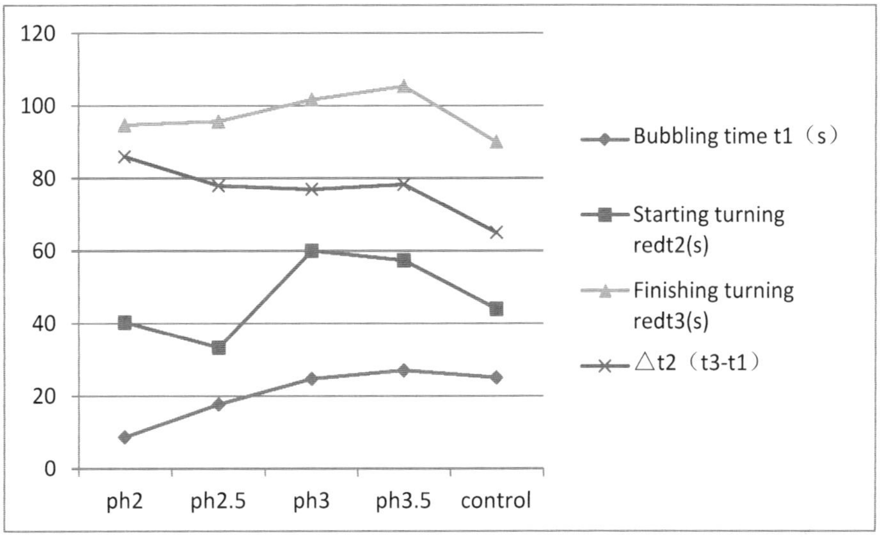

Figure 3. Effect of pH on the corrosion resistance of conversion coating

3.4. Effect of temperature

According to the basis of literature research, in the case of other conditions of unity, we tried 30,40 and 50 ℃ three temperatures. As can be seen from the data, 50 ℃ although the corrosion time is very long, but because the beginning of the replacement reaction time earlier. 30 and 40 ℃ of the three time parameters are very different, three temperatures in the results of the red time is almost the same. So a comprehensive analysis of view, 30-40 ℃ are reasonable temperature.

Table 8. The gradient of temperature and the corrosion resistance of conversion coating

28g/L, pH 2.0, 10min, 10g/L H_3PO_4	Starting turning red (s)	Finishing turning red (s)	Δt
30℃	28	58	30
40℃	29	60	31
50℃	16.33	60	43.67
Average control	23.22	75.38	52.16

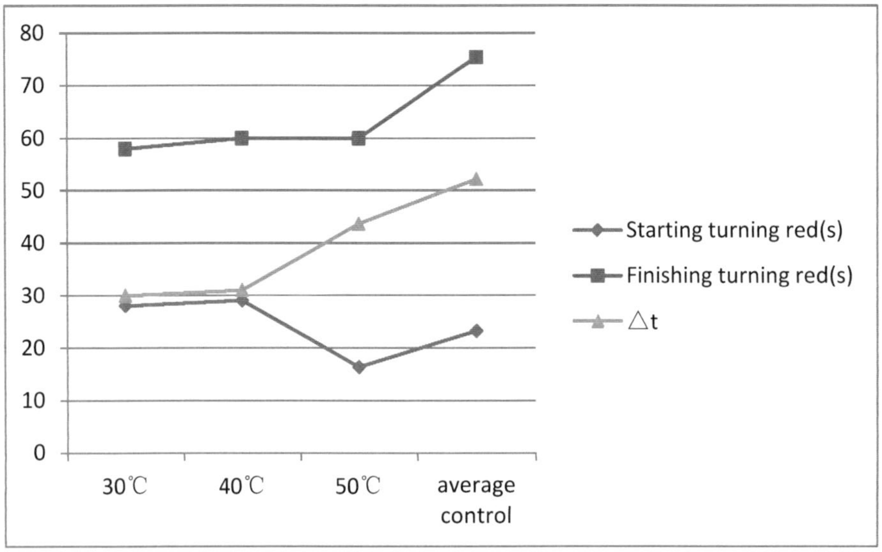

Figure 4. Effect of temperature on the corrosion resistance of conversion coating

3.5. Effect of K_2ZrF_6 concentration

The zirconium titanate oxide film formed by zirconium titanium has good protective properties and self-healing ability, but the concentration of zirconium and titanium in the conversion film is too high or too low to affect the quality of film formation. After the reaction conditions in the reference literature, we selected the process parameters of 20 g / l H_3PO_4, 10 g / l chromium sulphate solution at 40 ° C and pH 2.5 for 10 minutes, and the concentration of K_2ZrF_6 was 1, 2 , 3 and 4 g / L of the parallel test. It can be seen that the replacement time of the 3g / L group is the shortest and the corrosion time is the shortest, indicating that the conversion efficiency of the conversion film is the lowest. In contrast, the replacement reaction at 1 g / L group began to be the longest and the conversion time was the highest. In summary, the binary system of K_2ZrF_6 concentration should be 1g / L.

Table 9. The concentration gradient of K_2ZrF_6 and the corrosion resistance of conversion

ph2.5，40°C，10min，20g/l H_3PO_4，10g/l $Cr_2(SO_4)_3$	Starting turning red t1(s)	Finishing turning red t2(s)	Δt（t2-t1）
1	22.67	47	24.33
2	17.33	39.33	22
3	13	30.33	17.33
4	22	43	21

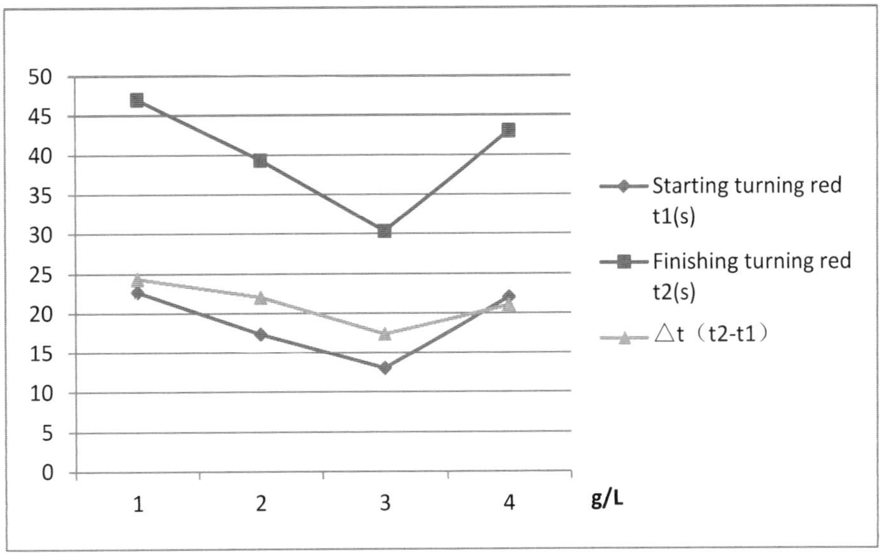

Figure 5. Effect of K_2ZrF_6 concentration on the corrosion resistance of conversion coating

3.6. Optimum reaction conditions

On the basis of above discussion, the optimum reaction conditions were obtained as follows: the concentration of chromium sulfate was 10 g / L, the concentration of phosphoric acid was 5 g / L, the reaction pH was 3.5, the reaction temperature was 40℃, and the concentration of potassium fluorozirconate was 1 g / L. At the same time, it can be seen from the very poor size that the effect of pH on the film formation is the largest, followed by $Cr_2(SO_4)_3$ concentration and phosphoric acid concentration.

4. Conclusion

(1) The optimized process condition for unitary conversion treatment are as follow: $Cr_2(SO_4)_3$ concentration 28g / L, phosphoric acid concentration 10g / L, reaction pH 2.5, reaction temperature 40 ℃.

(2) The optimized process condition for binary conversion treatment are as follow: $Cr_2(SO_4)_3$ concentration 28g / L, phosphoric acid concentration 10g / L, K_2ZrF_6 concentration 1g/L, reaction pH 2.5, reaction temperature 40 ℃.

Reference

[1] Wang D, Bierwagen G P 2009 Sol-gel coatings on metals for corrosion protection, *Prog. Org. Coat.* **64** 327

[2] Yu H C, Chen B Z and Shi X C 2009 EIS investigation of the deposition of trivalent chromium coatings on Al 6063 alloy *J. Appl. Electrochem.* **39** 303

[3] Yu H C, Chen B Z and Shi X 2008 Investigation of the trivalent-chrome coating on 6063 aluminum alloy *Mater. Lett.* **62** 2828

[4] Ramezani-Varzaneh H A, Allahkaram S R and Isakhani-Zakaria M 2014 Effects of phosphorus content on corrosion behavior of trivalent chromium coatings in 3.5wt.% NaCl solution *Surf. Coat. Technol.* **244** 158.

[5] Qi J, Hashimoto T and Thompson G 2016 Influence of Water Immersion Post-Treatment Parameters on Trivalent Chromium Conversion Coatings Formed on AA2024-T351

Alloy *J. Electrochem. Soc.* **163** 5
[6] Dardona S, Chen L and Kryzman M 2011 Polarization Controlled Kinetics and Composition of Trivalent Chromium Coatings on Aluminum *Anal. Chem.* **83** 6127
[7] Qi J T, Hashimoto T and Walton J R 2015 Trivalent chromium conversion coating formation on aluminium *Surf. Coat. Technol.* **280** 317
[8] George F O, Skeldon P and Thompson G E 2012 Formation of zirconium-based conversion coatings on aluminium and Al–Cu alloys *Corros. Sci.* **65** 231
[9] Cerezo J, Taheri P and Vandendael I 2014 Influence of surface hydroxyls on the formation of Zr-based conversion coatings on AA6014 aluminum alloy *Surf. Coat. Technol.* **254** 277
[10] Chang Y Z, Tsai P H and Li J B 2013, Zr-based metallic glass thin film coating for fatigue-properties improvement of 7075-T6 aluminum alloy *Thin Solid Films* **544** 331

IOP Conf. Series: Materials Science and Engineering **359** (2018) 012048 doi:10.1088/1757-899X/359/1/012048

Rheological Properties of Asphalt Mortar with High Viscosity

M L Xing[1,2], H X Chen[2], Z Z Li[2], X Y Ma[2], Y Sun[2] and Z H Liang[2]

[1]Guangxi Key Laboratory of Road Structure and Materials, Nanning, 530007,China
[2]Engineering Research Centre of Transportation Materials, Ministry of Education, Chang'an University, Xi'an710061,China

Abstract. In order to study the rheological and rotary viscosity characteristics of asphalt mortar with high viscosity containing different asphalt and filler, the component, filler-bitumen ratio and temperature are considered. The results show that among these factors, viscosity and specific surface area are the main factors that influence the ability of resistance to high temperature deformation. The temperature sensitivity and G*/sinδ increase with increasing filler-bitumen ratio.The growth rate changes corresponding to the filler-bitumen ratio of 1.2. A good exponential relationship exists between G*/sinδ and Brookfield rotary viscosity at 135°C and 150°C. Thus, the results of Brookfield rotary viscosity can be used to estimate the influence of filler-bitumen ratio to asphalt mortar characteristics at high temperature instead of filler-bitumen ratio to asphalt mortar characteristics at high temperature instead of G*/sinδ.

1. Introduction

Asphalt mortar combined with asphalt and mineral powder plays an important role in asphalt mixture. In the mixture of porous asphalt pavement, the strength lies on the character of asphalt mortar mainly because of its structure with skeleton space and large air void. E.Ray Brownfound that the use of the Dynamic Shear and Bending Beam Rheometers does seem to be viable with mortarsand the. Data obtained with the Dynamic Shear and Brookfield suggests the good correlation to Bending Beam Rheometers [1]. Drutaused the PG64-22 asphalt binder test data initially to predict mastics' performance parameters - shear modulus (G*) and accumulated permanent shear strainand then same properties for asphalt mixtures, in order to find a correlation between the three materials. Binder and mastics were tested at three temperatures (46°, 55°, and 64°C) using a dynamic shear rheometer (DSR), while the mixtures were tested at two temperatures (55° and 64°C) using the Superpave shear tester. Newly developed Hirsch model was used for estimating the shear moduli of asphalt mastics and mixtures. Zhang, Z.Q. and Guo, Panalysed the influence of fibre to performance of mortar at different temperature by Dynamic Shear and Bending Beam Rheometers [3-5]. Liu, L verified that the matrix asphalt of 90# is approximately Newtonian fluid when the temperature is higher than 90℃ [6-7]. And all of studies which had been done aim to the common asphalt mortar, but rarely to the characters of asphalt mortar with high viscosity.

In this paper, influences of asphalt, mineral powder and ratio of filler to bitumen to rheological behavior are studied by Dynamic Shear Rheometers and Brookfield. And the influences of asphalt, mineral powder, and asphalt mortar are analysed basing on the testing results.

2. Experimental

2.1. Bitumen
Two kinds of modified bitumen are made from matrix bitumen,ESSO70#. One(named B1) is modifiedwithTPS being 12% of bitumen in weight, and the other(named B2) is 5%SBS.

2.2. Mineral powder
Two types of limestone powder from different resourceare used in the study. And their specific surface areas and particle sizes are tested by different methods. The table 2 shows the results.

2.3.Asphalt mortar
Asphalt mortar is the mixture of bitumen and mineral powder. Firstly, mineral powder is heated and dried to a constant weight at 105±5°C. Then, a certain quality mineral powder is added into modified asphalt under 160°C. And the mixture is stirred until it isup to the best. So three kinds of asphalt mortar are chosen for studied from 2 types of bitumen and 2 kinds of mineral powder.

Table 1. Indexes of bitumen.

Properties	Test values	
	B1	B2
Softening point (°C)	90.2	79.2
Penetration(25°C,0.1mm)	50.3	56.4
Ductility(cm) (5°C)	44.1	37.9
Toughness（N·m）	29.6	30.2
Tenacity（N·m）	16.9	20.9
Dynamic viscosity(60°C,Pa·s)	41743	16713

Table 2. Indexes of mineral powder.

Properties		1#	2#
Apparent density（t/m^3）		2.64	2.68
Size range	<0.6mm（%）	100	100
	<0.15mm（%）	99.4	98.4
	<0.075mm（%）	86.4	79.4
Specific surface area（<0.075mm）（m^2/kg）		2392.5	3224.6
Average particle size（<0.075mm）（μm）		5.476	4.879

Table 3. Asphalt mortar.

Asphalt mortar	Modified asphalt	Mineral powder	Ratio of filler to bitumen
B1+1	B1	1#	
B1+2	B1	2#	0.4,0.6,0.8,1.0,1.2,1.4,1.6
B2+1	B2	1#	

2.4. Dynamic Shearing Rheologicaland Brookfield Testing
The Bohlin c-VOR dynamic shear rheometer produced by Malvern is used in this test. Strain control mode is adopted during the test. And the testing parameters are as follows: γ=10%,ω=10rad/s. The diameter of sample is 25mm and the thickness is 1mm. Temperature for testing changes from 64°Cto 84°C. While, Brookfield testing is carried out at the temperature of 135°C, 150°C, 165°C, 175°C separately.

3. Results and discussion

3.1. Dynamic Shearing Rheology

DSR for three kinds of asphalt mortar with high viscosity are done at different temperature and different ratio of filler to bitumen. Results of G*/sinδ are showed in Figure 1.

Fig.1. shows that G*/sinδ increaseswith the increase of the filler-bitumenratio at the same temperature.And the anti-rutting capability ofasphalt mortaris enhanced. This is because with increasing of the filler-bitumen ratio, the content of fillerwith larger specific surface area increased gradually. And more asphalt is need to form the structural asphalt which resulted in the reduction of free asphalt in asphalt mortar. Consequently the relative increase of solid content in asphalt mortar led to the decrease of flow deformationat high temperature.Therefore the anti-rutting ability of asphalt mortar is enhanced. Also G*/sinδ decreases with the increasing of temperature. Then the rate of descent is influenced by the characters of asphalt and mineral filler and the ratio of filler to asphalt. Figure 2 shows the differencerate of descent between the three asphalt mortars at the ratio of filler to asphalt being 1.0. It is obviously that the G*/sinδ is influenced by characters of asphalt mortar when the ratio of filler to asphalt being at the same. And from Fig.2, it can be seen that the value of G*/sinδfor B1+1and B1+2 is remarkably higher than that of B2+1. This result indicates that the character of bitumen is the most important factor which influences the value of G*/sinδ for asphalt mortar. And it also shows that the higher the viscosity of the asphalt is, the stronger the resistance to deformation at high temperature will be.The trend of G*/sinδwith temperature can be expressed as following formula [f1].

$$G^* / \sin \delta = Ae^{BT} \tag{f1}$$

Where, A and B are the regression coefficients, and T is the temperature. The smaller the absolute value of B is, the less temperature sensitivity of the asphalt mortar is. The regression results ofG*/sinδ to temperature are shown in Table 4.

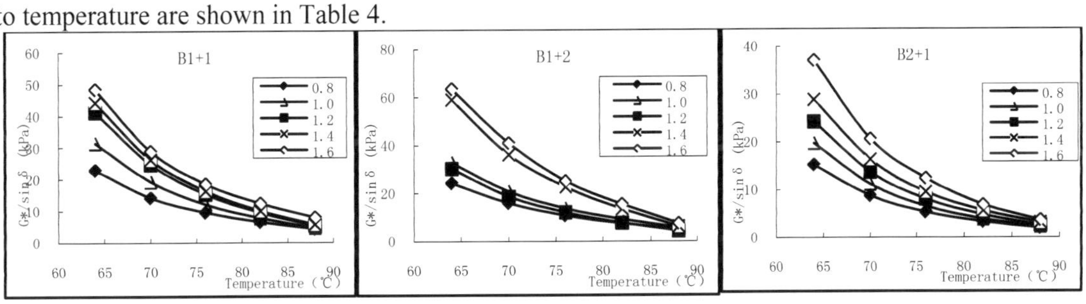

Fig.1 Results of G*/sinδ for three different asphalt mortar

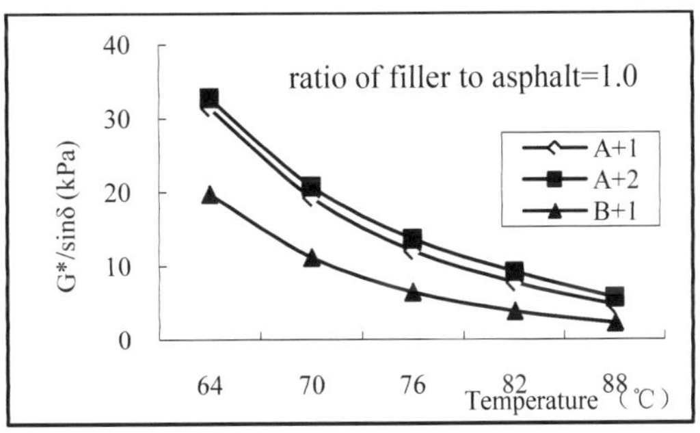

Fig.2. Curves of G*/sinδ with temperature and type of asphalt mortar

Table 4. Regressionresults of G*/sinδ to temperature

Type of mortar	Ratio of filler to bitumen	Equation formula	Correlation coefficient R^2	B
B1+2	0.8	$y = 1628.7e^{-0.066x}$	0.9985	-0.066
	1.0	$y = 3323e^{-0.0723x}$	0.9992	-0.0723
	1.2	$y = 5260.6e^{-0.0804x}$	0.9973	-0.08
	1.4	$y = 23128e^{-0.0923x}$	0.9869	-0.0923
	1.6	$y = 19015e^{-0.088x}$	0.9894	-0.088

Results in Table 4 show that the temperature sensitivity of asphalt mortar increaseswith the increasing of filler-bitumen ratio. Also the characters of asphalt and filler influence the temperature sensitivity of asphalt mortar, especially the dynamic viscosity of asphalt and the specific surface area.

3.2Brookfield RotaryViscosity

Figure3 shows the trend of Brookfield rotary viscosity changing with the temperature in different types of asphalt mortar. Brookfield rotary viscosity of asphalt increase with the increasing of filler-bitumen ratio. However, the growth of rotary viscosity is different between the asphalt mortars with different ratio of filler to bitumen. Among the testing range from 0.8 to 1.2, the growthof rotary viscosity is lower than that of ratio changing from 1.2 to 1.6. So the rotary viscosity of asphalt mortar withis closely related to the type of filler, temperature and the ratio of filler to bitumen. And the rotary viscosity decreases with the increasing of temperature which can be described by the equation of power function (f2).

$$y = Mt^N \quad \text{(f2)}$$

Where, y is Brookfield rotary viscosity, t is temperature, and M and N are the regression coefficients.

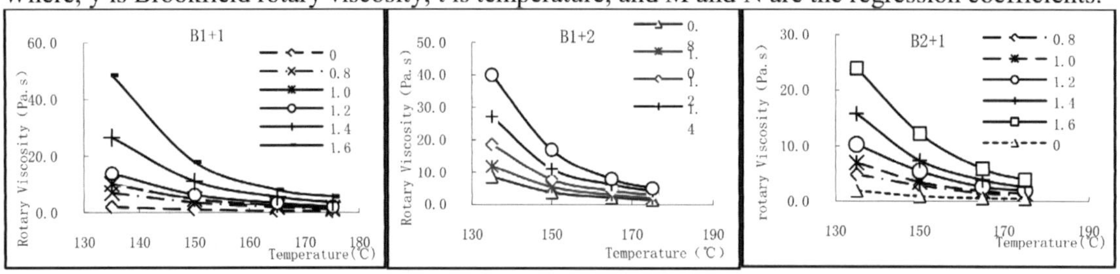

Fig.3. Curves of rotary viscosity and temperature

M stands for the value of rotary viscosity and |N|indicates the temperature sensitivity. Therefore, the large value of |N| means the mortar is more sensitive to temperature.

According to ASTM D2463-95a, the logarithm of viscosity is proportional to the temperature. Then, figure 4 shows the results of regression curves between Brookfield rotary viscosity and temperature.

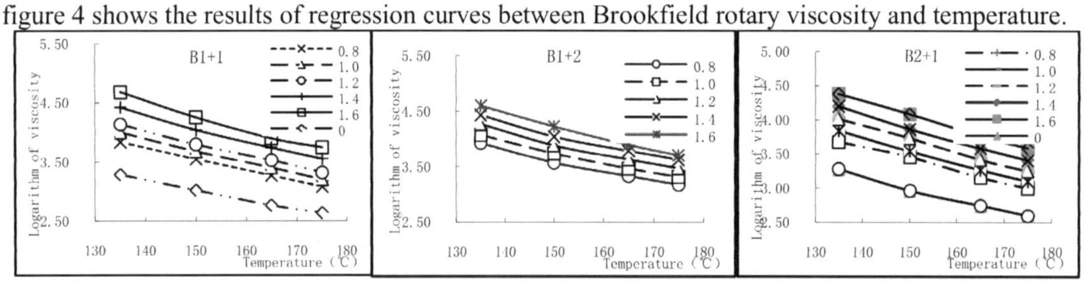

Fig.4. Curves of asphalt mortar and temperature

Curves in figure.4 also verifies the linear relationship between the logarithm of rotary viscosity and temperature. For one kind of asphalt mortar, the slope of the curves increase with the ratio of filler to bitumen. And this result also reveals the more temperature sensitivity. At the same ratio of filler to bitumen, $|N|_{B1+2}<|N|_{B1+1}$ means that the temperature sensitivity of asphalt named B1+2 is lower than that of B1+1.

3.3 Correlation Analysis of Different Rheological Results

Brookfield rotaryviscosity indicates the ability of asphalt mortar to resist shearing deformation under external force. And G*/sinδ is another parameter which can indicate the abilityof asphalt mortar to resist permanent deformation. Therefore, it is necessary to analysis the correlation between G*/sinδ at 64℃ and Brookfield rotary viscosity at different temperature（The viscosity of 135℃ and 150℃, for example）. The results of correlation betweenBrookfield rotaryviscosity and G*/sinδare showed in figure 5. It can be seen from the curves of figure 5 that there is a good exponential relationship betweenBrookfield rotaryviscosity and G*/sinδ. And these result indicate that the influence regularity of the filler-bitumen ratio to Brookfield rotaryviscosity is consistent with thattoG*/sinδ. Therefore, to some extent, results of Brookfield rotaryviscosity can be used to estimate the influence of filler-bitumen ratio to asphalt mortar character at high temperature instead of G*/sinδ.

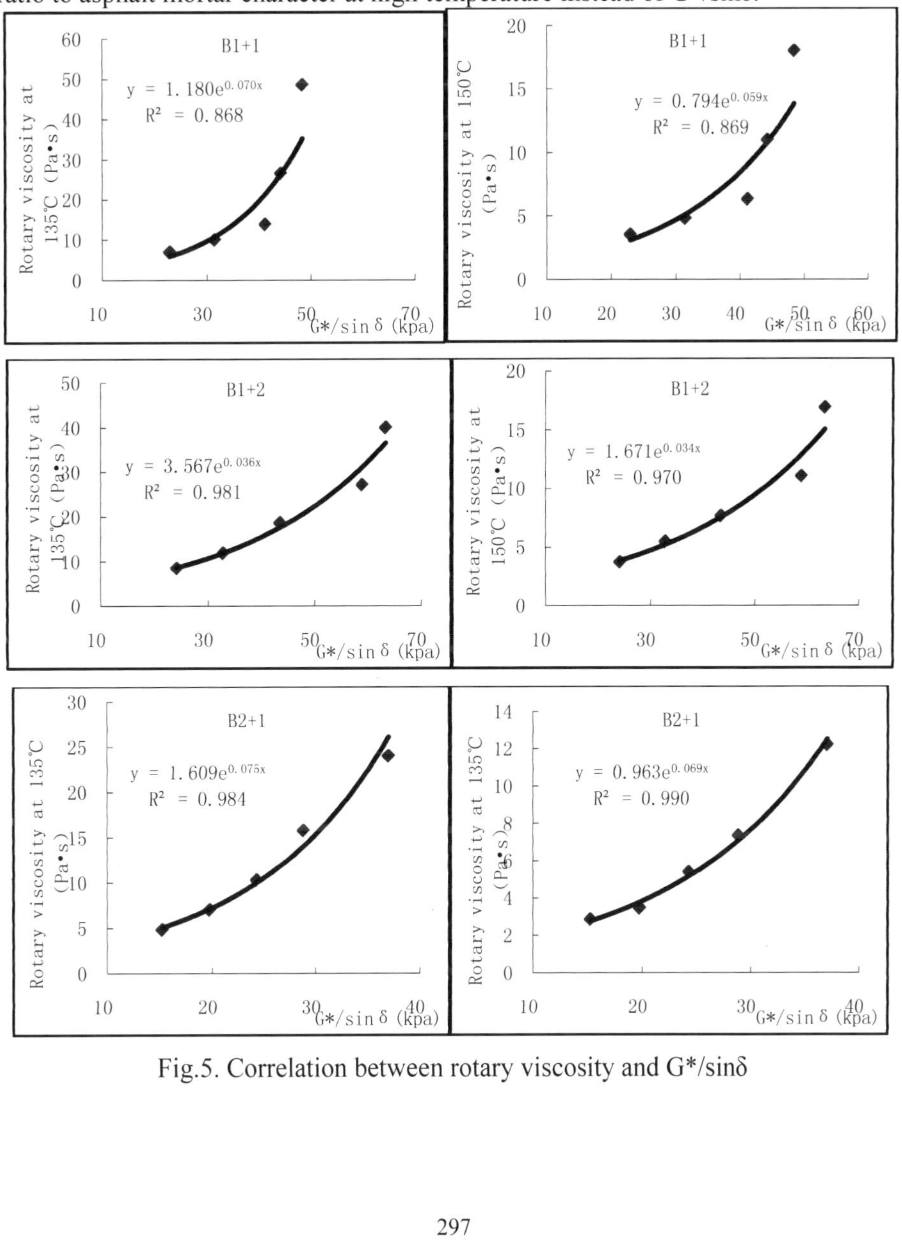

Fig.5. Correlation between rotary viscosity and G*/sinδ

4. Conclusions

This study investigated Dynamic shearing rheology and Brookfield rotaryviscosity of asphalt mortar with different ratio of filler to bitumen. The value of G*/sinδ is influence by character of asphalt and filler. Asphalt with higher viscosity and filler with larger specific surface area can both improve the value of G*/sinδ. Therefore, the G*/sinδ of mortar B1+2 is the highest, and the G*/sinδ of mortar B2+1 is the lowest. When the ratio of filler to bitumen changes from 0.8 to 1.2, *G*/sinδ*increases quickly with the increase of filler-bitumen ratio. But, it increases slowly when the ratio of filler to bitumen is higher than 1.2. Also the temperature sensitivity increases with the increase of the filler-bitumen ratio. To some extent, increasing the specific surface area of filler can decrease the temperature sensitivity of asphalt mortar. Meanwhile, there is a goodexponential relationship betweenG*/sinδ and Brookfield rotaryviscosityat 135℃and 150℃. So that, the results of Brookfield rotaryviscosity can be used to estimate the influence of filler-bitumen ratio to asphalt mortar character at high temperature instead of G*/sinδ.

Acknowledgments

This research work was sponsored by the Key Laboratory of Road Structure and Materials of Guangxi, China Postdoctoral Science Foundation（2015M582592）and Special Fund for Basic Scientific Research of Central Collegesin Chang'an University(310831151080，310831153409).

Reference

[1]E.Ray Brown, John E.Haddock(1997). Characterization of Stone Matrix Asphalt Mortars [R]. Eighth International Conference on Asphalt Pavements. Washington.
[2]Cristian Druta. A Micromechanical Approach for Predicting the Complex Shear Modulus and Accumulated Shear Strain of Asphalt Mixtures from Binder and Mastics[D]. LouisianaStateUniversity, 2006:41-58
[3]Zhang Z.Q. Li P(2005). Effect of fiber and mineral filler on asphalt mortar performance [J]. Journal of Chang'an University (Natural Science Edition), 25(5)15-18
[4]Guo P (2010).Multifactor influence on rheological property of fiber asphalt mortar [J]. Journal of Guangxi University (Natural Science Edition), 35(1):105-109
[5]Zhang Z.Q. Li P(2005). Ratio of filler bitumen of asphalt mixture [J]. Journal of Chang'an University (Natural Science Edition), 24(5):7-10
[6]Liu L. (2007). High temperature properties and evaluation method of asphalt mortar [J].Journal of Chang'an University (Natural Science Edition), 27(5):30-34
[7]Liu L. (2007).Properties and evaluation method of asphalt mortar [D]. Xi'an, Chang'an University
[8]Shen, J.A.(2001). Road performance of asphalt and mixture [M]. Beijing,China Communications Press.
[9]JTJ052-2011 Test procedure for asphalt and asphalt mixture of highway engineering [S].

IOP Conf. Series: Materials Science and Engineering **359** (2018) 012049 doi:10.1088/1757-899X/359/1/012049

Tribological Characteristics of Submicron SiC(p)-GR(p)/Zn-35Al-1Mg Composites in Semisolidification Casting process

S Liu*, Z W Fang and L X Li

Key Laboratory for Ferrous Metallurgy and Resources Utilization of Ministry of Education ,Wuhan University of Science and Technology, Wuhan 430081,China.

E-mail:liusheng@wust.edu.cn, Liu.sh2008@163.com

Abstract. Uniform SiC(p)-GR(p)/Zn-35Al-1Mg composites were prepared by powder pressing and semisolidification stirring-casting process by adding submicron silicon carbide and graphite reinforcement particles in an aluminum-zinc alloy matrix. Micro Vickers hardness and microstructures of the novel composites were studied, and their wear properties and wear temperature were measured for different load and friction conditions. The results show that silicon carbide and graphite particles homogeneously mix in the matrix, while contained silicon carbide particles improve the matrix hardness to 8.4%, graphite improves the matrix hardness to 16.8%, but two of them, combined, reduce the matrix hardness to 7.6%; the rate of temperature rise of the zinc-aluminum matrix alloy is the highest than the other three composites and is up to 48.5° C/s at 1.69MPa. At 0.56MPa and sliding 26.4km, the graphite composite anti-wear effect is optimal, while at 1.13MPa, the wear resistance of silicon carbide and graphite compound particles is the best; in the other case of only silicon carbide particles, the wear resistance is increased to 35% at 1.69MPa and 26.4km, and its anti-wear effects are excellent.

1. Introduction

Casting Zn-Al based alloy is widely used in Mechanical engineering for its favorable processing, mechanical, and wear-resisting properties, and it has been drawing attention from manufacturers and researchers because of its wide application on bearing bush,piston and sliding blocks[1-3]. However, the defects represented by the low hardness and unsatisfying under heavy load and high temperature have been limiting its range of application[4]. But with the addition of SiC and graphite and other particles enforcement into Zn-Al matrix alloy, the wear-resistance and friction decreasing properties can be acquired for the matrix alloy[5].SiC as a hard phase in Zn-Al matrix alloy can improve the wearproperty effectively, while graphite being soft phase, can increase friction-resisting property as well. The wear-resistance of the key component-bearing bush, which is damageable as a part of engine, has to be improved as much as possible while ensuring the safety of its spouse friction pair[6,7].

The research of improving wearresistance on matrix by adding SiC and graphite particles simultaneously is rarely reported, so this work involves the preparation of Zn-Al matrix composites containing SiC and graphite particles, the study of its wear characteristics and wear temperature principles under different load and wear distance, thus providing important parameters for the further application of new type Zn-Al matrix composites.Therefore, as a kind of light weight material, the new type casting $SiC_{(p)}$-$GR_{(p)}$/Zn-35Al-1Mgcomposites used as transmission components in mechanical engineering will have a wider ranger of application in the future.

2. Materials and experimental

Zn-35Al-1Mg matrix alloy was prepared by casting using pure Zn(99.9%), pure Al(99.9%) and pure Mg(99.9%) allocated in portion. The middle composites were produced by Aluminium particles with the particle size of 300~400 meshes, SiC particles and GR particles with the particle size of about 5000 meshes.First of all, Al,SiC and Gr particles put into the QM3SP4L planetary ball mill were ball milled for 8hoursat speed of 400rpm.After that, the mixed powder were allocated in rectangularforming mould, and then completely compressed on YES-2000 Digital hydraulic pressure machine, which the middle composites were produced after the finalization of the process.Second,Al ingots were placed in steel crucible with ZnO painted in its inner surface, which were put into SG2-3-10 resistance furnace with automatic temperature control system to be heated at 720℃ and kept for 8 minutes. ThenAr was conducted as protective atmosphere to the furnace while addedZn,Mg ingots and 10%$SiC_{(p)}$/Al, 10%$GR_{(p)}$/Al and 5%$SiC_{(p)}$+5%$GR_{(p)}$/Almiddle composites together wrappedto the bottom of Al liquids, after the addition for 5 mins, the mixed melting liquids were stirred at 500rpm and kept the furnace temperature cooled to 600℃ for 10 minutes until they were mixed well. Finally experimental ingots were prepared inΦ15mm×50mm metal mould.

As-casting composites were processed to Φ8mm specimens, then they were subjected to heat treatment,which was performed by solid-solution at 500℃ for 4 hours, aged naturally for 36hours. After heat treatment, the friction wear properties of the composites at the friction distance of 0.45km, 2.2km, 6.6km, 13.2km, 26.4km were measured under the loads of 0.56MPa, 1.13MPa and 1.69MPa on friction wear testing machine. In the meantime, the wear temperatures of materials under different conditions against friction distance were recorded using TP1000 temperature recorder.

Wear sections were embedded in conductive resin using ZXQ-2 automatic mosaic machine. then themicrostructural examinations from embedded ingots were measured by ZEISS metallographic microscope and PHILIPXL30TMP SEMequipments with standard metallographic techniques,that as rough-grinding,fine-grinding and polishing,thenthey were etched in a beakerwhich mixed etching solutionfor 4 minutes.Themixed etching solution contained two hydrofluoric acid,three muriatic acid,fivenitric acid and one hundred ninety distilled water,which was placed in a constant temperature mixing crucible. After the etching, the specimens were dried using hot air for following observation.

3. Results and discussion

The curve showing mass loss of different compositeswithsliding distance is given in Fig.1.In the view of these results,$SiC_{(p)}$ the single addition exhibited better effect of improving wear resistance of matrix alloy under heavy load than $SiC_{(p)}$+$GR_{(p)}$compositesaddition or $GR_{(p)}$ the single addition does. What's more,according to the appearance of wear surface and wear temperature,$GR_{(p)}$exhibited softness and possessed favorable lubricity and it was favorable for decreasing wear mass loss under lower load;on the contrary,a moderate amount of homogeneously dispersed $SiC_{(p)}$exhibited high hardness,strong carrying capacity under heavy load,which was beneficial for decreasing wear mass loss of materials,besides,they also possessed small thermal expansion index and high thermal conductivity,etc. that enabled$SiC_{(p)}$ to disperse the high heat on friction surface yielded by friction under heavy load, therefore protecting the matrix material from being further damaged.

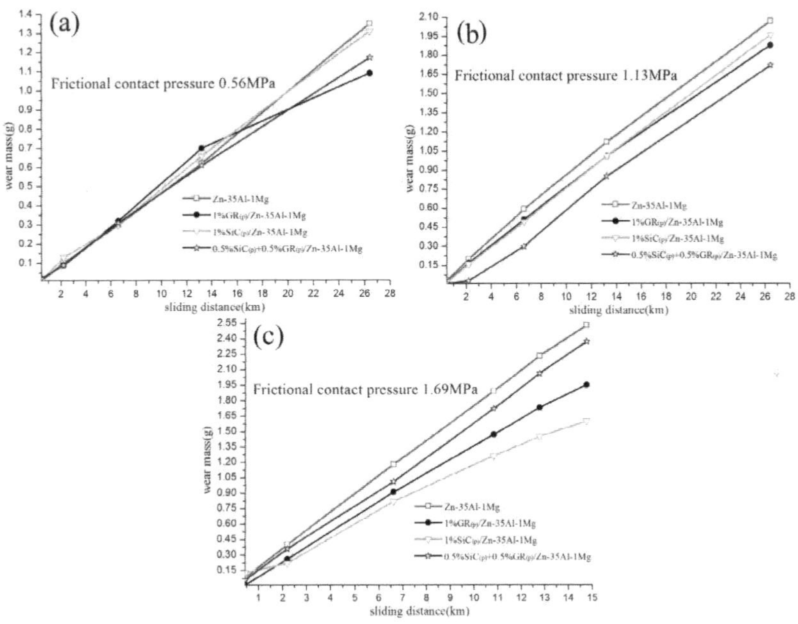

Fig.1 The relationship between wear mass and sliding distance of the composites:(a) 0.56MPa,(b)1.13MPa,(c)1.69MPa

The appearance of the wear stripping films of composites under load the 1.69MPa is shown in Fig.2. It can be seen that four compositess as-cast Zn-35Al-2Mg, composites single added $SiC_{(p)}$, composites single added $GR_{(p)}$ and composites added $SiC_{(p)}$+$GR_{(p)}$composite all yielded many striping films and these films accumulated on the margin of the wear section of wear specimens. As is shown in Fig.2(b), composites single added $SiC_{(p)}$, exhibited a different wear surface compared with three other composites, which can be explained by the fact that homogeneously dispersed $SiC_{(p)}$enabled partial Zn-Al solid solution to conduct and disperse the heat in time and that Zn-Al alloy remaining on the surface, which was oxygenized and melted, prompted the surface to form into paste.

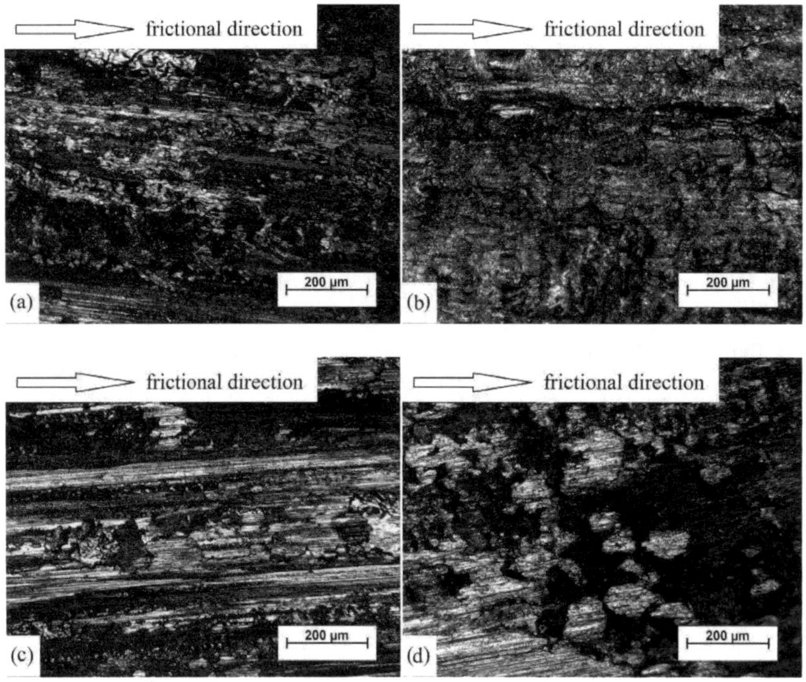

Fig.2 The wear morphology of composites in frictional contact pressure 1.69MPa:(a)Zn-35Al-1Mg,(b)1$SiC_{(p)}$+,(c)1$Gr_{(p)}$+,(c)0.5$SiC_{(p)}$+0.5$Gr_{(p)}$+

It can be seen from Fig.3that the friction started, the maximum temperature increasing rateoffour compositesin different frictional condition,which wascalculatedto 48.5℃/s;similarly, when the wear ended, the maximum temperature decreasing rates of these four composites can also be calculated as 51.4℃/s. Besides, it can be seen from Fig.3 thatduring friction wear process, composites added graphite particles, SiC particles or these two combined exhibited different wear temperature change. Specifically, the addition of SiC and graphite under load of 0.56MPa can yield a lower wear temperature after wearing at a sliding distance of 1.32km compared with composites single addedSiC, single added graphite or the matrix alloy; Also, not surprisingly, the maximum wear temperature under load of 1.13MPa was slightly higher than the wear temperature under the load of 0.56MPa, for the load of 0.56MPa, the maximum wear temperature was produced by the direct wear of Zn-Al matrix materials and can rise to 230℃ approximately during wear, while for the load of 1.13MPa, the maximum wear temperature was produced when single adding SiC particles and can rise to approximately260℃ during wear, and wear temperature of composites containing SiC particles also rose with the increase of the load. However, composites single added graphite all displayed lower wear temperature under different loads, because the distribution of graphite particles enabled the compositess to exhibit good friction reduction property on wear surface after it was stripped away from worn matrix alloy under low load; however, the weartemperature decreasing effect was not evident for the composites addedSiC and graphite mixed particlescompared with composites single added SiC particles or single graphite particles.

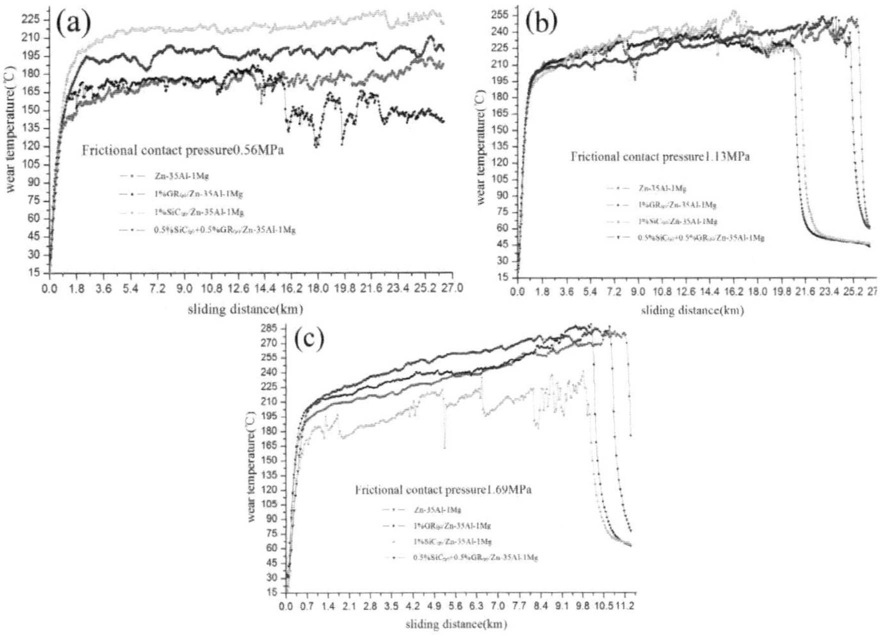

Fig.3 The relationship between wear temperature and sliding distance of the composites:(a) 0.56MPa,(b)1.13MPa,(c)1.69MPa

The mircrostructure of new type $SiC_{(p)}$-$GR_{(p)}$/Zn-35Al-2Mg composites wear surface under load of 1.69MPa was shown in Fig.4, it is known that the parent phase in the structure of composites was dendritic Zn-Al solid solution, consisting of primaryα-phase, and α+β mixed eutectoid phase; and the white dendritic α-phase was solid solution formed by Zn melting into Al; η-phase was the solid solution formed by Al melting into Zn, around the margin of dendrite of which, a small amount of Mg-Zn compound was distributed. It can be seen in Fig.4 that the silver gray flaky particles appearing in Fig.4(b)(d) wereSiC particles, while the black particles were graphite particles. Besides, the solid soulution in Fig.4(a)(b)(c)(d) exhibited different appearances, for example, it can be seen from Fig.4(a) that dendrites of Zn-Al solid solutions were slender and long; dendrites of Zn-Al solid solution; the Fig.4(b), however,were thick, short and small in Fig.4(b) affected $SiC_{(p)}$ particle; Fig.4(c) showed a decreasing amount of dendrites, which existed in the form of tiny particles dispersively distributed; as was shown in Fig.4(d), the addition of $SiC_{(p)}$ and $GR_{(p)}$ composite particles transformed dendritic Zn-Al solid solution intotiny petal clusterdispersively distributed. As a result of the description of the mircostructural appearance above, it is knownthat $SiC_{(p)}$ and $GR_{(p)}$ particles did have some influences on the mircostructure of high Zn-Al matrix alloy, more specifically, dendrites ofZn-Al solid solutions thickened after the introduction of $SiC_{(p)}$ and $GR_{(p)}$ particles, with previously slender dendrites transformed into short, thick petal shape and the blunting of sharp angles of former slender dendrites.

IOP Conf. Series: Materials Science and Engineering **359** (2018) 012049 doi:10.1088/1757-899X/359/1/012049

Fig.4 The microstrucure of worn surface in different composites during the same condition:(a)Zn-35Al-1Mg,(b)1$SiC_{(p)}$+,(c)1$Gr_{(p)}$+,(c)0.5$SiC_{(p)}$+0.5$Gr_{(p)}$+

The hardness of samples was measured by HV-1000/HV-1000A Vickers mircohardness tester with 1.961N set for the test stress series and hardness range was set within HV0.2. During the test, the mircohardness of 7 points along the radial direction were measured every 0.1mm. And the center of the specimens was chosen as the the benchmark for the determination of these 7 points. The average of hardness of them were determined as final Vickers mircohardness value.TheHV of Zn-35Al-2Mg,3%$SiC_{(p)}$/Zn-35Al-2Mg, 3%$GR_{(p)}$/Zn-35Al-2Mg and 1.5%$SiC_{(p)}$+1.5%$GR_{(p)}$/Zn-35Al-2Mg was 131,142, 109, 121 separately. It can be seen from the results that $SiC_{(p)}$ increased the microhardnessby 8.4%, on the contrary, the hardness of composites added $GR_{(p)}$ decreased by 16.8% compared with matrix material, however the introduction of $SiC_{(p)}$+ $GR_{(p)}$ caused mircohardness to decrease sharply, which was 7.6% less than the matrix alloy.

The friction surface of materials and its spouse mill during friction process were softened due to sharp temperature increase, thus diminishing the deformation resistance of the materials and the mill. So during the wear under the continuous effect of friction shearing strength, the wear surface broke apart from the matrix and peeled off, which also took away high heat on wear surface thus cooling the wear surface. Then temperature would rise again due to the frictionuntil reaching the stripping temperature and then the stripping films would take away the heat again, therefore the temperature during the process would circulate in this way until reaching stability. When there was relative sliding between composites and its spouse friction steel plate, the friction surface temperature roseand large plastic deformation occurred because of the friction on contact surface caused by outer load, the metals on the surface exhibited partial gelling effect under thermal compressing influence and therefore would be broken by molecular stress during the following relative sliding, during which the metal fragments were dragged away from the components thus causing abrasion on component surface, so this was the reason why stripping films formed on wear surface in the experiment and it also determined that it was the gelling wear form that distinguished the wear of compositess.

It can be seen from Fig.4 that $SiC_{(p)}$ and $GR_{(p)}$weredispersively distributed in the form of small particles in the matrix. The reciprocal compression between friction parts after the start of friction made$SiC_{(p)}$ and $GR_{(p)}$ on contact surface behave in different forms, for $SiC_{(p)}$, it was top distribution, the actual contact area was small, and plastic deformation of matrix would occur once under large contact stress effect. However, differently, $GR_{(p)}$ stripped away from matrix directly fell into friction contact area causing lubrication and friction-reducing effects, which effectively improved the wear-resisting property under heavy load but did not exhibit apparent effects under the heavy load. The reason why $GR_{(p)}$failed to play an important role under heavy load was that the large stress between contact surfaces forced material to go through both the plastic deformation and creep deformation, which in turn decreased deformation resistance of material sharply and stripped away Zn-Al solid solutions and graphite particles possessing soft phase from matrix alloy during friction, therefore leading to the failure of the $GR_{(p)}$friction-resisting effect. However, in the meantime the $SiC_{(p)}$with favorable microhardnessbegan to substitute $GR_{(p)}$and it also exhibited favorable heat conductivity, which lately can effectively prevent materials from going through creep deformation caused by increasing wear temperature; what's more, according to the comparison of Vickers hardness,the white dendriticαphase possessed higher hardness thanηphase solid solutionsformed by Al melting into Zn. Besides, it is known that forαphase, Zn can form infinite solid solution in Zn, while forηphase, Al can only form finite solid solution in Zn, therefore according to solid solution reinforcement effects, whenαandηphase coexist, the existence of moreαphase can lead to higher wear-resistance. Moreover, the coexistence ofαandηphasetransformed brittle dendritic structure (especially fragile under continuous high temperature sheer stress)to petal shaped structure, which bluntedthe sharpness of dendrites thus positively hindering the breakage of them, sobetter wear-resisting property was acquired by adding both $SiC_{(p)}$ and $GR_{(p)}$. Finally, the small amount of Mg in matrix alloy improved the creep strength of matrix and led to effective soakage of SiC particles into matrix, which in turn increased the dislocation density of grains therefore improving the wear-resistance of materials effectively.

4. Conclusions

1.Zn-Al matrix alloy composites with reinforcement particles mixed homogeneously can be produced by stirring casting process after powder-compacting,the mircrohardness increase of composites compared with the matrix differed when adding different particles, which was 8.4% for the signgle addition of SiC and 16.8% for graphite particles,however,the mircohardness exhibited a decrease by 7.6% when adding mixed SiC and graphite particles.

2.The maximum temperature rising rate of four composites was 48.5℃/s when the wear began, while the maximum temperature decreasing rate of these composites was 51.4℃/s in the end of the wear, the wear temperature change of materials single added SiC particles under heavy was not apparent with the increase of wear time.

3.The wear mass loss decreases compared with matrix alloy differed between different composites under low load at a sliding distance of 26.4km, among which, the highest was 18% produced by composites added graphite particles decreasing, followed by 12% decrease exhibited by adding graphite and SiC particles, however, the addition of SiC single did not show an evident effect.

4.Under the heavy load, different wearresisting indexes were tested for different composites, among which, the SiC particles increased the index by the largest percent to 35%, while the index increases of composites added Gr + SiC were7% separately compared with matrix alloy.

5. Acknowledgement

This work is supported by The Key Laboratory for Ferrous Metallurgy and Resources Utilization of Ministry of Education through its Open-Fund Innovation Program with item numberFMRU201305.

References

[1]Miroslav B, Slobodan M and Branislav J 2010The influence of heat treatment on the sliding wear behavior of ZA-27 alloy. *Tribology International*, 43(1-2): 16-21.

[2]MurphyS and SavaskanT 1984 Comparative wear behaviour of Zn-Al-based alloys in an automotive engine application, *Wear* 98: 151-161.

[3] BarnhurstRJ 1988 Guidelines for designing zinc alloy bearings-a technical manual, in: *Proceedings of the International Congress and Exposition, Detroit, Michigan.*

[4]Ranganath G,Sharma S C and Krishna M 2001 Dry sliding wear of garnet reinstressd zinc/aluminum metal matrix composites. *Wear*,251:1408-1413.

[5]IlijaB,JovanaR,BiljanaB,MiroslavB,AleksandarVand SlobodanM 2014 Microstructural characterization and artificial aging of compo-casted hybrid $A356/SiC_{(p)}/Gr_{(p)}$ composites with graphite macroparticles.*Materials Science & Engineering A*,612:7-15.

[6]MondalDP, DasS, RaoRN and SingM 2005 Effect of SiC addition and running-in-wear on the sliding wear behaviour of Al-Zn-Mg aluminium alloy.*Materials Science and Engineering A*, 402:307-319.

[7]SharmaS.C., GirishB.M., KamathR. and SatishB.M 1998 Graphite particles reinforced ZA-27 alloy compositess for journal bearing application, *Wear*, 219-2:162-168.

Comparative Analyses of Creep Models of a Solid Propellant

J B Zhang[1,2], B J Lu [1,2], S F Gong[1,2] and S P Zhao[1,2]

[1]713 Research Institute of China Shipbuilding Industry Corporation, Zhengzhou, 450015, China

[2]Henan Key Laboratory of Underwater Intelligent Equipment, Zhengzhou, 50015,China.

E-mail: zjb66688@163.com

Abstract. The creep experiments of a solid propellant samples under five different stresses are carried out at 293.15 K and 323.15 K. In order to express the creep properties of this solid propellant, the viscoelastic model i.e. three Parameters solid, three Parameters fluid, four Parameters solid, four Parameters fluid and exponential model are involved. On the basis of the principle of least squares fitting, and different stress of all the parameters for the models, the nonlinear fitting procedure can be used to analyze the creep properties. The study shows that the four Parameters solid model can best express the behavior of creep properties of the propellant samples. However, the three Parameters solid and exponential model cannot very well reflect the initial value of the creep process, while the modified four Parameters models are found to agree well with the acceleration characteristics of the creep process.

1.Introduction

Most solid propellant is generally considered to be a typical viscoelastic material [1-5].Because this materials exist obvious viscoelastic characteristics (the phenomenon such as creep, replies and stress relaxation), that strongly depend on the loads, environment temperature and time process [6]. Most solid propellant material belongs to high-molecular polymer, is also a typical viscoelastic material. Creep is one of the main mechanical properties indexes of solid propellant materials. However, there are relatively few researches on the creep characteristics of solid propellant. No creep model can be widely used in propellant materials. Choosing different creep model to describe the creep properties of the propellant materials, this can help to analyse its viscoelastic behaviour. With the help of creep test data can obtain the material Parameters of the models.

The common basic viscoelastic models are usually composed of the springs and the dampers, such as Kelvin and Maxwell model. A variety of complex fluid and solid models is widely applied to express the rheological characteristics of materials, such as bituminous sand [7-8], rocks [9] and asphalt mixtures [10], However, there are relatively few researches on the creep characteristics of solid propellant. To choose can be applied to the creep model of propellant material. The five models are used to express the propellant material creep properties under different stress level in this research, and compared with the experimental result analysis obtained a better creep model.

2.Experimental

The solid propellant materials mainly include: nitration cotton (55%), nitroglycerin (29.3%), the second nitro toluene (10%), other (5.7%).The test sample are derived from the solid propellant grain. The standard samples are obtained through the standard QJ924-85 of the aerospace sectors. The schematic diagram of the standard dumbbell drawing sample is shown in figure 1. In order to remove the residual stresses generated during machining, all test samples must be kept at 50 degrees for more than 48 hours. Before creep experiment, the sample and fixture will be installed in the attemperator of the QJ211B material universal testing equipment. Five kinds of constant stress were 4MPa, 4.5MPa, 5MPa, 5.5MPa and 6MPa, respectively; under the environment temperature 293.15 K. Five kinds of constant stress were 1MPa, 1.5MPa, 2MPa, 2.5MPa and 3MPa, respectively, under the environment temperature 323.15 K.

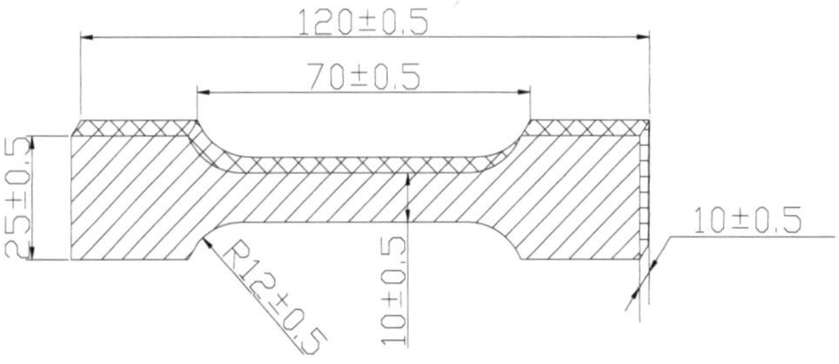

Figure 1. Solid propellant samples

Before creep experiment, the sample and fixture will be put in the attemperator for 10 hours. In order to eliminate the influence of the experimental system of mechanical clearance, the test equipment system will be offered preloading load of 2N. The specimen and fixtures were by thermal insulation in the temperature controller after 1 hour. The computer program was restarted, which the creep test data were recorded real-time by. The test results averaged from the repeated five times in each case.

3.Results and discussions

The creep curve of solid propellant material under ambient temperature 293.15 K and 323.15 K is shown in figure 2. The creep curves of each stress level in figure 2 show the creep curve under different stress levels contains only deceleration creep and constant creep at 293.15K; the creep curves under low stress level contains deceleration creep and constant creep, while under high stress will be accelerated creep stage, and the creep rate increased rapidly until the test sample was destroyed at 323.15k.

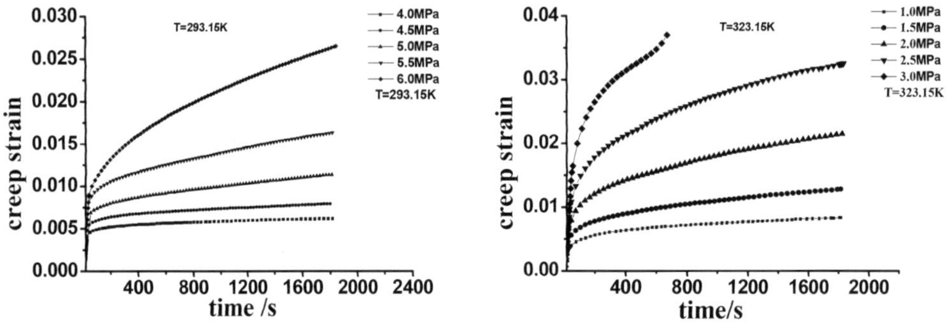

Figure 2. The creep curve of propellant material at different ambient temperature

The superposition of each curve using the time-temperature-stress equivalent principle in figure 2, it is concluded that solid propellant creep can be divided into three stages. The first stage is the deceleration creep process. The deformation is increasing rapidly and creep strain rate is decreasing. The time of this stage is about 5% of the creep life, and the main performance is transient elastic deformation. The next stage is the constant velocity creep process. The creep deformation nearly linear increase, the strain rate nearly remains the same. The time period of this stage is about 80% of the creep life, mainly manifested as viscous flow deformation. The third stage is the accelerated creep process, also known as the creep damage process. The creep deformation increases rapidly, the strain rate rise sharply, causing the failure of the sample in a short time. The time period of this stage is about 15% of the creep life, mainly the visco-plastic deformation.

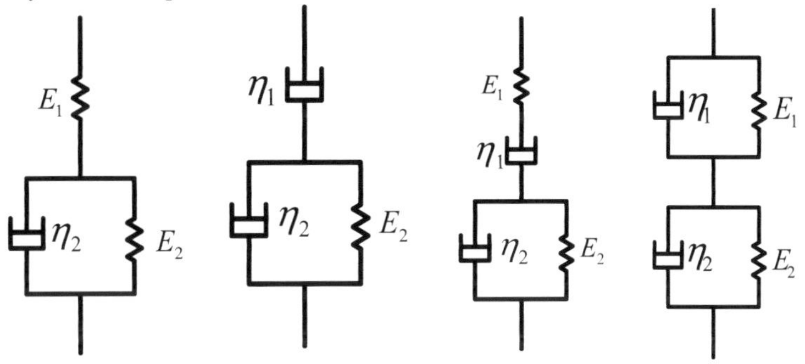

Figure 3. Four kinds of viscoelastic models. (a) Three Parameters Solid. (b) Three Parameters Fluid. (c)Four Parameters Fluid (d)Four Parameters Solid.

Table 1. All sorts of the differential and creep equations of models

models	differential equations	creep equations
three parameters solid	$\sigma+\frac{\eta_2}{E_1+E_2}\dot{\sigma}=\frac{E_1E_2}{E_1+E_2}\varepsilon+\frac{E_1\eta_2}{E_1+E_2}\dot{\varepsilon}$	$\varepsilon(t)=\left[\frac{1}{E_1}+\frac{\left(1-e^{-E_2t/\eta_2}\right)}{E_2}\right]\sigma_0$
three parameters fluid	$\sigma+\frac{\eta_1+\eta_2}{E_2}\dot{\sigma}=\eta_1\dot{\varepsilon}+\frac{\eta_1\eta_2}{E_2}\ddot{\varepsilon}$	$\varepsilon(t)=\left[\frac{t}{\eta_1}+\frac{\left(1-e^{-E_2t/\eta_2}\right)}{E_2}\right]\sigma_0$
four parameters fluid	$\sigma+\left(\frac{\eta_1}{E_1}+\frac{\eta_1+\eta_2}{E_2}\right)\dot{\sigma}+\frac{\eta_1\eta_2}{E_1E_2}\ddot{\sigma}=$ $\eta_1\varepsilon+\frac{\eta_1\eta_2}{E_2}\dot{\varepsilon}$	$\varepsilon(t)=\left[\frac{1}{E_1}+\frac{t}{\eta_1}+\frac{\left(1-e^{-E_2t/\eta_2}\right)}{E_2}\right]\sigma_0$
four parameters solid	$\sigma+\frac{\eta_1+\eta_2}{E_1+E_2}\dot{\sigma}=\frac{E_1E_2}{E_1+E_2}\varepsilon+$ $\frac{E_1\eta_2+E_2\eta_1}{E_1+E_2}\dot{\varepsilon}+\frac{\eta_1\eta_2}{E_1+E_2}\ddot{\varepsilon}$	$\varepsilon(t)=\left[\frac{\left(1-e^{-E_1t/\eta_1}\right)}{E_1}+\frac{\left(1-e^{-E_2t/\eta_2}\right)}{E_2}\right]\sigma_0$
exponential		$\varepsilon(\mathrm{t})=\left(A+Bt^n\right)\sigma_0$

The four viscoelastic models are shown in figure 3. The differential equations and creep equations of five models are shown in table 1. The fitting accuracy of different models is judged by correlation coefficient. According to the principle of least square method, the different creep models parameters and the correlation coefficients are obtained, as shown in table 2. As is known to all, the four Parameters solid model can best express creep properties curve of the propellant samples, followed by the four parameters fluid model, again is three parameters fluid and exponential model, and the fitting of three parameters solid model is the worst.

Table 2. Correlation coefficients of different creep models

model	293.15K					323.15K					mean value
	σ_0(MPa)					σ_0(MPa)					
	4.0	4.5	5.0	5.5	6.0	1.0	1.5	2.0	2.5	3.0	
three parameters solid	0.967	0.794	0.791	0.863	0.967	0.924	0.923	0.932	0.965	0.980	0.910
three parameters fluid	0.967	0.989	0.993	0.993	0.991	0.981	0.985	0.991	0.992	0.997	0.988
four parameters fluid	0.967	0.999	0.993	0.992	0.991	0.984	0.985	0.990	0.992	0.997	0.989
four parameters solid	0.967	0.997	0.998	0.998	0.998	0.995	0.997	0.998	0.999	0.997	0.995
exponential	0.962	0.998	0.990	0.989	0.995	0.979	0.985	0.986	0.991	0.967	0.984

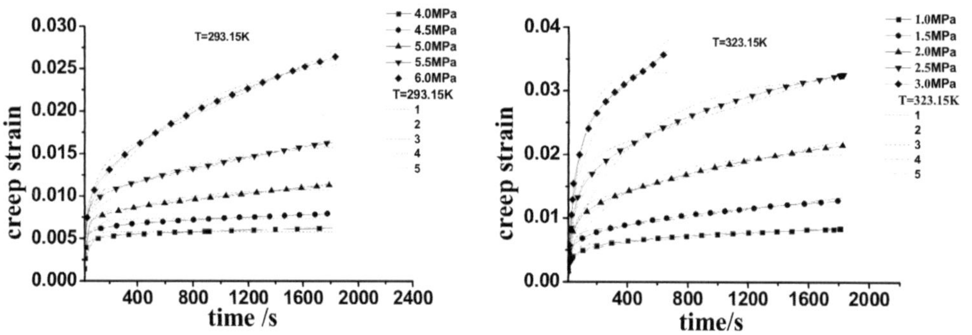

Figure 4. The fitting contrast of models (1-three Parameters solid; 2-three Parameters fluid; 3-four Parameters fluid; 4-four Parameters solid; 5- exponential).

The fitting results of creep curves of different models under different stress levels at 293.15K and 323.15K are shown in figure 4. By figure 4, particularly in the decelerating creep stages, four parameters fluid and four parameters solid model fitting precision is well which is consistent with the test curve under low stress, can better reflect the decelerating creep process, Followed by the three parameters fluid model. And the three parameters solid and exponential model fitting precision is the worst, which don't reflect the initial process of test curve. In constant creep phase, these five models fitting precision have little difference. But four parameters solid model fitting precision is well which is consistent with the test curve under high stress. Secondly is the four parameters fluid model. While the three parameters fluid model and the exponential model fitting precision is bad, which can't really describe the creep process of test curve. When the high temperature and high stress, four parameters solid model fitting precision is well that is consistent with the decelerating creep and the constant creep stage of the test curve, which has some deviation with the test curve in the accelerated creep stage, followed by the three parameters fluid and four parameters fluid model. While fitting precision of exponential model is the worst, which can't really describe the creep process of test curve.

4. Conclusions

Described above five creep models can reflect the second stage characteristics of the creep process under the same temperature and different stress level. Four parameters solid model not only can well fit the constant creep phase of creep curve under the different temperatures and different stress level, also can well reflect the first stage characteristics of the creep process, followed by three parameters fluid and four parameters fluid model. While the three parameters solid and exponential model can't reflect the initial value of the creep process of the solid propellant material. the modified four parameters models can better describe the acceleration characteristics of creep process.

References

[1] Shiang-Woei Chyuan 2003 Studies of poisson's ratio variation for solid propellant grain under ignition pressure loading *Internationan Journal of Pressure Vessels and Piping.* **80** 871-877.

[2] Nevière Robert 2006 An extension of the time-temperature superposition principle to non-linear viscoelastic solids *International Journal of Solids and Structures.* **43** 5295-5306.

[3] Xu,N Aravas,P Sofronis 2008 Constitutive modeling of solid propellant materials with evolving microstructural damage *Journal of the Mechanics and Physics of Solids* **56** 2050-2073.

[4] Ren P,Hou X,Rang He G 2010 ComParative Research of Tensile and Compressive Modulus of Composite Solid Propellant for Solid Rocket Motor *Journal of Astronautics* **31** 2354-2359.

[5] Yıldırım H C, Özüpek Ş 2011 Structural assessment of a solid propellant rocket motor: Effects of aging and damage *AerosPace Science and Technology* **15** 635-641.

[6] Li D,Zhou C S,Ju Y T 2009 Experimental Research of Constitutive Relationship of DB Solid Propellant Grain *Journal of Ballistics* **21** 6-9.

[7] DeshPande V S, Cebon D 2000 Uniaxial experiments on idealized asphalt mixes *Journal of Materials in Civil Engineering* **12** 262-271.

[8] Ye Y,Yang X H,Chen C Y 2008 Viscoelastoplastic constitutive model for creep deformation behavior of asphalt sand *Journal of Central South University of Technology* **15** 13-16.

[9] Yuan H P,Cao P,Xu W Z 2006 Visco-elastoplastic constitutive relationship of rock and modified Burgers creep model *Chinese Journal of Geotechnical Engineering* **28** 796-799.

[10] Ye Y 2009 ExPerimental researches on visco-elastoPastic constitutive model of asphalt mixture *The Huazhong University of Science and Technology.*

IOP Conf. Series: Materials Science and Engineering **359** (2018) 012051 doi:10.1088/1757-899X/359/1/012051

A Study on the Rural Residence in the Northern Area of Zhejiang Province from the Perspective of Green Living Environment

J Wang, W J Gao and C Wang
The University of Kitakyushu1-1-1 Hibikino, Wakamatsu-ku, Kitakyushu-city

ZheJiang university architectural design institute co. LTD NO.148 Tian mushan Road

Email: 389190731@qq.com

Abstract At present, owing to the rapid development of rural construction, it lacks corresponding theories and practices and damages the features of rural area, ignoring the geography, suitability and green living environment factors. The research selects rural residence as the object, defining "courtyard" as the basic unit for rural residence and . It utilizes the principle of topology as the expanding media, by the method of principle of cellular structure and green living environment design strategy. The essay establishes the design and construction system of "rural basic unit", combining functions and structures, prototype menu, chamber space and compound interface, from the perspective of green living environment. It aims to guide rural construction and protect the ruralliving environment.

1 Background and Subject

Since the ancient time, the northern Zhejiang Province has been an important area for China's economics, society and culture. In short, it has consistent ecological, environmental, resources, social and cultural features, and is indeed a subject for research and construction integrating natural terrain unit, economic and social unit and rural housing unit [1].

Currently, the area manifests features of industrialized and commercialized development of industry, networking of human society, and modernization of construction technology. At the same time, rural settlement construction blindly uses the mode of urban development, with too many demands, rapid development, and large cost. It makes the "high carbon" feature and tendency much more obvious with the time passing by, which requires corresponding solutions as soon as possible [2].

Rural settlement is just like thestructure of life system, from a cell to the biosphere. It can also be disintegrated into a system based on basic cells and many upper levels. "The rural courtyard" is the key component of the rural construction system, and also a basic vector for realizing a rural strategy of low carbon emissions. According to this, we can build a "basic unit" of rural area--"the rural courtyard", namely an independent living area combined with houses and their courtyard. We can learn about the essential relation of rural living by researching the "basic unit" of "rural courtyard", and study its composition and structure, gathering and growing, which are the foundation for analyzing and researching rural area.

Components	Functions	Examples	Cellular Structure
terrain limit	*limit, division and setting*	*homestead boundaries or terrain*	*cytoderm*
function prototypes	*respective indoor basic functions*	*modules of living room, bedroom, kitchen, bathroom and storeroom*	*organelle*
additional cavity	*bearing, adjusting and flowing*	*a chamber space enclosed by each subject and boundary, bearing light, air, wind and activities*	*cytoplasm*
multi-layered skin	*enclosing, protecting*	*roof, wall, floor, door and window*	*cytomembrane*
Courtyard--rural basic unit			**Biological cell**

Figure 2-1 Division for "Rural Basic Units" (self-made)

2 Theories and Framework

2.1The cognition on rural living unit based on the perspective of low carbon emissions is a systematic research integrating many subjects. According to the principle of cytoarchitecture, the author analyzes the courtyard unit and establishes a design method and construction system for "basic rural unit" of function prototypes + additional cavity + multi-layered skin + terrain limit. (Figure 2-1)

2.2Based on low carbon perspective and the previous analysis, the research applies a method of prototype theory, with the assistance of bioscience, and sticks to the strategy of passive energy conservation technology. It also uses the principle of topology as an expanding media to construct a menu for the living unit in the northern Zhejiang Province. (Figure 2-2)

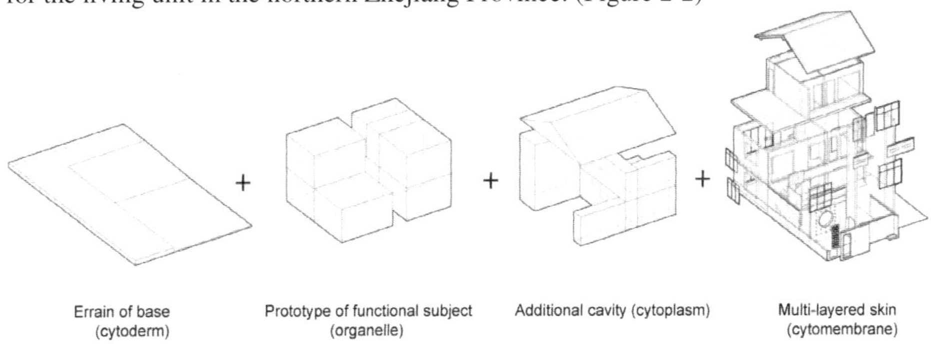

Figure 2-2 Diagram for Basic Courtyard Units(self-made)

3 Prototype and Menu

3.1 Main Function Prototype

3.1.1 The Cube Model

According to the instruction of low carbon emissions , we take the cube as the basic model ,with the length of 9.9 meters. Both the first and second floor are 3.3 meters high,and the rest is a 3.3-meter-high roof. This modle meets the requirements of various factors. (Figure 3-1)

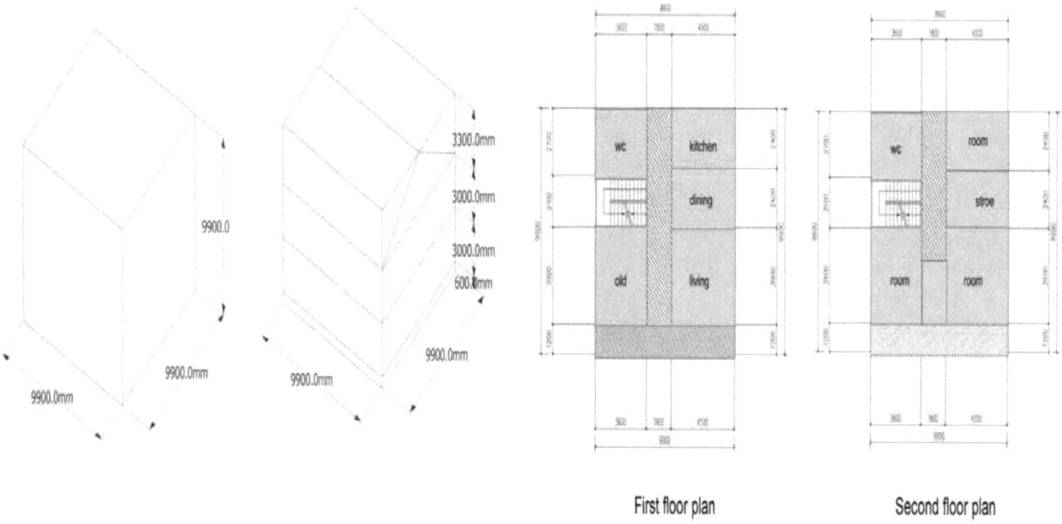

Figure 3-1 9.9-meter-length Cube Model (self-made)

- Optimum building shape coefficient ——A cube as a building model has the smallest building shape coefficient and he smallest energy consumption. Taking the cube as the prototype, and making small changes according to the , is the most energy efficient way.
- Suitable width and depth——9.9-meter-depth helps to ventilation and sunshine .(The depth of the building will not affect the sunshine requirement when it is less than 14 meters.)
- Meet the reality——Based on data analysis of many villages,The site area of rural residence is about 150 m^2~300 m^2 and the total areais about 200 m^2~300 m^2。 As the factors of the yard，we make the site area to be 9.9m x 9.9m，the total areais about 200 m^2 with 2 floors.
- Increasing Conversions——The plane of residence changes with base form,sunshine conditionand so on. Buta square provides the most convenient possibility for change.

3.1.2 Functions and Structures

Locate the plans according tolighting,ventilation,forthcomingdistrict and so on .(Figure 3-2)

- 1.2-meter-Lobby—For Staying and storage ,regulating the sunshine and air
- 4.5mx9.9m Kitchen- Dining room- Living room combinations for different needs.
- Bedrooms—Different types for hosts,children,old people according to their different habits .

Figure 3-2 The first and second floor plan (self-made)

3.1.3 The chamber space

The chamber space in building is a dynamic regulation system, regulating wind, air and sunshine and energy.(Figure 3-3)

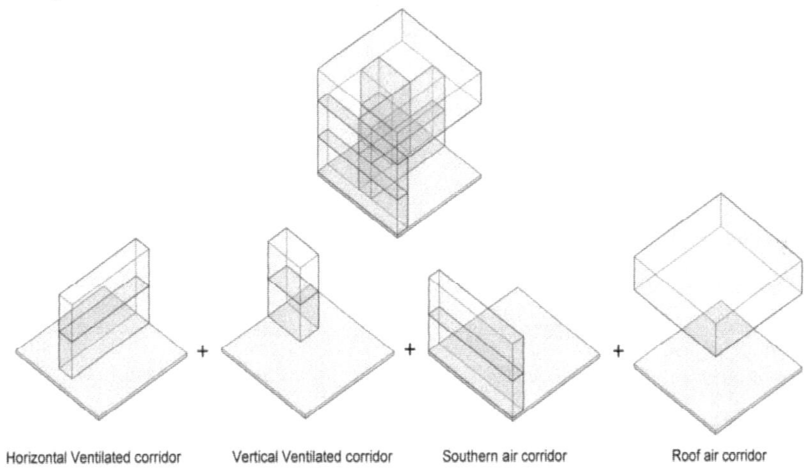

Figure 3-3 Cavity space schematic diagram (self-made)

3.2 Extension menu

According to the factors of yard, sunshine, ventilation, corridor, roof layer ,terrain of base , chamber space and compound interface , we construct a menu for the living unit in the northern Zhejiang Province.(Figure 3-4)

IOP Conf. Series: Materials Science and Engineering **359** (2018) 012051 doi:10.1088/1757-899X/359/1/012051

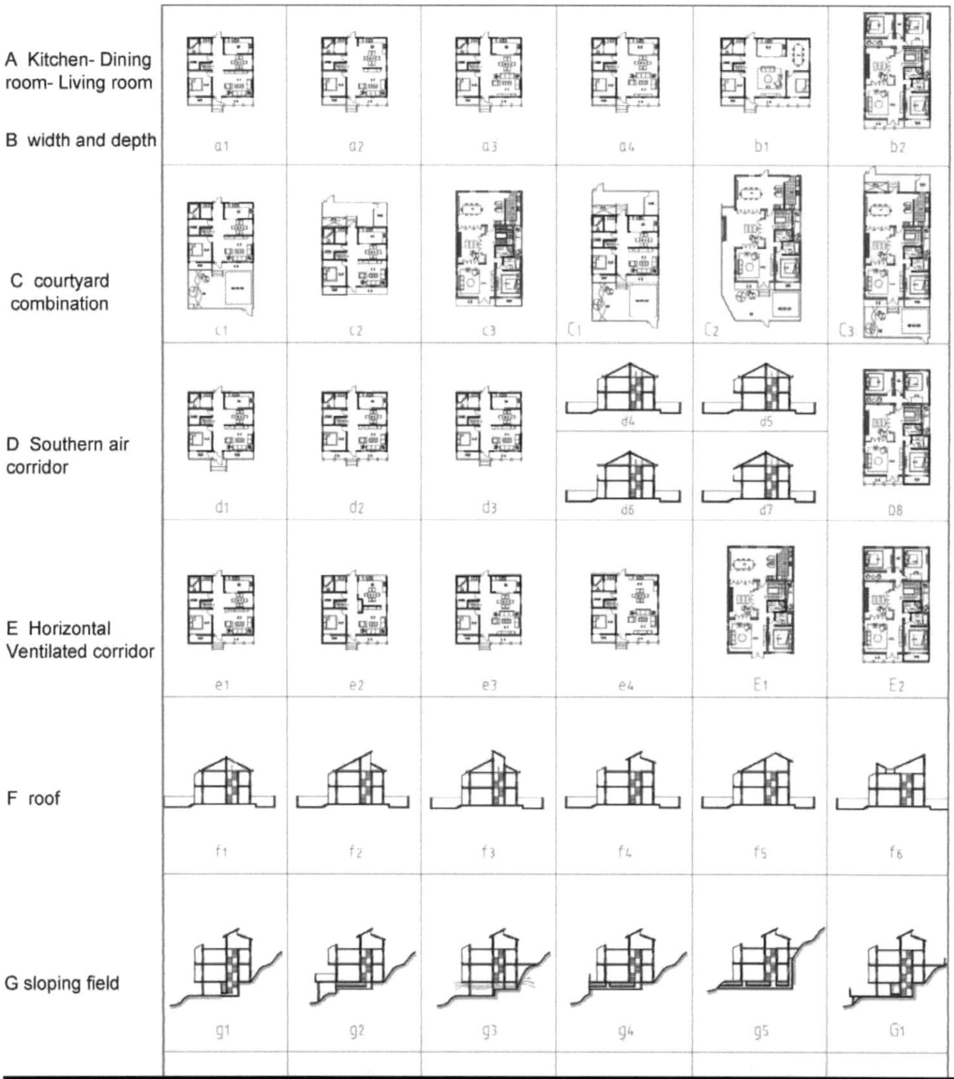

[redacted]

Figure 3-4 Extension menu (self-made)

References

[1]ZhuWei .Study of Village Settlement Space in Northern Zhejiang Basis on Geography[D].Zhejiang University,2009: 60-61.

[2] LinTao .Research on Rural Agglomeration and Settlement Space Evolution Mode of North Zhejiang.Zhejiang University,2012: 1-7.

IOP Conf. Series: Materials Science and Engineering **359** (2018) 012052 doi:10.1088/1757-899X/359/1/012052

The Reliability Estimation for the Open Function of Cabin Door Affected by the Imprecise Judgment Corresponding to Distribution Hypothesis

Z P Yu[1], Z F Yue[2], W Liu[3]

[1]PhD candidate, College of mechanics and civil engineering, Northwestern Polytechnical University, Xi'an, P. R. China

[2]Professor, College of mechanics and civil engineering, Northwestern Polytechnical University, Xi'an, P. R. China

[3]Associate Professor, College of mechanics and civil engineering, Northwestern Polytechnical University, Xi'an, P. R. China

E-mail:yuzhipu@163.com

Abstract. With the development of artificial intelligence, more and more reliability experts have noticed the roles of subjective information in the reliability design of complex system. Therefore, based on the certain numbers of experiment data and expert judgments, we have divided the reliability estimation based on distribution hypothesis into cognition process and reliability calculation. Consequently, for an illustration of this modification, we have taken the information fusion based on intuitional fuzzy belief functions as the diagnosis model of cognition process, and finished the reliability estimation for the open function of cabin door affected by the imprecise judgment corresponding to distribution hypothesis.

1. Introduction

In the reliability design of cabin door system, complicated working condition and expensive experiment cost have made experts to complete reliability estimations under the limited experiment data. Therefore, when the reliability estimation can be separated into the cognition process of distribution hypothesis and reliability calculation based on distribution hypothesis, the reliability estimation relied on expert judgments (e.g.,"It seems to obey Welbul distribution." , " It seems to obey exponential distribution.") have become an meaningful research for reliability design of engineering system.

For this kind of reliability estimation, some experts have separated expert judgments from other reliability data and put them in the group of informative prior for the reliability estimations under the Bayesian framework. For example, Rajabi etc. have has used approximate posterior distribution derived from surrogate modeling for an approach of accelerating the fuzzy Bayesian inference algorithm[1]. As the uncertainties of expert judgments, Coolean has also discussed the use of imprecise probability in reliability estimation, which have translated expert judgment into imprecise probability under Bayesian framework[2]. Moreover, Cizelj etc. have proposed a likelihood function for the Bayesian inference, which has aggregated the influence from component operating conditions on component failure rate[3].

However, as subjective judgments can be easily affected by environment factors and personal preference, rational judgment based on information fusion seems to be more convincible than

intuitional judgment based on personal preference [4–7]. Therefore, besides the quantification of single judgments, experts have also proposed subjective quantification methods based on D-S theory[8–12], Bayesian-belief net[13–15] or neuronal network[16]. Moreover, when Intuitionistic Fuzzy (IF) set is regarded as a better quantification method than fuzzy set for the subjective quantification, the information fusion based on IF sets has also attracted the experts' attention[17,18]. Consequently, different from the reliability estimation under Bayesian framework, some reliability experts have highlighted the importance of expert judgments and used Human Reliability Analysis (HRA) for the reliability estimation relied on expert judgments, which has aggregated cognition process in the calculation[19–21]. Owing this, three generations of HRA methods can be sorted by the performance shaping factor(PSF), static cognition process and dynamic cognition process[22]. Moreover, with the more engineering applications of HRA, experts have proposed more efficient cognition processes for HRA. In the reliability design of nuclear power plant disturbance, Pyy has proposed a dynamic cognitive method for HRA under the decision dimension, which have separated the cognition process of distribution hypothesis into nine different cognition stages[23]. Mkrtchyan etc. have discussed the feasibility of Bayesian belief networks in the HRA, which can aggregate the subjective information and some other types of information[24].
Therefore, to highlight the importance of expert judgments and complete the reliability estimation for the open function of cabin door, here we introduce the cognition process advised in SHARP approach in this reliability estimation process, and separate the reliability estimation into three parts: (1) Observation; (2) Diagnosis; (3) Manual actions[22]. Meanwhile, we have taken IF belief functions proposed by Song as reasoning model in the diagnosis stage, which can complete the information fusion of multiple judgments based on the IF set and D-S theory[25]. Consequently, the first part of this paper is the introductions of IF set theory and information fusion based on IF belief function. Then, the second part is used to introduce the reliability estimation for the open function of cabin door. At last, the discussion and conclusion are used to end this paper.

2. IF set

As a generation of fuzzy set theory, IF set theory proposed by Atanassov can express the hesitation within subjective judgment[26,27]. The main definitions of IF set theory can be expresses as follows:
Definition 1. Let $X=\{x_1,x_2,\cdots x_m\}$ be a finite universal set, the IF set is

$$A=\{<x_l,\mu_A(x_l),\nu_A(x_l)>| x_l\in X\}, \tag{1}$$

where the functions $\mu_A: X\mapsto[0,1]$ and $\nu_A: X\mapsto[0,1]$ are membership degree and non-membership degree of element x_l, respectively [28].
For each $x_l\in X$, if $0\le\mu_A(x_l)+\nu_A(x_l)\le 1$, let $\pi_A(x_l)=1-\mu_A(x_l)-\nu_A(x_l)$. $\pi_A(x_l)$ is the third member of IF set A, and called the intuitionistic index. Therefore, the simple form of IF set A can be written as $A=<\mu_A(x_l),\nu_A(x_l)>$.

3. The information fusion based on IF evidence functions

As a probability estimation method based on the IF set and D-S theory, Song's method can complete the information fusion of multiple imprecise judgments[25]. The core definitions of this method can be expressed as two parts:

3.1. The probability estimation

Definition 2. Let $X=\{x_1,x_2,\cdots x_n\}$ be a universe of discourse, and IF be the family of all fuzzy set correspondence with X. Then, the IF belief function of A_i is defined as

$$\{<A_i,m(A_i),\mu_{A_i}(x_j),\nu_{A_i}(x_j)>\}, \tag{2}$$

where A_i is an IF event within IF, $\mu_{A_i}(x_j)$ is the membership function of the IF set A_i, $\nu_{A_i}(x_j)$ is the non-membership function, $m(A_i)$ is the basic probability assignments (BPA) of A_i, and $j=1,2,\cdots,n$.

Definition 3. Define the probability estimation based on the different pieces of linguistic information as follows:

$$P(x_j)=[a_j,b_j], \tag{3}$$

where

$$\begin{cases} a_j=\sum_{A_i\in F}\dfrac{m(A_i)\ \mu_{A_i}(x_j)}{\sum_{j=1}^{n}(1-\nu_{A_i}(x_j))-\pi_{A_i}(x_j)} \\ b_j=\sum_{A_i\in F}\dfrac{m(A_i)(1-\nu_{A_i}(x_j))}{\sum_{j=1}^{n}\mu_{A_i}(x_j)+\pi_{A_i}(x_j)} \end{cases}, \tag{4}$$

3.2. The priority based on the probability estimations
Definition 6. Define the e_j as the priority value of $P(x_j)$ as

$$P(x_j)=[a_j,b_j], \tag{5}$$

$$e_j=(a_j+b_j)/2, \tag{6}$$

Therefore, for the judgment $P(x_{\tau_1})$ and $P(x_{\tau_2})$, if $e_{\tau_1}\geq e_{\tau_1}$, $P(x_{\tau_1})$ is better than $P(x_{\tau_2})$, where τ_1, $\tau_2\in[1,2,\cdots,N]$ and $\tau_1\neq\tau_2$.

4. The reliability estimation for the open function of cabin door system

In this paper, the cabin door system contains control system, hydraulic system and a high attitude environment. Since the reliability experiment for the open function in the high attitude environment is difficult to simulate with virtual prototype, here we choose the five parameters as the subsystem of the open function, and the reliability estimation for open function of cabin door system can be expressed as follows:

$$R_S=R_{P_1}\times R_{P_2}\times R_{P_3}\times R'_{P_4}\times R'_{P_5} \tag{7}$$

where R_S is the usage reliability of open function. R_{P_1}, R_{P_2}, R_{P_3}, R'_{P_4}, R'_{P_5} correspond to the usage reliability estimations of the viscosity coefficient of hydraulic fluid P_1, the noise of control signal P_2, the synchronous rate of actuator cylinder P_3, the turbulence environment P_4, and the gust environment P_5, respectively.

Meanwhile, to simple the calculation, define R_{P_1} as the usage reliability based on the unknown distribution hypotheses, and list the rest of subsystem reliability information in Table 1.

Table 1. Design information

Parameter	Reliability information
The viscosity coefficient of hydraulic fluid	Unknown distribution hypothesis
The noise of control signal	R_{P_2} ~Weibul(1, 1.1)
The synchronous rate of actuator cylinder	R_{P_3} ~Weibul(1, 1.2)
Random turbulence	R'_{P_4}=[0.9,0.95]
Random time of gust	R'_{P_5} =[0.9,0.95]

Therefore, based on these background, the reliability estimation for the open function of cabin door system can be expressed as

$$R_S = R_{P_1} \times R_{P_2} \times R_{P_3} \times R'_{P_4} \times R'_{P_5}, \tag{8}$$

where

$$\begin{cases} R_{P_1} = \bar{R}_{P_1} \times P_1 \\ R_{P_3} = 1.1x^{0.1}e^{-x^{1.1}} \\ R_{P_4} = 1.2x^{0.2}e^{-x^{1.2}} \\ R'_{P_4} = R'_{P_5} = [0.9, 0.95] \end{cases}, \tag{9}$$

$\bar{R}_{P_1}$ is the reliability estimation based on the distribution hypothesis, and P_1 is the probability estimation for the expert judgment corresponding to the hypothesis distribution.

Moreover, based on the cognition process advised in SHARP approach[22], the flow chart of this reliability estimation can be expressed as figure 1, and the calculation of this reliability estimation can be expressed as follows:

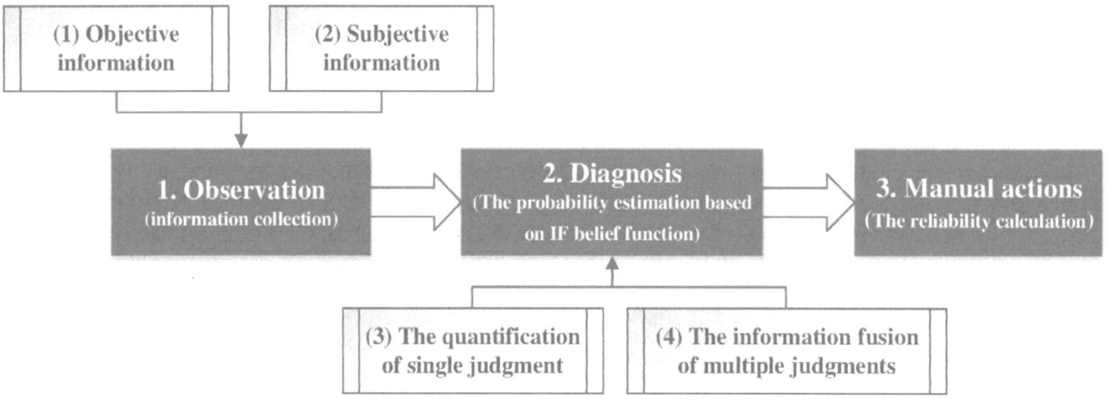

Figure 1. Flow chart of reliability estimation

4.1. Information collection

In figure 2, expert needs to find the most appropriate Pareto distribution hypotheses for P_1, where has presented the five experiment data and fitting results with the time interval (30 days) as abscissa and the reliability estimation the corresponding absorbency.

4.2. The quantification for the basic subjective judgment

Based on figure 2 and expert experience, the subjective information corresponding to each distribution hypothesis has been calculated with equation 3, which have been aggregated in table 2.

4.3. The probability estimation based on IF belief functions

With equations (3-6), the probability estimation and priority value of distribution hypothesis can be calculated in table 3. As the priority value of Weibul (1,1) is the biggest, the distribution hypothesis of P_1 can be regarded as Weibul (1,1), and the probability estimation of expert judgments is [0.34,0.41].

4.4. The reliability estimation at $x = 0.4$

With the cognition process above, it is easy to calculate the usage reliability of open function. Therefore, when usage time $x = 0.4$, the reliability estimation can be calculated as

$$R_S = [a^1, b^1] \times \overline{R}_{P_1} \times 1.32x^{0.3}e^{-x^{1.1}-x^{1.2}} \times R_4^{'} \times R_5^{'} = 1.32 \times [a^1, b^1]\overline{R}_{P_1} \times x^{0.3}e^{-x-x^{1.1}-x^{1.2}} \times R_4^{'} \times R_5^{'} \quad (10)$$

where

$$\begin{cases} [a^1, b^1] = [0.34, 0.41] \\ R_{P_4}^{'} = R_{P_5}^{'} = [0.9, 0.95] \end{cases}, \quad (11)$$

Consequently, the usage reliability of open function at $x = 0.4$ is R_S=[0.0162,0.049].

Table 2. The expert information

Expert judgment	BPA	Frame of discernment	Expert preference	Expert judgment in the form of IF set
It may be Weibul(1,1) or Weibul(0.9,0.7).	m_1^1=0.6	Weibul（1，1）	0.75	[0.80,0.10]
		Gamma（0.642,0.333）	0.35	[0.40,0.50]
		Weibul（0.9，0.7）	0.65	[0.70,0.20]
It may be Weibul(0.9,0.7).	m_2^1=0.3	Weibul（1，1）	0.85	[0.90,0]
		Gamma（0.642,0.333）	0.35	[0.40,0.50]
		Weibul（0.9，0.7）	0.75	[0.80,0.10]
It may be Gamma(0.642,0.333).	m_3^1=0.1	Weibul（1，1）	0.25	[0.30,0.60]
		Gamma（0.642,0.333）	0.45	[0.50,0.40]
		Weibul（0.9，0.7）	0.55	[0.60,0.30]

Table 3. The probability estimations

Index	The probability estimation of expert judgment	The priority value
Weibul(0.9,0.7)	[0.34,0.41]	0.375
Gamma(0.642,0.333)	[0.20,0.26]	0.23
Weibul(1,1)	[0.4,0.56]	0.48

5. Discussion

5.1. The feasibility

In figure 3, two reliabilities of the viscosity coefficient of hydraulic fluid have been used as a comparison. The reliability ignored the influence from expert judgments have supplied an acceptable estimation result at $x = 0.4$, and the reliability under IF belief function and IF set has supplied a dangerous estimation result. As the core of this calculation is just a prediction based on the limited experiment data, a dangerous estimation result seems to be convincible and acceptable. Therefore, to emphasize the influence from expert judgment to reliability estimation, this modification can be regarded as feasible.

5.2. The limitation

As the flow chart has described, although the dynamic cognition process may be more appropriate for the human cognition process, we have used the static cognition process to simulate the cognition process in the reliability estimation. This is because the limit experiment data may cause not only the uncertainties if fitting results but also the short of expert experience. Therefore, this static cognition process is more realistic than the dynamic cognition process for the reliability estimation under the limited experiment data.

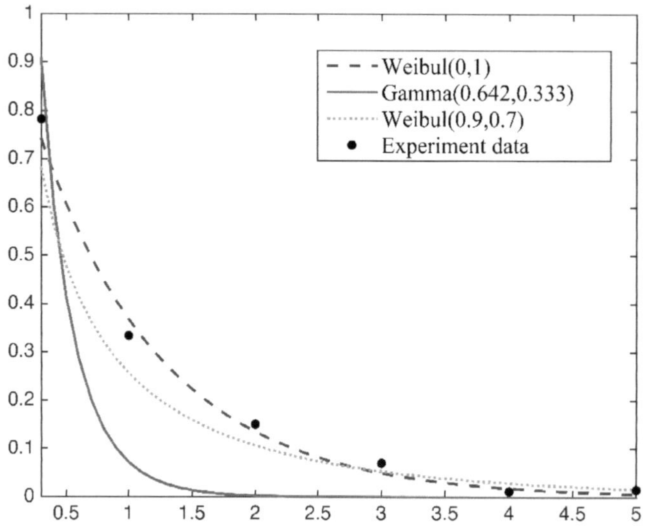

Figure 2. Fitting results based on experiment data

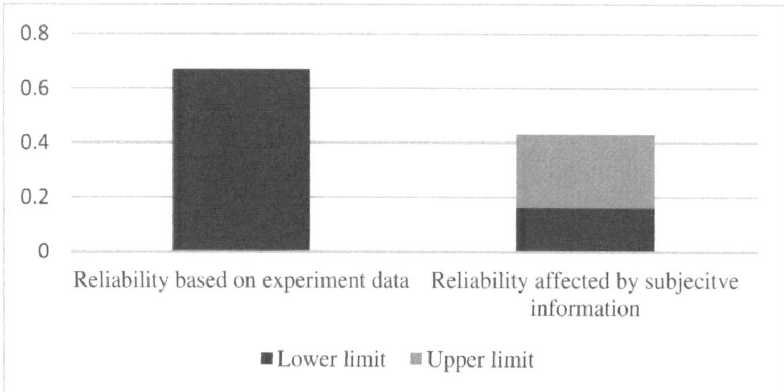

Figure 3. The comparison of reliabilites

6. Conclusion

This paper has completed the reliability estimation for the open function of cabin door system, which is affected by the imprecise judgment corresponding to distribution hypothesis. As this reliability estimation is based on the joint probability of expert judgment and traditional reliability, it is not only a reliability estimation for an engineering design but also a meaningful application of HRA, which is meaningful for the engineering design with limited data and less intelligent cognition process.

References

[1] Rajabi M M and Ataie-Ashtiani B 2016 Efficient fuzzy Bayesian inference algorithms for incorporating expert knowledge in parameter estimation *J. Hydrol.* **536** 255–72

[2] Coolen F P A 2004 On the Use of Imprecise Probabilities in Reliability *Qual. Reliab. Eng. Int.* **20** 193–202

[3] Cizelj R J, Mavko B and Kljenak I 2001 Component reliability assessment using quantitative and qualitative data *Reliab. Eng. Syst. Saf.* **71** 81–95

[4] LeBoeuf R A, Shafir E and Bayuk J B 2010 The conflicting choices of alternating selves *Organ. Behav. Hum. Decis. Process.* **111** 48–61

[5] Lieberman M D 2000 Intuition: A social cognitive neuroscience approach. *Psychol. Bull.* **126** 109–37

[6] Thompson V A, Prowse Turner J A and Pennycook G 2011 Intuition, reason, and metacognition *Cognit. Psychol.* **63** 107–40
[7] Wang P 1996 Heuristics and normative models of judgment under uncertainty *Int. J. Approx. Reason.* **14** 221–35
[8] Elouedi Z, Mellouli K and Smets P 2004 Assessing sensor reliability for multisensor data fusion within the transferable belief model *IEEE Trans. Syst. Man Cybern. Part B Cybern.* **34** 782–7
[9] Shafer G 2016 Dempster's rule of combination *Int. J. Approx. Reason.* **79** 26–40
[10] Smets P 2005 Belief functions on real numbers *Int. J. Approx. Reason.* **40** 181–223
[11] Xiao W, Cheng W, Zi Y, Zhao C, Sun C, Liu Z, Chen J and He Z 2015 Support evidence statistics for operation reliability assessment using running state information and its application to rolling bearing *Mech. Syst. Signal Process.* **60–61** 344–57
[12] Zouhal L M and Denoeux T 1998 An evidence-theoretic k-NN rule with parameter optimization *Ieee Trans. Syst. Man Cybern. Part C-Appl. Rev.* **28** 263–71
[13] Cadini F and Gioletta A 2016 A Bayesian Monte Carlo-based algorithm for the estimation of small failure probabilities of systems affected by uncertainties *Reliab. Eng. Syst. Saf.* **153** 15–27
[14] Jaynes E T 1977 Confidence intervals vs. Bayesian intervals *Found. Probab. Theory Stat. Inference Stat. Theor. Sci. Proc Intern. Res Colloq Univ West. Ont. Lond. Ont 1973 Vol II* 175–257. Univ. Western Ontario Ser. Philos. Sci., Vol. 6
[15] Sun Z, Gong E, Li Z, Jiang Y and Xie H 2013 Bayesian estimator of human error probability based on human performance data *J. Syst. Eng. Electron.* **24** 242–9
[16] Nahar J, Imam T, Tickle K S and Chen Y-P P 2013 Computational intelligence for heart disease diagnosis: A medical knowledge driven approach *Expert Syst. Appl.* **40** 96–104
[17] Feng T, Mi J-S and Zhang S-P 2014 Belief functions on general intuitionistic fuzzy information systems *Inf. Sci.* **271** 143–58
[18] Grzegorzewski P 2013 On some basic concepts in probability of IF-events *Inf. Sci.* **232** 411–8
[19] Dai L, Li Z and Li P 2011 HRA in China: Model and data *Saf. Sci.* **49** 468–72
[20] Reason J 2000 Human error *West. J. Med.* **172** 393–6
[21] Swain A D 1990 Human reliability analysis: Need, status, trends and limitations *Reliab. Eng. Syst. Saf.* **29** 301–13
[22] Hannaman G W and Spurgin A J 1984 Systematic Human Action Reliability Procedure (SHARP)
[23] Pyy P 2000 An approach for assessing human decision reliability *Reliab. Eng. Syst. Saf.* **68** 17–28
[24] Mkrtchyan L, Podofillini L and Dang V N 2015 Bayesian belief networks for human reliability analysis: A review of applications and gaps *Reliab. Eng. Syst. Saf.* **139** 1–16
[25] Song Y and Wang X 2015 Probability Estimation in the Framework of Intuitionistic Fuzzy Evidence Theory *Math. Probl. Eng.* **2015** 1–10
[26] Atanassov K T 2012 *On Intuitionistic Fuzzy Sets Theory* vol 283 (Berlin, Heidelberg: Springer Berlin Heidelberg)
[27] Atanassov K T 1999 Intuitionistic Fuzzy Sets *Intuitionistic Fuzzy Sets: Theory and Applications* ed K T Atanassov (Heidelberg: Physica-Verlag HD) pp 1–137
[28] Zadeh L A 1968 Probability measures of Fuzzy events *J. Math. Anal. Appl.* **23** 421–7

Acknowledgments
We would like to thank all referees for their valuable comments and suggestions to improve the readability and quality of this paper.

IOP Conf. Series: Materials Science and Engineering **359** (2018) 012053 doi:10.1088/1757-899X/359/1/012053

Extended Importance Sampling for Reliability Analysis under Evidence Theory

X K Yuan[1], B Chen[1], B Q Zhang[1]

[1]School of Aerospace and Engineering, Xiamen University, Xian 710072, China

Abstract. In early engineering practice, the lack of data and information makes uncertainty difficult to deal with. However, evidence theory has been proposed to handle uncertainty with limited information as an alternative way to traditional probability theory. In this contribution, a simulation-based approach, called 'Extended importance sampling', is proposed based on evidence theory to handle problems with epistemic uncertainty. The proposed approach stems from the traditional importance sampling for reliability analysis under probability theory, and is developed to handle the problem with epistemic uncertainty. It first introduces a nominal instrumental probability density function (PDF) for every epistemic uncertainty variable, and thus an 'equivalent' reliability problem under probability theory is obtained. Then the samples of these variables are generated in a way of importance sampling. Based on these samples, the plausibility and belief (upper and lower bounds of probability) can be estimated. It is more efficient than direct Monte Carlo simulation. Numerical and engineering examples are given to illustrate the efficiency and feasible of the proposed approach.

1. Introduction

The aim of conducting structural reliability analysis and optimization design is to make the optimum choice between the economy and reliability, considering various inevitable uncertainty factors of the structure, like boundary conditions, material properties, random load and so on. Most uncertainties can be divided into two kinds, objective uncertainty and epistemic uncertainty [1]. Objective uncertainty can be studied under probability model after a large number of data statistics. Thus reliability analysis [2][3] and optimization design [4] methods for random uncertainty has been well developed. Epistemic uncertainty is due to the lack of data and information, this kind of problems, it is usually more difficult to handle. There is a few theories can be used to study epistemic uncertainty based on non-probabilistic description, like evidence theory (Dempster-Shafer theory)[5]-[7], possibility theory[8][9], fuzzy theory[10], convex models [11][12] and so on.

Evidence theory can be used to effectively deal with uncertainty information, like interval, fuzziness, random and so on. In addition, reliability problem based on probability theory can be a particular case of evidence theory. Thus, evidence theory is an effective way to handle discontinuous variables. However, due to the cost of calculation [13], research achievement in the area of reliability analysis under evidence theory is rare, although some progress has been made in this research orientation. Bae et al. [13][14]built multi-point approximate surface based on a certain point at the limit state surface, in order to solve reliability analysis under the epistemic uncertainty. Du [15] et al proposed a new reliability-analysis method that probability and evidence theory are combined. Jiang et al. [16] proposed a structural reliability method using evidence theory by introducing a non-probabilistic reliability index approach. Nevertheless, high efficient method for reliability analysis still need to be developed.

2. Reliability analysis based on evidence theory

For simplicity, a two-dimensional case is used to describe the propagation process of evidence theory

$$y = g(\boldsymbol{x}) = g(x_1, x_2) \tag{1}$$

where $\boldsymbol{x} = [x_1, x_2]^{\mathrm{T}}$, $x_1 \in A, x_2 \in B$ are two independent variables of the problem(A, B are two sets); y is the output; $g(\boldsymbol{x})$ is the limit state function. Let $m(x_1), m(x_2)$ be the basic probability assignment (BPA) of variables x_1 and x_2, respectively. The combined BPAs of both x_1 and x_2 can be calculated by Dempster's rule. Defining a vector $c_{ij} = [x_{1i}, x_{2j}]$ where i and j indicate focal elements, the BPA of C can be defined as:

$$C = A \times B = \{c_{ij} = [x_{1i}, x_{2j}], x_{1i} \in A, x_{2j} \in B\} \tag{2}$$

Considering the independence of x_1 and x_2, the BPA of C can be calculated by:

$$m(c_{ij}) = m(x_{1i})m(x_{2j}) \tag{3}$$

In reliability analysis, suppose the safe domain G is defined as

$$G = \{y : y = g(x_1, x_2) > 0, c_{ij} = [x_{1i}, x_{2j}] \subset C\} \tag{4}$$

Then $Bel(G)$ and $Pl(G)$ can be calculated by:

$$Bel(G) = \sum_{c_{ij} \subseteq G} m(c_{ij}) \tag{5}$$

$$Pl(G) = \sum_{c \cap G \neq 0} m\left(c_{ij}\right) \tag{6}$$

where $c \subseteq G$ indicates that the joint focal element c_{ij} must be entirely within the domain G, while $c_{ij} \cap G \neq 0$ means that c_{ij} entirely or partly within G. However, it's difficult to deal with $c_{ij} \subseteq G$ and $c_{ij} \cap G \neq 0$. In order to determine whether c_{ij} satisfies $c_{ij} \subseteq G$ or $c_{ij} \cap G \neq 0$, the following equation should be calculated:

$$[y_{\min}, y_{\max}] = [\min_{\boldsymbol{x} \in c_{ij}} g(\boldsymbol{x}), \max_{\boldsymbol{x} \in c_{ij}}(\boldsymbol{x})] \tag{7}$$

If $y_{\min}$ and $y_{\max}$ are both greater than zero, then it means that at this focal element interval, the joint BPA $m(c_{ij})$ should be reckoned in both $Bel(G)$ and $Pl(G)$. However, if $y_{\min}$ is less than zero while $y_{\max}$ is greater than zero, $m(c_{ij})$ is only reckoned in $Pl(G)$. If $y_{\min}$ and $y_{\max}$ are both less than zero, $m(c_{ij})$ should not reckoned in both $Bel(G)$ and $Pl(G)$. Through the above analysis, it can be seen that the calculation should be carried out at all focal element intervals. This is the difficulty of the reliability analysis under evidence theory. Approximate method [17]-[19] can be used to calculate the linear or approximate linear problem. However, for highly non-linear or complicated cases, simulation based method should be adopted.

3. Extended Importance sampling for reliability analysis under evidence theory

In order to achieve compute efficiency, traditional importance sampling method can be extended to solve the problem with epistemic uncertainty. The key of importance sampling for reliability analysis lies in the identifying of the important region and setting sampling centre on it. In this contribution, the 'importance region' is also identified which has larger joint BPA and is in the vicinity of limit state function. This can be done by introducing an 'equivalent' auxiliary sampling density according to the joint BPA, and then solve the design point, and the region around the design point indicates the region contributing most to *Bel* and *Pl*. This is called 'extended importance sampling' in this paper, and it is illustrated as follows.

(1) Construct auxiliary sampling density

In order to apply traditional reliability analysis to solve problem with epistemic uncertainty, it is necessary to introduce an auxiliary sampling density (ASD) for each epistemic uncertainty variable. Generally, The ASD should be chosen according to the BPA of each variable. The principle of determine the ASD is that it reflects the amount of probability of BPA. For example, one optional way is given as

$$BPA([x_a, x_b]) = p(x_b) - p(x_a) \tag{8}$$

where $p(x)$ is the introduced ASD of variable x; x_a and x_b are interval endpoints. Eq.(8) guarantees that the ASD can reflect the probability assignment of the BPA.

(2) Solve the design point

The design point is solved in 'equivalent' reliability problem with variables are distributed as auxiliary sampling density $p(\boldsymbol{x})$. This can be done by using advance first order second moment method (AFOSM) or other optimization algorithm.

(3) Carry out importance sampling

Construct the importance sampling density $H(\boldsymbol{x})$ based on the design point, and generate the samples of variables according to $H(\boldsymbol{x})$.

(4) Calculate the $y_{\min}$ and $y_{\max}$ in each focal element

For every generated sample, say, $\boldsymbol{x}^{(j)}$, first determine which focal element c' it locates in, and then decide whether or not to calculate the limit state function *G* by judging which case the focal element c' is.

Case 1: $0 < y_{\min} < y_{\max}$, compute $y^{(j)} = g(\boldsymbol{x}^{(j)})$, update the $y_{\min}$ and $y_{\max}$ in current focal element;

Case 2: $y_{\min} < 0 < y_{\max}$, no more compute;

Case 3: $y_{\min} < y_{\max} < 0$, compute $y^{(j)} = g(\boldsymbol{x}^{(j)})$, update the $y_{\min}$ and $y_{\max}$ in current focal element.

Note that the $y_{\min}$ and $y_{\max}$ in current focal element is calculated according to the samples falling in this element.

(5) Estimate the $Bel(G)$ and $Pl(G)$

By using eqs. (5) and (6), the measures $Bel(G)$ and $Pl(G)$ can be obtained. They can be viewed as the lower and upper bounds of the true reliability P_r, which satisfies

$$Bel(G) \le P_r \le Pl(G) \tag{9}$$

Note that if we use auxiliary sampling density $p(\boldsymbol{x})$ to directly generate samples in step (2) and (3), it can be seen as direct Monte Carlo simulation (MCS) method.

4. Examples

In this section, example is given to demonstrate the effectiveness of the proposed EIS method. Direct Monte Carlo and other methods are also used to compare with the proposed method.

4.1. A ten-bar truss

Consider a ten-bar truss structure shown in Fig. 1. The length L=360 in, and the modulus of elasticity E= 15,000 ksi. A_j ($j = 1,2,\ldots,10$) denotes the cross-section area, and $A_j = 10$ in ($j = 7,8,9,10$). The truss is subjected to three forces, F_1=100kip, F_2=120kip and , F_3=400kip. The limit state function is given by

$$g = d_y - \delta_2\left(A_1, A_2, A_3, A_4, A_5, A_6\right) \tag{10}$$

where $d_y = 1.5$ in is the allowable displacement; δ_2 is the vertical displacement of joint 2 which is the function of $A_1, \dots, A_6$ and

$$\delta_2 = \sum_{i=1}^{6} \frac{N_i^0 N_i}{A_i} + \sqrt{2} \sum_{i=7}^{10} \frac{N_i^0 N_i}{A_i} \frac{L}{E} \tag{11}$$

where $N_j, j = 1,2, \dots, 10$ are forces and can be obtained by solving the following equations.

$$\begin{aligned} & N_1 = F_2 - \sqrt{2}/2N_8, N_2 = -\sqrt{2}/2N_{10}, N_3 = -F_1 - 2F_2 + F_3 - \sqrt{2}/2N_8, \\ & N_4 = -F_2 + F_3 - \sqrt{2}/2N_{10}, N_5 = -F_2 - \sqrt{2}/2N_8 - \sqrt{2}/2N_{10}, \\ & N_6 = -\sqrt{2}/2N_{10}, N_7 = \sqrt{2}(F_1 + F_2) + N_8, N_8 = (a_{22}b_1 - a_{12}b_2)/(a_{11}a_{22} - a_{12}a_{21}), \\ & N_9 = \sqrt{2}F_2 + N_{10}, N_{10} = (a_{11}b_2 - a_{21}b_1)/(a_{11}a_{22} - a_{12}a_{21}) \end{aligned} \tag{12}$$

And

$$\begin{aligned} & a_{11} = \left(1/A_1 + 1/A_3 + 1/A_5 + 2\sqrt{2}/A_7 + 2\sqrt{2}/A_8\right)L/2E, a_{12} = a_{21} = L/2A_5E, \\ & a_{22} = \left(1/A_2 + 1/A_4 + 1/A_6 + 2\sqrt{2}/A_9 + 2\sqrt{2}/A_{10}\right)L/2E, \\ & b_1 = (F_2/A_1 - (F_1 + 2F_2 - F_3)/A_3 - F_2/A_5 - 2\sqrt{2}(F_1 + F_2)/A_7)\sqrt{2L}/2E \\ & b_2 = (\sqrt{2}(F_3 - F_2)/A_4 - \sqrt{2}F_2/A_5 - 4F_2/A_9)L/2E \end{aligned} \tag{13}$$

The cross-section areas $A_1, \dots, A_6$ are treated as evidence variables, and the BPA structures are given in Table 1.

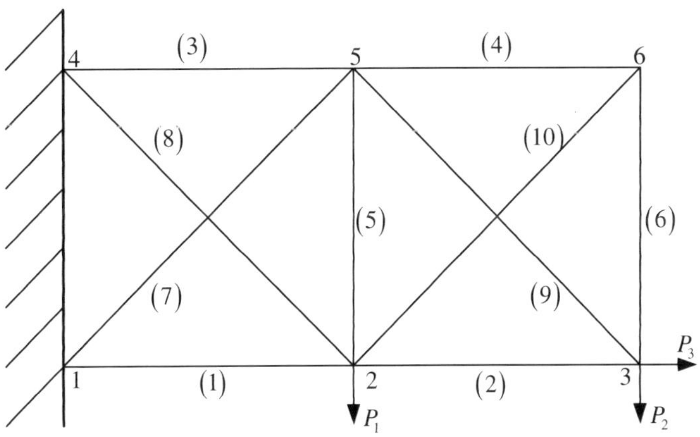

Figure 1. Ten-bar structure

Table 1. The BPA structure of evidence variables

A_1、A_2、A_3 、A_4、A_5、A_6	
Interval	BPA
[2.0 6.0]	0.09
[6.0 10.0]	0.41
[10.0 14.0]	0.41
[14.0 18.0]	0.09

The proposed EIS and direct MCS are used to calculate the *Bel*(*G*) and *Pl*(*G*). The computed results are is summarized in Table 2. In order to compare the method more clearly, the mean values and standard deviations of *Bel*(*G*) and *Pl*(*G*) estimates are also given, which are obtained by carrying out the analysis method for a number of times, say 10, in this example. Also, the optimization methods, i.e., Active Set and Genetic Algorithm, are applied to solve this problem. The results of different methods are listed in Table 2.

Table 2. The statistical results of running 10 times by using different methods

	Bel(*G*)		*Pl*(*G*)		Averaged No. of calls
	Mean	Standard deviation	Mean	Standard deviation	
EIS	0.6072	0.0082	0.9686	0.0018	10^4
MCS	0.6701	0.0045	0.9722	0.0010	10^5
Active Set	0.6064	---	0.9926	---	4.68×10^5
Genetic Algorithm	0.6059	---	0.9926	---	1.94×10^8

It can be seen from Table 2 that the results of different methods are approximately consistent with each other. Among which, the proposed obtains the results at the least computed cost which uses 10^4 function calls. By contrast, Direct MCS method uses 10^4 function calls. Note that as the number of focal elements in this example is 4^6= 4096, the optimization approach should be carried out at each focal element twice to find the minima and maximum value of the limit state function, thus the computational cost is very huge. As seen in Table 2, Active Set approach uses 4.68×10^5 function calls, while Genetic Algorithm uses more, 1.94×10^8. The proposed method can obtains the satisfied results at moderate computational cost, thus the feasibility and efficiency of the proposed method is shown.

5. References

[1] Agarwal H, Renaud J E and Preston E L 2004 Uncertainty quantification using evidence theory in multidisciplinary design optimization *J. Reliability Engineering and System Safety*. **85** 281-294

[2] Hasofer AM and Lind NC 1974 Exact and invariant second-moment code format *J. ASME J Eng Mech Div*. **100** 111-121

[3] Breitung KW 1984 Asymptotic approximation for multi normal integrals *J. ASCE J Eng Mech*. **110(3)** 357-366

[4] Younes Aoues and Alaa Chateauneuf 2010 Benchmark study of numerical methods for reliability-based design optimization *J.Struct Multidisc Optim* **41** 277–294

[5] Shafer G. A 1976 Mathematical Theory of Evidence NJ: Princeton

[6] Oberkampf WL, Helton JC and Sentz K 2001 Mathematical representation of uncertainty *AIAA* 2001-1645

[7] Sentz K and Ferson S 2002 Combination of evidence in Dempster-Shafer theory *SAND2002-0835 Report* Sandia National Laboratories,

[8] Klir GJ and Wierman MJ 1999 Uncertainty-based Information-Elements of Generalized Information Theory Heidelberg: Physica-Verlag

[9] Klir GJ 2004 Generalized information theory: Aims, results, and open problems *J. Reliab Eng Syst Safe* **85(1-3)** 21-38

[10] Zadeh L 1965 Fuzzy sets *Inf Control* **8** 338-353

[11] Benhaim Y and ElishakoI 1990 Convex models of uncertainties in applied mechanics Amsterdam: Elsevier Science Publisher

[12] Guo X, Bai W, Zhang WS, et al 2009 Confidence structural robust design and optimization under stiness and load uncertainties *J. Comput Method in Appl M* **198** 3378-3399

[13] Bae HR, Grandhi RV and Canfield RA 2004 Epistemic uncertainty quantification techniques including evidence theory for large-scale structures *J. Comput Struct* **82** 1101-1112

[14] Bae HR, Grandhi RV and Canfield RA 2004 An approximation approach for uncertainty quantification using evidence theory *J. Reliab Eng Syst Safe* **86** 215-225

[15] Du X P 2006 Uncertainty analysis with probability and evidence theories In: Proceedings of DETC/CIE International Design Engineering Technical Conference & Computers and Information in Engineering Conference, September 10–13, Philadelphia, Pennsylvania, USA

[16] Jiang C, Zhang Z, Han X. 2013 An evidence-based reliability analysis method for uncertain structures. *Chin J Theor Appl Mech* **45(1)** 103–15.
[17] Dong, W. M., and Shah and H. C 1987 Vertex Method for Computing Functions of Fuzzy Variables *Fuzzy Sets Syst* 24 pp 65–78.
[18] Zissimos P M and Zhou J 2006 A design optimization method using evidence theory *J.ASEM Journal of Mechanical Design* **128(7)** 901 ～ 908
[19] Agarwal H， Renaud J E and Preston E L 2002 Trust Region Managed Reliability Based Design Optimization Using Evidence Theory In: 44th Structures， Structural Dynamics， and Materials Conference

Acknowledgments
The authors would like to acknowledge financial support from the National Natural Science Foundation of China (Grant No. U1530122 and No. 51505398).

IOP Conf. Series: Materials Science and Engineering **359** (2018) 012054 doi:10.1088/1757-899X/359/1/012054

Reliability Analysis of Sealing Structure of Electromechanical System Based on Kriging Model

F Zhang[1]*, Y M Wang[1], R W Chen[2], W W Deng[1], Y Gao[1]

1.School of Mechanics, Civil Engineering and Architecture, Northwestern Polytechnical University, Xi'an 710129, China

2. Broad Homes Industrial International Co., Ltd, Changsha, China

Abstract. The sealing performance of aircraft electromechanical system has a great influence on flight safety, and the reliability of its typical seal structure is analyzed by researcher. In this paper, we regard reciprocating seal structure as a research object to study structural reliability. Having been based on the finite element numerical simulation method, the contact stress between the rubber sealing ring and the cylinder wall is calculated, and the relationship between the contact stress and the pressure of the hydraulic medium is built, and the friction force on different working conditions are compared. Through the co-simulation, the adaptive Kriging model obtained by EFF learning mechanism is used to describe the failure probability of the seal ring, so as to evaluate the reliability of the sealing structure. This article proposes a new idea of numerical evaluation for the reliability analysis of sealing structure, and also provides a theoretical basis for the optimal design of sealing structure.
Keywords: Y-type seal ring, Kriging model, reliability analysis, failure probability

1. Introduction

With the rapid development of the aerospace industry, the reliability requirements of the aircraft are getting higher and higher than any time. Sealing structure, a more critical part of the aircraft, its failure would cause the incredible malfunction for aircraft, affecting the completion of the task, and even crash. In June 10, 1996, when the flight 8105 was flying at an altitude of 8000 meters from Xiamen to Beijing in China, Cockpit seal failure suddenly occurred due to mechanical failure, which result the extremely dangerous situation in acute high-altitude anoxia and high-altitude decompression. Statistics has shown that in the typical incidents regarding to the failure of typical components of aircraft hydraulic system, over 45% of them are mainly caused by non-sealing. Therefore, more researchers has paid their attention to the problem of failure of the sealing structure. In order to improve flight safety in large, there is a necessary to evaluate the reliability of the sealing structure for the aircraft electromechanical system in the field[1-3].

Lots of literatures on the seal structure has been delivered, and the main research view focus on the study of constitutive model and structural parameters, and the discussion of failure criteria and so on[4-6]. And also, some research have explored numerical simulation using the relatively complete reciprocating seal theory, predicted the life value of the sealing structure, and got some referenced results[7-9]. Wang [10] has studied the status of rubber seal, mechanical seal and packing seal technology and then realized the numerical solution of the model through finite element and numerical iteration technology. The thermal deformation rules of the sealing face under the thermo-mechanical coupling effects were studied, and the structure of the sealing gasket and the loading analysis were analyzed as well. Liu [11] has explored the sealing performance of the rectangular rubber seal ring based on the finite element analysis software Abaqus, from which the researcher found both material characteristics and the friction factor are the important factors that could affect the sealing performance. In addition, Xing [12] analyzed the deformation and force condition of Y-type hydraulic seal ring under different working pressures, and obtained the relationship between the maximum contact pressure and oil pressure from the research of sealing performance of Y-ring using finite element analysis software. Thus, the pressure distribution and deformation between seal coupling under different working pressures has already been analyzed by research. According to the sealing principle of Y-ring, the relationship between structural design and working environment was been studied by former researchers.

Based on the discussion above, it not difficulty to see the sealing technology has conducted many research conclusions in this field[13]. There are, however, not too much research literatures about the failure probability of the seal ring, especially for the Y-ring which is an important aspect of reciprocating seal structure [14,15]. The Y-ring has good performance in long service life, and dynamic pressure seal in it is 5-10 times, which is higher than the traditional rubber sealing products. Under certain conditions, it has the same life with sealing matrix. And the dynamic and static friction are equal, which are smaller than the frictional resistance of O-ring. Thus, the research on the sealing performance of Y-ring has a great significance meaning in the field.

In this paper, the contact stress between the rubber sealing ring and the hydraulic cylinder wall is calculated by the finite element method, and the relationship between the contact stress and the pressure of the hydraulic medium is obtained, and then compare the friction force under different working conditions. What's more, the adaptive Kriging model has obtained by EFF learning mechanism through the co-simulation, which is used to calculate the failure probability of the seal ring. And finally, it can be seen that it is feasible and effective to analyze the reliability of the seal ring using this kind of method.

2. Finite element model analysis of seal ring

2.1. The establishment of geometric model

The sealing structure described below is composed of a Y-shaped sealing ring, a shaft and a sealing groove, and the three-dimensional model is shown in Fig. 1. In the process of sealing, the model needs to consider the squeezing action between the seal ring and the piston rod and the seal groove, which involves the nonlinear deformation simulation of the rubber material and the contact analysis between the seal ring and the rigid body. In this research, the material in this model is NBR, and the parameters Mooney-Rivlin function are used as the strain energy function of the rubber structure (C1 and C2 are 1.87 and 0.47, respectively), and finally shafts and seal grooves can be regarded as rigid bodies.

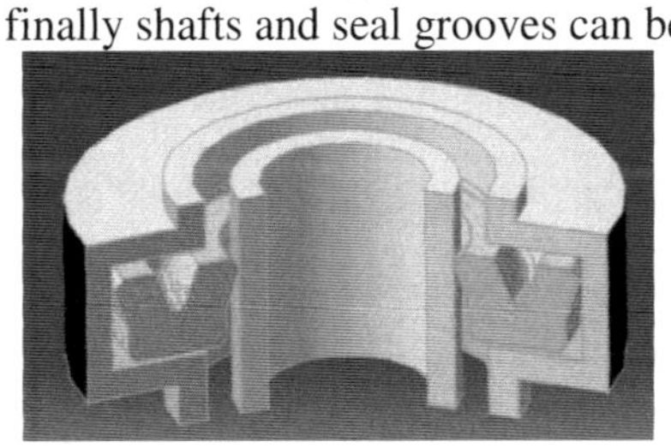

Fig. 1 Three-dimensional model of the sealing structure

2.2. The definition of material and boundary condition

This study intends to analyze the sealing performance of the sealing structure using the finite element analysis software which is called ANSYS. Based on the facts of geometry, material and boundary conditions of the seal structure, the three-dimensional model can be simplify analyzed into the axisymmetric model. For one thing, the elastic modulus of the sealing ring is 13.8 MPa and the Poisson's ratio is 0.498. For another, the modulus of elasticity E1 of the shaft and the sealing groove are 210 GPA, and the Poisson's ratio of the shaft and the sealing groove is taken as 0.33, and the sealing groove chamfer R is 1.5 mm. Therefore, the elastic modulus of them is much larger than the sealing rubber ring, and then its deformation can be neglected. The two-dimensional meshing model is shown in Fig. 2.

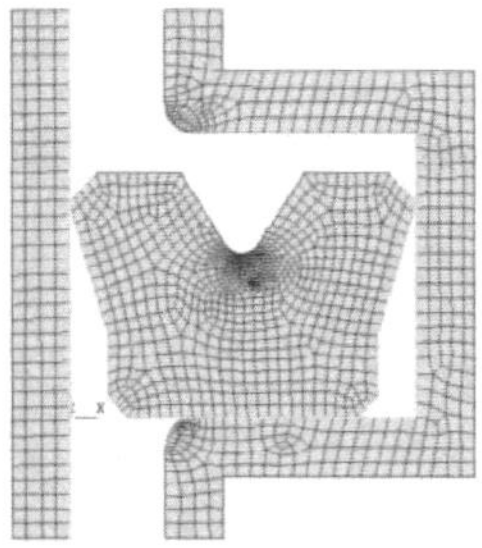

Fig. 2 Two-dimensional finite element model of sealing structure

2.3. The deterministic results analysis of finite element method

In the deterministic analysis, the X-direction displacement is 2mm and the applied medium pressure is 2MPa. It can be seen from the contact stress cloud chart that the easily area adapting to gap bites is located in the upper left and upper right of the sealing ring, exactly lip.

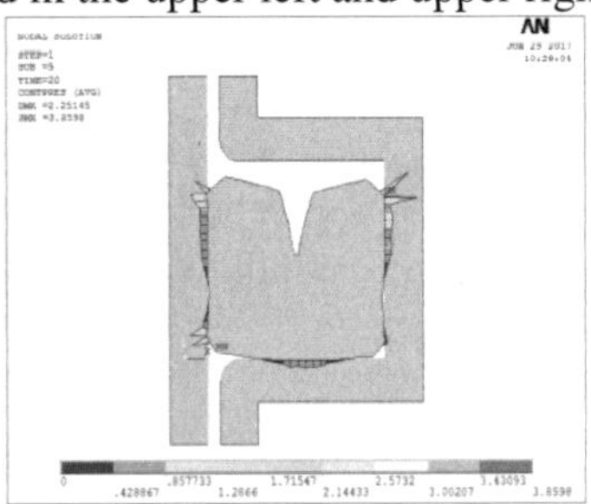

Fig. 3 Von Mises stress cloud chart in the installed state

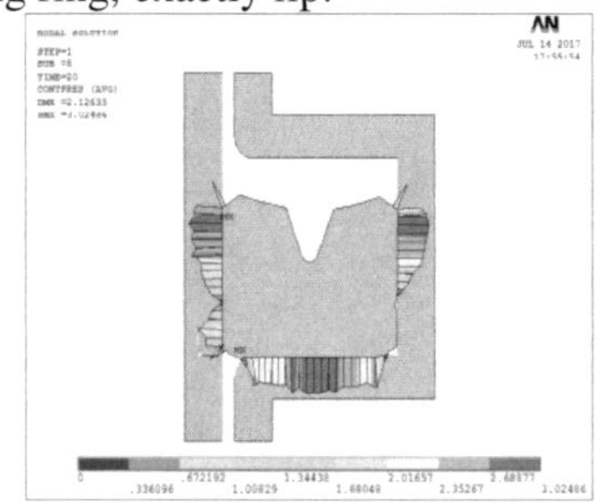

Fig. 4 Contact stress cloud diagram in static pressure working state

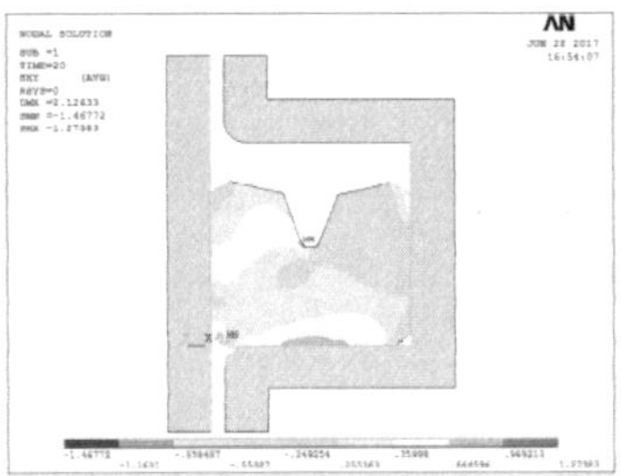

Fig. 5 XY shear diagram under 2 MPa medium pressure

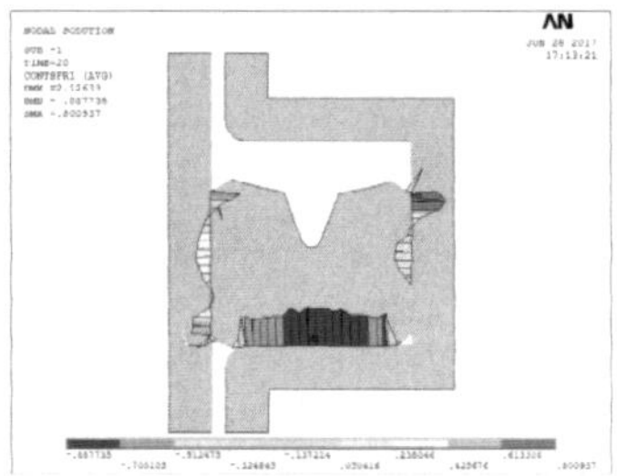

Fig. 6 Friction force diagram under 2 MPa pressure

From Fig. 3 and Fig. 4, we could see that there is a larger contact area between the upper and lower lips of the sealing ring and the shaft and the sealing groove and more frictional damage will be appeared in the upper lip than the lower lip of the sealing ring than the lower lip. The maximum shear force shown in Fig. 5 usually occurs in the tiny middle are of the seal ring. In addition, the largest shear force is appeared in the two areas where the lower lip of the sealing rings with piston.

If the amount of compression is 1.5mm, uniform load is applied to the sealing ring, and 0.5, 1, 1.5, 2, 2.5 and 3 MPa are selected as the working oil pressure for studying the contact stress respectively, the maximum friction value under different media pressure can been seen in following Fig. 7 based on the above working conditions.

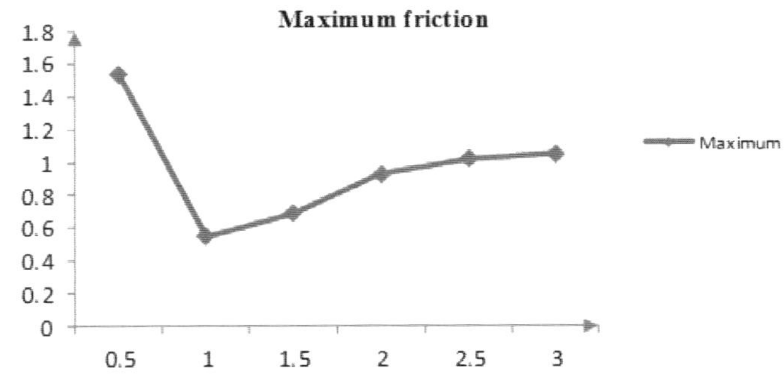

Fig. 7 The maximum friction force of the sealing ring under different oil pressure

Compared the maximum friction force value of the sealing ring under different working pressures, it can be seen from Fig. 7 that the maximum friction force decreases first and then increases, and finally tends to be stable in the pre-compression amount of 1.5mm case with the suffered hydraulic medium pressure of Y-ring lip is getting bigger and bigger. In the processing, the numerical value is defined in a normal range, which indicates that the wear of sealing ring is getting small. So, under this situation, it could not be taken into account when considering the failure mode.

If the pre-compression volume are 1.5mm and 2mm, applying uniform load to the sealing ring, still under the above working conditions, 0.5, 1, 1.5, 2, 2.5, 3MPa are selected as the working oil pressure for studying contact stress respectively, the value of contact stress are different under different media pressure (see Fig. 8).

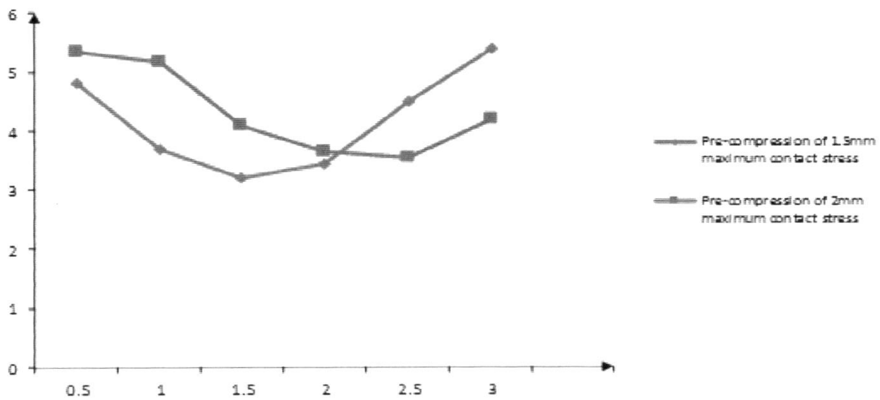

Fig. 8 Relationship between maximum contact stress and oil pressure

when pre-compression amount is 15%

In a deterministic calculation, analyzing of Fig. 8, it shows that when the medium pressure increases, the contact stress decreases first and then increases. As the amount of pre-compression increases, the difference between the contact stress and the applied medium pressure is increasing as well. According to the sealing criterion, if the maximum contact stress is greater than the medium stress, the structure can be safe and reliable.

3. Reliability analysis

3.1. Reliability analysis of sealing ring structure based on Kriging agent model

In the reliability calculation of Y-ring process, if we directly calculate using the traditional Monte Carlo method, it will take a lot of time to co-simulation Ansys and MATLAB. Therefore, in order to improve the computational efficiency, this paper adapts the Kriging model to calculate the failure probability of the sealing ring. Also, we extract a small number of design point samples in the known distribution variables applying Latin hypercube sampling (LHS) to, so as to the initial fitted Kriging model can get better fitted results. Generally, the Kriging model consists of two parts. The first part is the parameter part, which is in the form of linear regression. The second part is the non-parametric part, which is achieved through a random process. Assuming that $\boldsymbol{x}=[x_1,x_2,\cdots,x_7]^T$ is the input variable of the model and $y(\boldsymbol{x})$ is the output response of the model, then the Kriging model can be expressed in the following equation:

$$y(\boldsymbol{x})=F(\beta,\boldsymbol{x})+z(\boldsymbol{x})=f^T(\boldsymbol{x})\beta+z(\boldsymbol{x}) \tag{1}$$

The Eq.(1) represents the two parts of Kriging model, the polynomial parameter part and the random distribution non-parametric part. Where β is the regression coefficient, $f^T(\boldsymbol{x})$ is the basis function vector of polynomial regression. $z(\boldsymbol{x})$ is a stochastic process function, which obeys the normal distribution.

In conclusion, $f(\boldsymbol{x})$ represents the fitting of the whole global model, which can make an approximate estimation for the entire undetermined model and then determine and obtain the overall trend of response. Based on the Eq. (1), unknown samples $\boldsymbol{x}_{new}$ can be used to predict its response value $y(\boldsymbol{x}_{new})$. The optimal estimation of the Kriging model is shown as follows:

$$\hat{y}(\boldsymbol{x}_{new})=f^T(\boldsymbol{x}_{new})\hat{\beta}+\boldsymbol{r}^T(\boldsymbol{x}_{new})\boldsymbol{R}^{-1}(\boldsymbol{Y}-\boldsymbol{F}\hat{\beta}) \tag{2}$$

where $f^T(\boldsymbol{x}_{new})$ is the base function value of all unknown sample points $\boldsymbol{x}_{new}$. $\boldsymbol{r}^T(\boldsymbol{x}_{new})$ is a 7-dimensional column vector, representing the relationship between design samples $\boldsymbol{x}=[x_1,x_2,\cdots,x_7]^T$ and unknown samples $\boldsymbol{x}_{new}$, which can be expressed as:

$$\boldsymbol{r}^T(\boldsymbol{x}_{new})=[R(\boldsymbol{x}_{new},x_1),R(\boldsymbol{x}_{new},x_2),\cdots,R(\boldsymbol{x}_{new},x_7)]^T \tag{3}$$

At the same time, to solve the variance $\sigma_{\hat{y}}^2(\boldsymbol{x}_{new})$ obtained by Kriging fitting to the unknown sample $\boldsymbol{x}_{new}$.

$$\sigma_{\hat{y}}^2(\boldsymbol{x}_{new})=\sigma^2(1+\boldsymbol{u}^T(\boldsymbol{F}^T\boldsymbol{R}^{-1}\boldsymbol{F})^{-1}\boldsymbol{u}-\boldsymbol{r}^T(\boldsymbol{x}_{new})\boldsymbol{R}^{-1}\boldsymbol{r}(\boldsymbol{x}_{new})) \tag{4}$$

where $\boldsymbol{u}=\boldsymbol{F}^T\boldsymbol{R}^{-1}\boldsymbol{r}(\boldsymbol{x}_{new})-f(\boldsymbol{x}_{new})$.

The Kriging variance $\sigma_{\hat{y}}^2(\boldsymbol{x}_{new})$ describes the least square difference of the estimated value $\hat{y}(\boldsymbol{x}_{new})$ obtained by the Kriging model and the value $y(\boldsymbol{x}_{new})$ obtained from the real function at the unknown sample.

The EFF learning mechanism is evaluating the degree that the true function value meet equality constraint $y(\boldsymbol{x})=a$ in the $a\pm\varepsilon$ range though the reliability of nonlinear implicit state equation, adopting the EFF function index. The function $y(\boldsymbol{x})$ of the seal structure indicates whether the sealing structure is failure or not. If $y(\boldsymbol{x})>0$, the seal structure is safe; if $y(\boldsymbol{x})<0$, the seal structure is failure; if $y(\boldsymbol{x})=0$, it shows that the sealing structure is in a critical state.

In this study, we could conduct the biggest sample point that could result in the most influence on the last fitted Kriging model based on the EFF learning mechanism. For any unknown sample points $\boldsymbol{x}_n$, the predicted value of Kriging model obeys the normal distribution of $N(\hat{y}(\boldsymbol{x}_n),\sigma_{\hat{y}}^2(\boldsymbol{x}_n))$. Meanwhile, the mean value $\hat{y}(\boldsymbol{x}_n)$ and the variance $\sigma_{\hat{y}}^2(\boldsymbol{x}_n)$ can be obtained by Eq. (2) and Eq. (4). According to study these theories, we can get the EFF learning function as follows:

$$EFF(\boldsymbol{x}) = (\hat{y}(\boldsymbol{x}) - a)\left[2\Phi\left(\frac{a\text{-}\hat{y}(\boldsymbol{x})}{\sigma_{\hat{y}}(\boldsymbol{x})}\right)\text{-}\Phi\left(\frac{(a\text{-}\varepsilon)\text{-}\hat{y}(\boldsymbol{x})}{\sigma_{\hat{y}}(\boldsymbol{x})}\right)\text{-}\Phi\left(\frac{(a+\varepsilon)\text{-}\hat{y}(\boldsymbol{x})}{\sigma_{\hat{y}}(\boldsymbol{x})}\right)\right]$$
$$-\sigma_{\hat{y}}(\boldsymbol{x})\left[2\phi\left(\frac{a\text{-}\hat{y}(\boldsymbol{x})}{\sigma_{\hat{y}}(\boldsymbol{x})}\right)\text{-}\phi\left(\frac{(a\text{-}\varepsilon)\text{-}\hat{y}(\boldsymbol{x})}{\sigma_{\hat{y}}(\boldsymbol{x})}\right)\text{-}\phi\left(\frac{(a+\varepsilon)\text{-}\hat{y}(\boldsymbol{x})}{\sigma_{\hat{y}}(\boldsymbol{x})}\right)\right]$$
$$+\varepsilon\left[\Phi\left(\frac{(a+\varepsilon)\text{-}\hat{y}(\boldsymbol{x})}{\sigma_{\hat{y}}(\boldsymbol{x})}\right)\text{-}\Phi\left(\frac{(a\text{-}\varepsilon)\text{-}\hat{y}(\boldsymbol{x})}{\sigma_{\hat{y}}(\boldsymbol{x})}\right)\right]$$

$\Phi(\bullet)$ represents the standard normal distribution cumulative density function,
$\phi(\bullet)$ represents the standard normal distribution probability density function,
a represents the realized value of y specific fitting ability.

When calculating the reliability of Y-type sealing structure, it is necessary to focus on the critical state of function $y(\boldsymbol{x})=0$. So in this structure, $a=0$; $\varepsilon=2\sigma_{\hat{y}}(\boldsymbol{x})$. We define the point that the maximum $EFF(\boldsymbol{x})$ corresponding as the optimal sample point, which is $\boldsymbol{x}^* = \arg(\max(EFF(\boldsymbol{x})))$.

When the maximum learning function value satisfies the learning stop condition $\max(EFF(\boldsymbol{x})) \le \delta$, (δ is threshold. For the reliability calculation, let δ=0.001), which means the active learning process stop. It is considered that the Kriging model has reached the requirement of the optimal model at present.

In this paper, when we analyze the reliability of Y-type sealing structure, we select a series of alternative points through LHS (Latin hypercube sampling) first. And then, we calculate $EFF(\boldsymbol{x})$ and find the maximum value. If $\max(EFF(\boldsymbol{x})) \le \delta$, the algorithm is finished and the reliability is solved. If it cannot satisfy the values, we need to select another sample point as the new one, adding to the original alternative point, and then recalculate the value of $EFF(\boldsymbol{x})$ until it meet the condition $\max(EFF(\boldsymbol{x})) \le \delta$ .At that time, learning stop, from which the Kriging model with the best fitting effect and high accuracy can be obtained. So, we calculate the reliability through the adaptive Kriging model obtained from this EFF learning mechanism, which not only guarantee the accuracy of the calculation, also improve the computational efficiency.

3.2. Result analysis

In the model, the relevant parameters of the sealing structure are shown in Table 1.

Table 1 Related parameters of random variables

Parameter	Identifier	Distribution type	Mean	Standard deviation
The length of the lip	x_1	Normal	4	0.04
Pre compression quantity	x_2	Normal	1	0.01
The coefficient of friction	x_3	Normal	0.3	0.003
Oil pressure (MPa)	x_4	Normal	2	0.02
Parameter C_1	x_5	Normal	1.47	0.0147
Parameter C_2	x_6	Normal	0.47	0.0047
Poisson's ratio	x_7	Normal	0.498	0.00498

The research plans to delivery active learning through EFF learning mechanism. Firstly, 100 sample points are selected by LHS sampling. Then, the relationship between the number of new points and the value of learning function is shown in Fig. 9. As can be seen that with the increase of new sample points, the learning function value calculated by Kriging model satisfies the specified stopping criterion when the new samples points reached 163. As shown in Table 2, the failure probability of the sealing ring is calculated, 0.00035, by the Kriging model of the EFF learning mechanism.

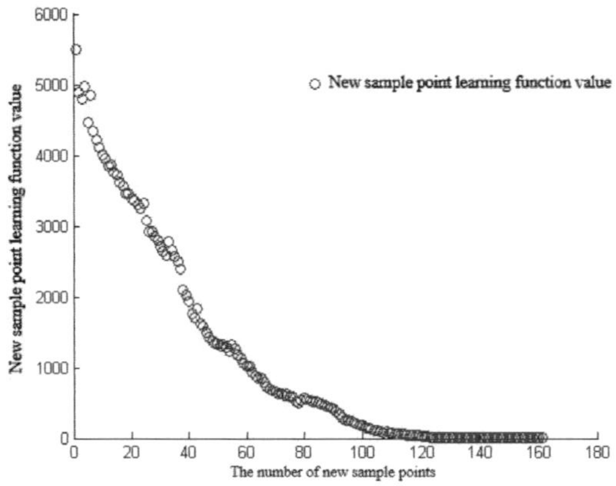

Fig. 9 Relation diagram of the number of new sample points and learning function values

Table 2 Failure probability of sealing structure calculation results

Reliability method	Sampling number	Failure probability
LHS-AK-EFF	100+163	0.00035

The calculation results show that the Kriging model using EFF active learning is effective. If the Monte Carlo method is directly used to analyze the reliability of the Y-type sealing structure of electromechanical system, the finite element model of Y-type sealing structure needs to be invoked about $100/P_f$ times, which can maintain the convergence result of estimated failure probability in the sealing structure. So, the huge amount of calculation limits the direct application of Monte Carlo method in the reliability analysis of sealing structure. However, the finite element is invoke only 263times if the Kriging model with EFF active learning, which could greatly saves the calculation time and meets the actual engineering calculation requirements.

4. Conclusion

The Kriging agent model is considered as a semi-parametric interpolation that need not a specific established mathematical model. In the paper, we apply it in the reliability calculation and analysis and conduct an effectively numerical evaluation. It can be a fundamental way for the effectively numerical evaluation reliability of the sealing structure as well as a theoretical basis for the optimal design for the sealing structure design. Compared with the Monte Carlo method in reliability analysis, the Kriging agent model obtained by EFF active learning can effectively select the optimum sample points according to the EFF learning function sequence and eventually improve the accuracy of the model and operational efficiency. Therefore, it can be effectively apply for the reliability analysis and the evaluation in practical engineering.

Acknowledgements

Authors gratefully acknowledge the support of China Scholarship Council, the fundamental research funds for the central universities (NPU-FFR-3102015BJ(Ⅱ) JL04), natural science foundation of Shaanxi province (2016JQ5109), and the seed foundation of innovation and creation for graduate students in Northwestern Polytechnical University (Z2017116).

Reference:

[1] Grenier A, Gutierrez AGP, Groult H, et al. Modified coin cells to evaluate the electrochemical properties of solid-state fluoride-ion batteries at 150°C[J], Journal of Fluorine Chemistry, 2016,191:23-28.

[2] Hao XH, Ju YL, Lu YJ. Experimental study on the sealing clearance between the labyrinth sealing displacer and cylinder in the 10 K G-M refrigerator[J], Cryogenics,2017, 57:203-208.
[3] Wang ZB, Fan JJ, Xiao SH, et al. Discussion on failure analysis methods for rubber seal rings[J], Failure Analysis and Prevention, 2015, 10(5): 314-319.
[4] Sun D, Wang S, Ai TT, et al. Numerical analysis and experimental research on the influence of swirl brakes on sealed static and dynamic characteristics[J]. Chinese Journal of Aeronautics. 2015, 36(9), 3002-3011.
[5] Liao CJ, Man M, Wang HR, et al. Study of mechanical characteristics of flange joint with metallic gasket under radial Loads[J]. Pressure Vessel Technology, 2014, 31(3):40-44.
[6] Gao BC, Meng XK, Li JY, et al. Thermal-mechanical coupled finite element model and seal performance analysis of mechanical seals[J]. Tribology, 2015, 35(5): 550-556.
[7] Zhou SM, Chen P, Shi Y. Analysis on sealing performance for a new type of rubber saddle-shaped sealing ring based on AQAQUS[J], Procedia Engineering,2015,130:1000-1009.
[8] Zhang K, Huang H, Duan ML, et al. Theoretical investigation of the compression limits of sealing structures in complex load transferring between subsea connector components[J], Journal of Natural Gas Science and Engineering, 2017,44:202-213.
[9] Shabani M, Carrapichano JM, Oliveira FJ, et al. Multilayered diamond mechanical seal rings under biodiesel lubrication and the full sealing conditions of pressurized water[J], Wear,2017,(384-385):178-184..
[10] Wang D Research on high-speed train service life of rubber seals evaluation techniques[D]. Qingdao University of Science & Technology, 2014.
[11] Liu M, Lu J, Duan D. Study on the sealing principle and structural optimization design of Y-type seal[J]. Special Purpose Rubber Products, 2012, 33(2): 57-59.
[12] Xing MJ. Study on sealing performance and leakage of metal W-ring in the aircraft engine[D]. Beijing Institute of Technology, 2015.
[13] Wang B, Jiao GQ, Lai DF, et al. Applying Roth Theory to Calculating Leak Rate of Rubber Gasket of Structure in Vacuum[J]. Journal of Northwestern Polytechnical University, 2010, 28(1): 129-133.
[14] Xue ZQ. Analysis on the reciprocating seal mechanism of the aircraft hydraulic cylinder[D], Zhe Jiang university, 2016.
[15] Cao YK, Wang HX, Wang X, et al. Application of domestic large pneumatic y-rings in mechanical press[J]. Lubrication Engineering, 2016, 41(5): 139-145.